The Essentials of

Technical Communication

The Essentials of

Technical Communication

FOURTH EDITION

ELIZABETH TEBEAUX
TEXAS A&M UNIVERSITY

SAM DRAGGA
TEXAS TECH UNIVERSITY

New York   Oxford
Oxford University Press

Oxford University Press is a department of the University of Oxford.
It furthers the University's objective of excellence in research,
scholarship, and education by publishing worldwide.
Oxford is a registered trade mark of Oxford University Press
in the UK and certain other countries.

Published in the United States of America by Oxford University Press
198 Madison Avenue, New York, NY 10016, United States of America.

For titles covered by Section 112 of the US Higher Education
Opportunity Act, please visit www.oup.com/us/he for the latest
information about pricing and alternate formats.

Library of Congress Cataloging-in-Publication Data

Names: Tebeaux, Elizabeth, author. | Dragga, Sam, author.
Title: The essentials of technical communication / Elizabeth Tebeaux, Texas
    A&M University, Sam Dragga, Texas Tech University.
Description: Fourth edition. | New York : Oxford University Press, [2018] |
    Includes bibliographical references and index.
Identifiers: LCCN 2017034974 | ISBN 9780190856144
Subjects: LCSH: Technical writing. | Communication of technical information.
Classification: LCC T11 .T295 2018 | DDC 808.06/66—dc23 LC record available at
    https://lccn.loc.gov/2017034974

9 8 7 6 5 4

Printed by LSC Communications, Inc.
Printed in the United States of America

Dedicated to
David H. Stewart
1926–2009

Our first English Department head, a scholar fully committed to the teaching of both writing and great literature. Without his support at the beginning of our careers, neither of us would have enjoyed as productive a life in technical communication as we have had. *The Essentials of Technical Communication* emerges from that support.

Brief Contents

Part One: **Principles**

Part Two: **Applications**

Detailed Contents

Part One: **Principles**

Part Two: **Applications**

Checklists

Preface

As we have emphasized in the first three editions of *The Essentials of Technical Communication,* no one wants to read what you write. In fact, they will read as little of what you write as they possibly can. Even if your document will be of value to your readers, you better make that point clear immediately, otherwise your document will likely be ignored or discarded. Your supervisor and colleagues may not have the necessary background, time, or inclination to sift through your e-mail, memo, or report and find the information they need.

We have developed *The Essentials of Technical Communication* as a practical introduction to all aspects of effective professional communication—a handbook to help you get your message across on the job, where time equals money and poorly crafted documents can produce a host of unwelcome consequences. We are teachers and researchers of technical writing, each of us with 40 years of experience, and we know that following a few simple guidelines leads to more efficient and effective communications. In this book we want to provide the guidelines you need as you plan, draft, and revise documents. Understanding these guidelines will help you manage your writing anxiety and enable you to write effectively and quickly—both requirements of employees who write for their jobs. *The Essentials of Technical Communication* will give you the information to help you design clear, concise, readable materials. From this foundation, you can learn to develop more complex documents as you advance in your career.

Approach

We have a simple rationale for our approach: we believe that the effective writer in a work situation must learn and internalize basic concepts of rhetoric and then apply these in developing documents. We've filled this brief book with memorable, concise guidelines. Each chapter in Part One focuses on basic rhetorical principles, and Part Two applies those principles to the planning and writing of particular types of documents.

A brief book enables instructors to adapt the book to their own uses. Many teachers want to build on principles by adding their unique approaches. This book provides the flexibility to allow for that possibility. In addition, many employees who did not study technical or business communication in college will find this book useful in learning how to write effectively in the workplace.

Organization

We have organized the book in two parts.

Part One (Chapters 1 through 6) lays out essential communication principles:

✦ Chapter 1, "Characteristics of Writing at Work," describes technical writing, or writing in the workplace, to show how it differs from academic writing. We stress the important privacy and security risks involved in using social media, e-mail, and texting.

✦ Chapter 2, "Writing for Your Readers," presents the essential elements of analyzing readers and then choosing the appropriate information, format, and style to meet the needs of the intended readers. We include a discussion of the composing process in this chapter.

✦ Chapter 3, "Writing Ethically," discusses the ethics of technical documents. While most professionals have standards of good practice, writers must also consider principles of communication ethics. We focus on standards of practice and include cases and exercises based on these standards.

✦ Chapter 4, "Achieving a Readable Style," explains how to write concise, pristine sentences and paragraphs.

✦ Chapter 5, "Designing Documents," covers basic principles for creating accessible and inviting documents. In a world of excess information, readers often miss or ignore important messages unless quickly available and easy to read.

✦ Chapter 6, "Designing Illustrations," provides guidelines for developing effective tables and figures. Graphics software offers infinite possibilities for data display, photo editing applications allow innumerable visual effects, and trillions of images are readily available online, but effective illustrations require an understanding of fundamental design principles and a thoroughgoing sensitivtiy to audience and purpose.

Part Two (Chapters 7 through 12) applies the principles from Part One to the types of documents most commonly prepared in the workplace:

✦ Chapter 7, "E-mails, Texts, Memos, and Letters," presents the basics of correspondence and demonstrates how to ensure that these routine messages are clear, readable, and effective. We also again emphasize ethics and accessibility in this chapter.

✦ Chapter 8, "Technical Reports," presents the elements of report development along with examples from abstracts and executive summaries to conclusions and recommendations.

✦ Chapter 9, "Proposals and Progress Reports," explains the purpose and design of each. In this chapter we use several student examples of proposals, as these respond to real situations in a university setting. We also include examples of online progress reports as these provide public information about the status of research and transportation construction projects.

✦ Chapter 10, "Instructions, Procedures, and Policies," describes how to develop clear instructions for a variety of situations to assure safety and efficiency of operation.

✦ Chapter 11, "Oral Reports," provides a short guide to developing and then presenting a concise, effective slide presentation.

✦ Chapter 12, "Résumés and Job Applications," describes how to prepare job application documents and prepare for interviews.

Key Features

- **Quick Tips:** Every chapter offers a brief section of concisely stated essential advice to jump-start your study and practice of effective technical communication.
- **Sample Documents:** This text, although concise, includes a range of sample documents covering the essential types and styles you will encounter in the workplace. Many of these documents are available for download on the book's companion website, www.oup.com/us/tebeaux, along with links to documentation resources.
- **Case Studies:** In most chapters case studies show how different types of documents function in different situations. These cases contextualize the documents to give you a sense of how and when the techniques we outline can and should be applied.
- **Checklists:** At chapter ends, we have included checklists—lists of questions you can use to ensure that your professional documents achieve your purpose. We hope you find these a handy reference tool.
- **Exercises:** Exercises at the end of each chapter guide practice in the techniques outlined in the text. Some of the exercises are designed to be done in class—individually or in in small groups—while others could be out-of-class assignments.
- **Appendices:** Appendix A offers a brief guide to grammar, punctuation, and usage. Appendix B gives a synopsis of information literacy and briefly explains three widely used systems for citing sources of information: APA, Chicago, and IEEE. Appendix C includes a sample report.
- **Companion Website:** The book's Companion Website at www.oup.com/us/tebeaux offers additional resources for students, including chapter overviews, self-quizzes, downloadable versions of the checklists from the book, helpful links, annotated document pages, and downloadable sample documents, including those from the exercises at chapter ends. The site also includes an Instructor's Manual, featuring downloadable PowerPoint files for use as lecture aids, chapter objectives, teaching strategies, workshop activities, writing projects, worksheets, and discussion questions. The Companion Website also includes revision assignments, multimodal writing assignments, and multilingual writing assignments.

New to This Edition

While improving upon our first three editions, we did not change those aspects of the book that have made it consistently popular with professors and students of technical communication. This new edition maintains the concise and practical nature of the original. We have, however, made several important changes based on the excellent suggestions from our expert panel of reviewers. We made each change to prepare students (1) to write in an increasingly dynamic, digital age and (2) to write for an increasingly diverse audience—both in the classroom and in the workplace.

Changes made for this fourth edition include the following:

- Chapter 1: Includes updated material on the need for information security as the most important difference between writing at school and writing at work. We believe that teachers have an ethical obligation to advise students about the risks of social media, texting, and e-mail, all of which in personal and business use carry legal liability.
- Chapter 2: Offers simplified and clarified wording and phrasing to make this chapter easier to read.
- Chapter 3: Includes accessibility as a key ethical consideration. We believe communicators must make their documents equally available to people regardless of their physical abilities.
- Chapter 4: Offers a revised organization and updated discussion of style with several new examples and exercises.
- Chapter 5: Includes advice on making documents accessible and on usability testing the design of your documents.
- Chapter 6: Includes new sample illustrations and added advice on usability testing your illustrations.
- Chapter 7: Includes advice on writing for social media and new exercises.
- Chapter 8: Offers a simplified and updated discussion with revised advice about writing reports in teams.
- Chapter 9: Includes new sample documents and new examples of online interactive reports.
- Chapter 10: Offers advice about usability testing your instructions.
- Chapter 11: Includes new sample documents to illustrate effective and ineffective slide presentations.
- Chapter 12: Includes advice about managing a social media profile to reinforce your professional reputation.

Finally, the Companion Website and Instructor's Resources have been updated with new examples, exercises, and materials. Of particular note is the revised Instructor's Manual, which now contains sections in each chapter on multimodal and multilingual writing, as well as new links, writing projects, and teaching strategies. The test bank has been updated as well and now offers a revised and expanded selection of test questions.

Acknowledgments

We are grateful to the dedicated book publishers of Oxford University Press for their conscientious efforts to make this book eloquent, elegant, concise, and cogent. We extend our thanks to the reviewers commissioned by Oxford for the earlier editions of this text: Susan Aylworth, California State University, Chico; Latonia Bailey, Crowder College; Elizabeth Childs, Auburn University; Cathy Corr, University of Montana-Missoula; Ed Cottrill, University of Massachusetts-Amherst; Richie Crider, University of Maryland; Melody DeMeritt, California Polytechnic State University; Scott Downing, Kenai Peninsula College, University of Alaska Anchorage; Leslie Fife, Oklahoma State University; Maureen Fitzsimmons, Syracuse University; Elizabeth Holtzinger-Jennings, Pennsylvania State University; Danica Hubbard, College of DuPage; Kendall Kelly, Southwest Texas State University; Kevin LaGrandeur, New York Institute of Technology; Elizabeth Lopez, Georgia State University; Lisa McClure, Southern Illinois University, Carbondale; Raynette Meyer, Aiken Technical College; Elizabeth Monske, Northern Michigan University; Brenda Moore, New Jersey Institute of Technology; Marguerite Newcomb, University of Texas at San Antonio; Mark Noe, University of Texas-Pan American; Roxanna Pisiak, Morrisville State College; Liza Potts, Old Dominion University; Ritu Raju, Houston Community College; Leslie St. Martin, College of the Canyons; Denise Stodola, Kettering University; Dawn Taylor, South Texas College; Aaron Toscano, University of North Carolina at Charlotte; Michelle Weisman, College of the Ozarks; and Linda Young, Oregon Institute of Technology. And we add our thanks to those who reviewed for this new edition: Paul M. Dombrowski, University of Central Florida; Jennifer Haber, St. Petersburg College; Helena Halmari, Sam Houston State University; Kevin LaGrandeur, New York Institute of Technology; David L. Major, Austin Peay State University; Richard Jeffrey Newman, Nassau Community College; Casey J Rudkin, Kenai Peninsula College; Michael Shuman, University of South Florida; William Clay Kinchen Smith, Santa Fe College; and Sonia Stephens, University of Central Florida.

We also thank the innumerable colleagues and students who have challenged and inspired us in the teaching of technical communication. And, as always, special thanks to Jene and Linda for their love and support.

Elizabeth Tebeaux, Professor Emerita, Texas A&M University
Sam Dragga, Professor Emeritus, Texas Tech University

PART **ONE**

Principles

Characteristics of Writing at Work

Technical writing (sometimes called business or professional writing) describes writing that occurs in a business or work setting. University offices, corporations, research centers, hospitals, businesses of all sizes, even nonprofit organizations produce large quantities of technical writing, which differs from academic writing in several important ways. These differences mean that you cannot write on the job the way you have written in school. Writing in school and writing at work differ because the purposes and the context of each differ. Thus, the products of each contrast sharply.

 Quick Tips

On the job, keep in mind that **no one wants to read anything you write. Most of the time they will not read all of what you write.** They will read because they need to, not because they want to. They will read because you have information they need to take actions or make decisions. They don't get paid to read: they get paid to take actions and make decisions. The more time they need to read your document, the less productive time they have.

Make sure everything you write is clear, correct, necessary, and polite. And never assume that anything you write is confidential.

Modern organizations must keep their information secure, whether it exists in paper or virtual form. Organizations that lose information to cyber thieves often face severe consequences.

Writing at Work Versus Writing at School

Workplace writing requires that you continue to apply what you have learned about effective paragraph development, correct sentence structure, punctuation, and usage. As an educated adult, your writing should exemplify correctness. Beyond these fundamental principles, business or technical writing will differ from writing you have done as a student in five important ways.

Writing at work:

1. Requires acute awareness of security and legal liability

2. Requires awareness that documents may be read by unknown readers, inside and outside the organization, for an infinite time

3. Achieves job goals

4. Addresses a variety of readers who have different perspectives from those of the writer

5. Requires a variety of written documents

Requires acute awareness of security and legal liability. The most fundamental characteristic of technical writing rests in the legal liability associated with workplace information.

Chief information officers in educational, business, government, and research organizations work diligently to protect the privacy of information about their employees and the knowledge generated by these employees by following both federal and state privacy laws. Identity and information theft can occur at any time, despite the best efforts of any chief information officer's staff and security team. People throughout the world continue to attack computing systems to gain access to credit card numbers, personal and medical information, and transcripts of academic work, creative work, and research data—essentially whatever hackers can access, either for their own use or to sell to crime cartels.

Electronic communication has become a blessing and a curse. Today's workplace requires extensive technology. Research organizations, hospitals, banks, financial organizations, law firms, physicians, and even small, locally owned businesses have to pursue strict security on all information they have about customers, clients, and patients. Organizations, like architectural firms, computer companies, engineering companies, and manufacturers, must protect their intellectual property from theft. The knowledge they produce for clients becomes the value of the organization. When you begin a job, you need to learn the security rules of your employer and follow them. For example, you will likely not be allowed to use your company e-mail for any purpose other than company business. Your company telephones will likely have the same restrictions. You should never access your personal blogs or social networking sites from your employer's computer.

To avoid potential security breaches:

- Remember that any text message you send will not be secure and may be legally accessed, whether the cell phone you use belongs to you or your employer. Company e-mail can be viewed by the company webmaster. Once you begin working for an organization, use caution in what you discuss via text message and e-mail.

- Avoid blogs, unless your company uses secure blogs for creating collaborative reports, for example. Remember that others can see what you have written. Be sure that your comments exemplify tasteful, helpful, and accurate tone and content.

- Any electronic communication—texts, e-mails, and social media messages—can be subject to subpoena. Your Internet provider must comply with "good cause" subpoenas. Again, what you say in cyberspace never goes away.

- Avoid using browsers available on company computers to locate information on any topic not related to your work.

- Use social media carefully. Your company may have a page on one of the social media sites, but do not use it or respond to it. Ask the purpose of the site and the rules for its use by employees. Note: Many students have been expelled from their universities for inappropriate use of social media. A business organization, because of concerns for information security, will watch how employees use social media. You can lose your job if your comments on blogs, wikis, and other forms of social media disparage the organization and perhaps divulge proprietary information.

- Many organizations, before they hire new employees, will check social media to see what potential employees have said about themselves. Again, criminals across the world also check. Divulging confidential information, personal or professional, can have major consequences for you and organizations for which you work, have worked, or will work. Because virtual messages never go away, ask yourself, "If I decided to run for public office in 20 years, would I want people I don't even know to see what I said about myself today?"

- If you have a personal web page, be sure that what you place on the page makes a positive statement about you and does not discredit your employer in any way.

- Guard your external storage drives carefully. Never leave one in your computer when you work in a public place, even for a few minutes. When you purchase a flash drive, be sure it has been manufactured by a reputable company. Never buy nonpackaged flash drives. Never use a flash drive given to you as a gift from an advertiser. You do not know what material, malware, or viruses have been placed on the drive.

- Never forget that everything you write can be accessed by others. Tip: Always write as if someone you do not know might be reading over your shoulder. And follow all rules your employer stipulates. Accepting and agreeing to follow rules of confidentiality of company information may be a condition of employment with that organization. When you interview for a job, ask about the company's website, all social media sites, and management of those sites.

In school, your primary obligation is to avoid plagiarism. But what you write at work can be used against you in lawsuits. Once you sign your name to a report or letter, your signature makes you responsible for the content. Hostile readers can use what you say to support claims against you and the organization you represent. Because we live in an increasingly litigious society, designing documents that will prevent their misuse should be one of your primary goals.

Requires awareness that documents may be read by unknown readers. Always anticipate unknown readers who may receive copies of your reports or e-mail. Ask yourself this important question: "Does my report or e-mail contain any information that could be misconstrued and affect me or the organization adversely if unknown readers see my communication?" While academic writing responds to assignments, applicable only for a specific semester, course, and professor, workplace communications have no specific life span. They can be read and then used in ways you never intended or envisioned.

You cannot underestimate the problem that unknown readers present. Nearly everything you write for an organization will remain in the organization's archive indefinitely. Copies of your reports and letters will be placed in files accessible to readers who may not know anything about you or the situation you discuss in your document. These documents will often be used in assessing your performance and in determining your promotion potential. What you say suggests how well you have done your job. Unknown readers may also use your reports to gain understanding of a work situation they have inherited with a new job assignment. On the job, what you write becomes much more than a knowledge indicator for a grade.

Achieves job goals. In school, you write to show your professor that you know the subject matter and to make a good grade. But in the workplace, writing is the major way that people achieve their job goals and document their work. Writing becomes documentation that you have done your work and how you have done it. How well you write will suggest how well you have done your work. It will become part of the organization's permanent archives.

Addresses a variety of readers who have different perspectives. In college, you write your assignments for a single reader, a professor, a specialist in a subject area. But in a work setting, you can expect to write to readers who have varied educational and technical backgrounds, readers who have different roles inside and outside the organization, and readers who may know less about a topic than you do. Your supervisor, for example, may have majored in a field of study very different from yours, or your supervisor's responsibilities may have channeled his or her technical knowledge into other areas. For example, you may report directly to a person whose educational background has been in physical chemistry or electrical engineering but whose responsibilities may now be in personnel management, database administration, quality control, or financial analysis. Many technical people earn a master's degree in business administration to assist them in management roles.

In a work context, these readers will feel no commitment to read what you write unless your messages help them as they do their own work. They will generally not read all documents completely. Each will be interested in how your message affects his or her job goals. What seems clear and important to you may lack clarity and importance to others. Because e-mail has become a common way of communicating within organizations, you really have no idea who will read what you write as any message and its attachments may be forwarded. Documents posted online on an organization's website have no security from prying eyes and hackers.

We live and work in an information age where the quantity of information grows rapidly, where people have more to read than they can ever hope to read. As they see your report or your e-mail subject line, they will immediately ask themselves questions such as "What is this?" "Why should I read it?" "How does it affect me?" "What am I going to have to do?" Without a carefully stated subject line, your readers may delete your message before opening it. If they do open your e-mail, they will want to find the main points and ideas quickly, and they will become impatient if they are unable to find them by glancing at the page. They will not usually read any document completely or bother to respond to it unless, at the beginning, the message indicates that reading it serves their best interests. How they respond to the first few sentences of your writing will often determine how much more of it they read.

On the job, your readers are not a captive audience, as your teachers have been. They do not have to read what you write. If you want your writing read, make your message clear and easy to read; make your message as interesting, relevant, and concise as possible. Because your readers often read selectively, conciseness and clarity are basic ingredients of effective business communication. Mechanical correctness remains a desirable quality, but correct writing that cannot be read easily and quickly will not be read.

Requires a variety of written documents. Most academic writing includes essays, essay examinations, research papers, and laboratory reports. You direct your writing to your teachers. At work, however, employees can expect to write a variety of documents not relevant to academic writing assignments: letters, e-mails, information and procedure memos, proposals, progress reports, project reports, feasibility studies, economic justification reports, policy statements, travel reports, news releases, speeches, training procedures, budget forecasts, employee evaluations, user documentation, and perhaps articles for publication in trade journals. What you write will change with your responsibilities, the kind of job you have, and your position in the organization. How you write each document will depend on the topic discussed, the situation leading to the document, your readers' needs and perspectives, and your purpose in writing.

The Foundations of Effective Writing at Work

Developing effective documents requires a process involving at least six stages:

1. Planning the document
2. Determining content

3. Arranging ideas

4. Drafting

5. Revising

6. Editing

While you may do each of these steps as a separate activity, when writing you will more than likely move back and forth from one activity to the other as you develop your document. Following this process will help ensure that the information is appropriate as well as correctly and effectively presented.

The Qualities of Good Technical Writing

Surveys show that organizations rank writing skills in this order of importance:

1. Accuracy

2. Clarity

3. Conciseness

4. Readability

5. Usability

6. Correctness

These qualities mean that a document

- makes a good impression when readers first interact with it: the document is neat, readable, well organized, and inviting
- can be read selectively—for instance, by some users, only the summary; by other users, only the introduction and conclusions; by still other users, the entire report
- shows a plan that reveals the purpose and value of the document
- makes sense: ideas appear in a logical sequence immediately evident from the document design
- uses visuals, if necessary, to help readers understand ideas or data
- conveys an overall impression of authority, thoroughness, soundness, and honest work
- makes sense to people who were not part of the initial readership
- makes a positive statement about the writer and the organization
- enables people who need to use your writing to perform a task to do so

Beyond all these basic characteristics, good technical writing has no typographical errors, grammatical slips, or misspelled words. Little flaws distract attention from the writer's main points and call into question the writer's literacy.

As you study and practice writing for a workplace setting, keep in mind these qualities as well as the differences between the writing you do as an employee and the writing you do as a student.

CASE 1–1

Bradshaw Engineering, LLP, understands the changes and challenges occurring in the field of chemical engineering. Matt Lunsford, one of the research engineers, tells Jerry Bradshaw, the senior principal, about a journal article he has read about carbon capture. Jerry has an established practice of posting summaries and articles on the company website to help employees remain informed. Matt provides a memo of transmittal to Jerry (Case Document 1–1A) and two short summaries of the article he has read and shared with Jerry (Case Documents 1–1B and 1–1C). Note how Matt develops the summaries with the knowledge level of both engineers and nonengineers in mind. He also uses basic principles of document design to create readable documents. How do his three documents reflect the qualities of good technical writing (see "The Qualities of Good Technical Writing")?

Matt "transmits" his summaries with a memo of transmittal to Jerry. At work, you will need transmittal documents (discussed in Chapter 7 for sending documents) as they introduce the documents to readers. Avoid the habit of using Post-it notes to transmit information. If the note is lost, the report may not reach its intended reader.

CASE DOCUMENT 1–1A

DATE: September 3, 2016
TO: Jerry Bradshaw
FROM: Matthew Lunsford
SUBJECT: **Technical and General Summaries of Article about Carbon Capture
            Technology**

I have completed and attached two summaries of a journal article describing technologies for carbon capture. The first, a technical summary, I have prepared for engineers here at BE. The second, a general summary, targets other employees. As you requested, all BE employees need to understand current research that relates to our work here at BE.

Subject and Purpose of Summaries

The summaries inform our employees about technological developments involving carbon capture. With the growing awareness of carbon dioxide emissions and their effect on the environment, reducing carbon dioxide emissions has been the subject of several important projects that BE has worked on in the last few years. Technology discussed in this article, such as power generation, could soon be standard in industries with which we work, and all of us need to remain aware of advances in carbon capture technology.

Reason for Summaries

Your request for summaries of state-of-the art research helps all of us. Thus, BE encourages employees to stay up to date on current engineering developments relevant to our projects and planning.

CASE DOCUMENT 1–1B

C. Jones, "CO_2 capture from dilute gases as a component of modern global carbon management," *Annual Review of Chemical and Biomolecular Engineering*, Vol. 2, 31–52, July 2011.

Technical Summary

Article Purpose
This article describes methods for carbon capture. It also describes the need for more detailed research and analysis before these methods can be implemented on an industrial scale.

Rationale for the Topic
With the increasing concentration of carbon dioxide in the atmosphere, the well-being of society in the future largely depends on carbon management.

Major Points Discussed
- Carbon management includes two different processes: (1) reducing carbon dioxide emissions from large sources such as coal-fired power plants (carbon capture and sequestration) and (2) capturing carbon dioxide directly from the atmosphere (air capture).
- Researchers face a major challenge in developing methods of carbon capture that are both effective and energy-efficient. Currently developed methods have large energy penalties of ~30 percent, meaning 30 percent of power produced by a plant would go toward powering carbon capture.
- A relatively mature technology for carbon capture and sequestration (CCS) uses absorption of carbon dioxide by aqueous amine solution. This method will most likely dominate first-generation CCS technologies. However, other absorption methods under development involve use of room-temperature ionic liquids (RTILs) to dissolve carbon dioxide.
- Carbon dioxide–selective membrane technology represents another strategy that, when developed, will enhance CCS. Many other types of membranes have been studied. These include membranes composed of polymers only and those that also have amine solutions, enzymes, or RTILs as active media. Research has achieved sufficient selectivities for carbon dioxide versus nitrogen, but practitioners need increased throughputs for viable large-scale use. Also, further research is needed on the stability of these membranes in the presence of contaminants contained in plant flue gas.
- A final method being developed for CCS is adsorption of carbon dioxide using either supported amine adsorbents or metal–organic frameworks. Metal–organic frameworks are attractive because they can be easily tuned to obtain the exact adsorption properties desired. However, more research must show the stability of these materials in the presence of contaminants contained in plant flue gas.

- Air capture technology, which would actually decrease the concentration of carbon dioxide in the atmosphere, is also under development. Current methods typically involve an aqueous basic absorption process using metal hydroxides. In the future this process could complement CCS, which only limits increases in atmospheric carbon dioxide concentration. However, costs associated with air capture are currently much higher than those with CCS. Research also lacks quantitative analyses of hypothetical air capture processes.

Conclusion

The technology most likely to be applied in the near future for CCS is aqueous amine absorption. Other methods that involve the use of RTILs, membranes, and adsorbents require further research and development before their use on a reasonable scale. Air capture also requires further research and detailed descriptions of process designs before its viability can be determined.

CASE DOCUMENT 1–1C

C. Jones, "CO_2 capture from dilute gases as a component of modern global carbon management," *Annual Review of Chemical and Biomolecular Engineering*, Vol. 2, 31–52, July 2011.

General Summary

Article Purpose
This article describes advances in a technology called "carbon capture," which involves the removal of carbon dioxide from plant emission sources and from the atmosphere.

Rationale for the Topic
With the increasing concentration of carbon dioxide in the atmosphere, the well-being of society in the future largely depends on our ability to manage the amount of carbon in the atmosphere.

Major Points Discussed
- Carbon management includes two different processes: (1) reducing carbon dioxide emissions from large sources such as coal-fired power plants (carbon capture and sequestration) and (2) capturing carbon dioxide directly from the atmosphere (air capture).
- Researchers face a major challenge in developing methods of carbon capture that provide both effective carbon capture and energy efficiency.
- Some relatively mature technology exists for carbon capture and sequestration (CCS). This technology involves the use of solvents to absorb carbon dioxide escaping from emission sources such as power plants. Other methods still in the development stage include use of membranes and adsorbents. These substances also selectively capture carbon dioxide while allowing other gases to pass through to the atmosphere.
- Air capture technology, which would actually decrease the concentration of carbon dioxide in the atmosphere, is also in development. In the future this process could complement CCS, which now only limits increases in atmospheric carbon dioxide concentration. However, the costs associated with air capture are currently much higher than those with CCS, and further research must continue.

Conclusion
The technology most likely to be applied in the near future for CCS is absorption of carbon dioxide. Other methods that involve the use of membranes and adsorbents require further research and development before they can be scaled to industry size. Air capture also requires further research before researchers can determine its viability.

 **EXERCISES**

1. Visit the website of one of the employers listed below. What kinds of writing do employees of these organizations do?

   - Virginia Department of Transportation (www.vdot.virginia.gov/projects)
   - McKinsey (www.mckinsey.com)
   - Deloitte (www.deloitte.com)
   - Chevron (www.chevron.com)
   - National Institutes of Health (www.nih.gov)

   Prepare a memo to your instructor explaining the kinds of writing you find on the website. For format, use the memo in Figure 1–1 as an example.

2. Based on the qualities of effective technical writing discussed in this chapter, write a memo to your instructor evaluating the memo in Figure 1–1. What does it do right and wrong? How would you make it more effective?

U.S. Department
of Transportation
Federal Highway
administration

Memorandum

Subject: **ACTION:** Request for 2016 Bridge   Date: January 17, 2017
Replacement Costs

From: Joseph L. Hartmann, Ph.D., P.E.   In Reply Refer To: HIBS-30
Director, Office of Bridges and Structures

To: Division Administrators

We hereby request that each Division office submit to the Office of Bridges and Structures by April 3, 2017, replacement costs for all highway bridges constructed in their State with Federal funds during fiscal year 2016. Collection of costs related to bridge construction is required by the National Bridge and Tunnel Inspection Standards (23 U.S.C.144). This memorandum provides guidance in collecting the required bridge costs.

Division offices are expected to review these costs in sufficient detail prior to submittal to ensure that States have followed the criteria defined in the link below. Please work with your State to resolve any inconsistencies and provide a discussion and recommended adjustments in conjunction with the submitted cost data to explain any anomalous data.

Cost data should be submitted electronically through the web-enabled NBI system by logging on to FHWA User Profile & Access Control System (UPACS). Use the Submittals tab in the NBI system. Data collected include the following for bridges on the National Highway System (NHS) and bridges off the NHS:

- Number of bridges replaced
- Total area of replaced bridges
- Total replacement cost of all bridges
- Bridge replacement unit cost

There is also a comment field available for additional information or discussion.

Additional guidance may be found at http://www.fhwa.dot.gov/bridge/nbi/uc_criteria.cfm .

Please direct questions to Fernando Luna at (202) 366-4621 or Fernando.Luna@dot.gov. Alternately, questions can be directed to Samantha Lubkin at (202) 366-1575 or Samantha.Lubkin@dot.gov.

cc:
Directors of Field Services
HIBS-30

● **FIGURE 1–1**

Writing for Your Readers

As you begin to plan your document, think first about your readers and then your purpose. Every decision you make in developing your document should reflect your audience, their needs, and your purpose.

Avoid becoming absorbed in ideas and information you plan to include. Never forget that the person or group who will read the document may have a very different perspective about the content. Your readers cannot climb into your mind and know your thoughts. When you carefully analyze your readers, you can often design and write your report in a way that helps them understand what's in your mind.

 Quick Tips

To develop any communication, you have three main goals that connect reader, purpose, and context:

1. You want your readers to understand your meaning exactly in the way you intend.
2. You want your writing to achieve its goal with the designated readers.
3. You want to keep the goodwill of those with whom you communicate.

Understand Your Readers—The Heart of the Planning Process

To achieve the three goals just listed, you must pursue the following four tasks, both before you begin to write and while you are actually composing your document:

1. Determine as fully as possible who will read what you write.

2. Know the goals you want your writing to achieve and the business context in which you need to communicate.

3. Understand your role in the organization as a writer and how your role should be reflected in what you write.

4. Determine the content by considering your readers' frame of reference and your purpose in writing.

Keep in mind that business readers want answers now. Employees in most organizations, particularly large ones, have more to read than they can and will read. If you want your letter or report to be read, be sure that important information occurs on the first page of a report and in the first paragraph of an e-mail, memo, or letter and that you answer the following questions your readers will ask:

- What is this?
- Do I have to read it?
- How does it affect me?
- What will I have to do?
- What are the main ideas?

If you answer their questions at the beginning of your document, readers are more likely to continue reading.

To answer these questions in the minds of your readers, you will have to answer three questions yourself about the document you are writing:

- Who will read what I write?
- Who will act on what I write?
- Who else may read what I write?

In many cases, your primary reader will transmit your document to someone else for action. Perhaps this individual is one of your secondary readers or someone unknown to you.

Determine your readers and their perspectives. When you consider your readers, determine as much as you can about them.

- How much do your readers know about your topic?
- Do your readers have expertise in this area?

Readers with technical expertise in the area you discuss have different needs (and often different perspectives) from readers who lack technical expertise.

- How much do your readers know about your topic?
- Do you know or can you estimate your readers' educational levels?
- Do you know your readers' cultural backgrounds?

If you work in an organization that does business with readers from other cultures, plan to do background reading on these cultures.

- Will your readers have interest in what you write? If not, how could you present your message to make it appealing?
- What kind of relationship do you have with these readers? Do you know the readers' attitude toward you, the subject matter you need to communicate, the job you have, and your area within the organization? Do you have credibility with these readers?

A host of factors determine your readers' perception: education, family, geographical and cultural background, job responsibilities, rank in the organization, age, life experiences, and gender—just to name a few demographics that define how people see the world. How much your readers know about your topic determines what you say and the technical level of your presentation.

- How well do you know your readers?

You may not know your readers personally. However, if you know an individual's level in the organization, the responsibilities associated with that level, and the kind of technical expertise your reader has, this information will help you decide what you need to include and how to present your information. Knowing your readers' responsibilities in the organization can help you anticipate their attitude—if your subject will interest them. Because people tend to read only what they can use, try to relate your message to your readers' job. Knowing the readers' attitude toward the topic addressed in your message will help you determine how to present your information.

- Who else might read what you write?

Most reports and letters have distribution lists: the names of those who receive copies. A person on the distribution list may be the person who will ultimately act on what you write. Thus, the needs and perceptions of those who receive copies should be considered.

- Why is each person on the distribution list receiving a copy?
- How much does each person on this list know about your topic?

Sometimes your primary reader may know the situation you are discussing, and the purpose of the report may be to inform others within the organization by going through proper channels.

- What situation led to the need for this document?

Often, you can better understand your reader's perspective if you understand the situation that requires you to write the document. The need for written communications develops from interactions of people involved in a work environment. To select the appropriate information, level of language (technical or general), and amount of explanation needed in a business context, a writer must carefully determine the needs of each reader.

Determine your purpose.

- Why do you need to write this document?
- What do you want to achieve with your document?

Determining why you are writing is as important as determining who your readers are. *Purpose always relates to readers.* And you may have more than one purpose. For example, you may be writing to provide information and to recommend action. In addition, what you say may serve as documentation—proof of your efforts to provide the information requested. Written messages that document employees' activities serve a major function in today's business organizations. Without documentation, you may have difficulty proving that you performed specific tasks.

Understand your role as a writer.

- What position do you have in the organization?

As an employee, you will be hired to perform the duties that define a particular job. As the one responsible for performing specific tasks, you will be communicating with employees above you, below you, and on your own level. In writing to individuals in any group, you will communicate not as you would with a friend or family member but as the person responsible for the work associated with that position. When you write, you create a personality that should fit the position you hold.

　　To have credibility as a writer in an organization, the image that you project should be appropriate to your position. What you write and how you write it should reflect your level of responsibility in the organization—the power relationship that exists between you and the reader. The image you project will change, depending on your readers. You will project the image of a subordinate when you write to those higher than you, but you will transmit the image of a supervisor to those who work directly under you. When you communicate with others on your own job level, you will convey the image of a colleague. Effective writers fit their message to each reader.

Plan the content.

- What ideas should you use to achieve the goals of the message?
- What ideas should you omit?
- How should you arrange your ideas?

　　Once you have analyzed your readers and your purpose, you can decide what you want and need to include and how you will phrase and arrange your ideas.

- How do you want your message to sound?

　　Knowing how your message should sound will always be critical. Always try to convey a respectful tone appropriate to your position in the organization. How a message is presented may often be as important as the information itself.

　　Case 2–1 shows how a writer's assessment of audience and purpose changes the information and presentation in the e-mails for each reader involved in the situation.

Anticipate the context in which your writing will be received.

- How will readers use your writing?

Once your document has reached its primary destination, it may be placed in a stack for later reading; it may be skimmed and then routed to the person who will be responsible for acting on it; it may be read, copied, and distributed to readers unknown to you; it may be read and used as an agenda item for discussing a particular point; or it may be read carefully and later used as a reference. Knowing how readers will use the documents they receive can often guide you in deciding not only what to include but also how to organize the information and arrange it on the page.

The Basic Parts of the Composing Process

The composing process, integral to your analysis of audience, has six main stages:

1. **Analyzing** the situation
2. **Choosing/discovering** information
3. **Arranging** information
4. **Drafting**
5. **Revising**
6. **Editing** the finished draft

A writer who tries to do all stages at once usually creates a document that will fail. Research has shown that good writers usually follow a standard process—one that will make your writing tasks easier and the results more effective.

Analyzing the writing situation—purpose, readers, and context. The first step in composing is the most critical. In this step, you need to know *why* you need to write: what you want to achieve with your document, what situation or problem has led to the necessity of your writing this document. Then, you need to consider your readers—those who will or may read your document.

Every technical or workplace document responds to a specific situation. Each document has a targeted audience. Writing responds to both—the situation and the readers in that situation. When you write, you do not simply compile information about a subject.

Choosing/discovering information. You select information for your document based on your purpose, *what* your reader needs and *how* you think your reader perceives the subject.

As you search for information, remember your purpose, what you want your reader to know and do with what you write. Then, begin to list ideas you can use to develop your topic. Based on these ideas, ask yourself what additional information you will need to locate. Don't like what you wrote? Delete it. You may want to begin your document by writing your purpose at the beginning to help you stay on track.

CASE 2–1

Running Brook Neighborhood needs money to maintain the entrance to the subdivision. As Case Document 2–1A indicates, Charles Fields, treasurer of the property owners' association, has decided to write a flyer to place in each mailbox of the Running Brook homes. He has received very little money for the required upkeep. He drafts his request, but before he makes copies for the project, he asks a neighbor for her opinion of his flyer. She offers to revise the flyer (Case Document 2–1B) because she sees the difference between his version and what she thinks should be written. He makes the required number of copies, places one in each mailbox, and receives more than the amount RB needs for its entrance maintenance.

CASE DOCUMENT 2–1A

Subject: Running Brook Subdivision Marker (RBSM) Fund

I have been remiss in reporting the status of the RBSM fund. In the past, an oral report was delivered at the annual subdivision picnic. I did not make such a report at the last picnic since I felt the information should be provided to all RB residents and not just to picnic attendees. The report for 2016–2017 will be distributed with the flyers that announce the 2018 Running Brook picnic. Since the subdivision has several new families, a historical element was added to this report.

The POA supplies each subdivision with a marker, which generally is located in the middle of a small landscaped area. Maintenance, decoration, and any embellishment of the marker area is the responsibility of the residents of the subdivision. Resident involvement in the care of the RBSM began in December 2014 when a waterline and meter were installed at a cost of $450.00 and obligation to pay an annual water usage fee was assumed. The fee allows for consumption of 4,000 gallons of water each month. Water consumption above 4,000 gallons incurs an additional charge, an event that has occurred during the hot summer months of the past 3 years. The fee has increased from $106.98 to $123.30 over the past 3 years, and I pay it from my checking account every 2 months throughout the entire year.

As of October 2015, cost of care of the marker was $712.34 (waterline and meter, water usage, plants, and landscaping material). This cost was defrayed by donations amounting to $705.00 from 70 percent of the residents and money ($123.78) from the 2017 picnic. The account had a positive balance of $116.44 ($828.78 minus $712.34).

Expenditures for the period October 2015 through October 2016 included water usage ($126.72) and plants ($88.76) for a total of $215.48. For a time, the account had a negative balance of $99.04. Donations amounting to $245.00 were received from only 16 of the 62 families (25 percent) in the subdivision. The account was healthy again with a positive balance of $145.96.

Expenditures for the period October 2016 through October 2017 were $230.12 (water usage $372) and plants, fertilizer, hat, pumpkins, etc. $93. Once again, the fund had a negative balance ($84.16). Recent donations from 9 residents totaled $135.00. The current fund balance is $50.84.

Projected costs from now until October 2018 include water ($123.30) and plants in the spring and fall ($90.00–$100.00), or approximately $250.00. Clearly, the present fund balance will not support these costs. Accordingly, contributions are needed. The treasurer for the fund is Joann Fields. Hopefully, every RB family will participate. I sense we all believe that participation provides a measure of the RB community spirit.

Seasons Greetings,

Charles Fields

2 Roaring Brook Court, phone 933-3314, e-mail clfields@earthlink.com

---

CASE DOCUMENT 2–1B

Running Brook Marker Fund Needs Your Donations

The Problem

- Our marker maintenance fund currently has a balance of **$50.84**. Our projected maintenance cost this year is $250.00.
- Since 2015, the number of residents contributing has dropped from 70 percent of the residents to only 9 residents. Without contributions, maintenance cannot occur.

How You Can Help

Please contact Joann Fields, treasurer, at 922-3314 or clfields@earthlink.com. Or you can mail her a check, made out to the Running Brook Marker Fund, for $10. This amount will cover the cost of water, plants, and maintenance. Any money left over will be applied to next year's maintenance.

Expenditures and Balances to Date			
	Costs	Donations	Account Balance
Oct. 2015	$712.34	$705	$116.44
Oct. 2016	$215.48	$245	$145.96
Oct. 2017	$230.12	$135	$50.84

Information for New Running Brook Residents

Each subdivision in Hot Springs Village has a marker that denotes the entrance to the subdivision. The Property Owners' Associations provide the markers, but residents must maintain and decorate the area surrounding the marker.

A few years ago, we installed a waterline and meter. Water costs have increased, particularly during the summer. Other expenses include plants, fertilizer, hay, and pumpkins.

Keeping our marker area well maintained provides a positive first impression of our neighborhood.

Contact Information—Give us a call if you have questions

Charles and Joann Fields

2 Roaring Brook Court

Hot Springs Village

phone 933-3314, e-mail clfields@earthlink.com

Why is Case Document 2–1B better than Case Document 2–1A? Why would you be more inclined to read and respond to Case Document 2–1B? Charles also decides to paper-clip an empty envelope with his name and address to the request. Why would this decision help Charles achieve his goal?

CASE 2–2

The director of online sales at Pine Avenue Books would like to perform a customer survey regarding the store's website in order to improve the efficiency of the online shopping experience. She drafts the following e-mail that she will send to all customers who made online or on-site purchases at the store in the last 2 years (Case Document 2–2A). The store manager, however, tells her that her draft, if sent by e-mail, will likely not be read: (1) The subject line does not encourage busy readers to open the e-mail. (2) The purpose of the message does not appear until the second paragraph. (3) The e-mail as a whole appears dense and difficult for readers to skim. The online sales director decides to revise the original to respond to the three issues.

CASE DOCUMENT 2–2A

Subject: Please take a 5-minute, multiple-choice survey; details follow.

I am the director of online sales at Pine Avenue Books and am conducting a study of the online shopping sections of the store's website. If you have purchased books online in the last 2 years, please consider participating and passing the attached invitation on to your friends for their potential participation.

This study wants to determine how customers view the store's website and, in particular, the sections related to online shopping. Your participation may provide useful information about online shopping and help us to improve your online shopping experience in the future. I am looking for customers who have bought books online at Pine Avenue Books in the last 2 years.

If you participate, your obligations will be low. You will complete a short, anonymous survey via the Internet that will require approximately 5 minutes of your time and will be returned to me via Survey Monkey, an online data collection service. If you complete a survey, your responses will be returned to me anonymously (I will not be able to identify your e-mail address, your IP address, or any other information that would inform me as to your identity or your location). If you agree to do so, you will also participate in a 15- to 20-minute follow-up session (this meeting will occur by telephone or e-mail as you choose). All data for surveys and follow-up interviews will be strictly confidential. Your identity will never be revealed in any results, discussions, or presentation of the research. In addition, all information will be destroyed after I have analyzed the data.

Completion of the survey and postsurvey interviews is voluntary; you may skip questions and can quit any portion of the study at any time. If you are willing to participate in this study, please click on the following link and complete the survey: http://www.surveymonkey.com/PineAvenueBooks. If you wish to participate in a follow-up interview, please include contact information at the end of the survey.

You may contact me via e-mail (PineAvenueBooks@spectrum.com) or phone (512-823-9235).

CASE DOCUMENT 2–2B

Subject: Please take a 5-minute survey about online shopping at Pine Avenue Books.

Survey Purpose

This study seeks to gauge your opinion of the online shopping sections of the Pine Avenue Books website (www.pineavenuebooks.com). Your participation will help us to improve your future online shopping experience.

If you have purchased books online in the last 2 years, please consider participating and passing the attached invitation on to your friends for their potential participation.

Survey Requirements

If you participate, your obligations will be minimal.

1. You will complete a short, anonymous survey via the Internet that will require approximately 5 minutes of your time and will be returned to me via Survey Monkey, an online data collection service.
2. If you complete a survey, your responses will be returned to me anonymously. (I will not be able to identify your e-mail address, your IP address, or any other information that would inform me as to your identity or your location.)
3. If you agree to do so, you will also participate in a 15- to 20-minute follow-up session (this meeting will occur by telephone or e-mail as you choose).
4. All data for surveys and follow-up interviews will be strictly confidential. Your identity will never be revealed in any results, discussions, or presentation of the research. In addition, all information will be destroyed after I have analyzed the data.

5. Completion of the survey and follow-up interviews is voluntary; you may skip questions and can quit any portion of the study at any time.

6. If you are willing to participate in this study, please click on the link http://www.surveymonkey.com/PineAvenueBooks and complete the survey. If you wish to participate in a follow-up interview, please include contact information at the end of the survey.

You may contact me via e-mail (PineAvenueBooks@spectrum.com) or phone (512-823-9235)

Arranging information. As you collect and begin summarizing information, consider how to arrange the material. In what order should you present your ideas? Memos, for example, need to begin with the news or essential information to ensure that readers at least read what's most important before they start skimming the document or stop reading it altogether. Most reports begin with an introduction and a summary of the report. The discussion section follows with supporting information. Most reports adopt some version of this arrangement. Many business organizations have templates for reports.

If you know what arrangement you want or need to use, sort material by creating folders for each segment of your report. Then arrange material within each folder before you begin drafting. Add copies of pertinent information sources into these same folders for later access when you begin to draft your document. This method allows you to track material you use and insert appropriate citations when you use material from a specific source.

Drafting. Every individual drafts differently. Most writers work on a document in a start/stop fashion. When you begin your draft, open your file and save it with the name of your report. Then, begin typing ideas or sections. (You may wish to move/paste material you listed, arranged, and then developed in step 2.) You may wish to type the names of your main segments, boldface those, and insert information beneath the appropriate segment. This method helps you keep track of the information that you are using to develop your draft. Note that some of the ideas in your list become headings. Some may be combined with other ideas. You can arrange, delete, and add ideas as you need to.

As you continue to draft, you will revise. But during the drafting stage you should revise only to improve the meaning. Try to avoid worrying about sentences that don't sound "quite right." If the sentence you write captures your core idea, even clumsily, don't stop to revise. You can "clean up" these sentences later. Don't attempt to correct mechanical problems unless you feel you can do so without slowing your ability to transfer your ideas from your mind to the screen. Focus on presenting your material to your readers; then you can begin a formal revision process once you believe you have your basic ideas on the screen.

Revising. During the formal revision process you need to revise several times and focus on different issues:

- **Logic.** Does your presentation make sense? Try reading paragraphs aloud that seem to be "scrambled." Hearing what you have written often tells you if/where problems in logic occur. Does your material occur in the appropriate order for your purpose and for your readers?

- **Completeness.** Does your presentation seem complete in terms of your purpose and your readers' needs and requirements? Have you checked all your information for correctness? Does your document contain all requested information?

- **Style.** Examine each paragraph and each sentence. Are your paragraphs really paragraphs? Do they have topic sentences? Do all the sentences in the paragraph pertain to the meaning you are building in the paragraph? Start each paragraph with a topic sentence. Eliminate or recast sentences that provide little support for the topic sentence. Today's readers usually dislike wordy, dense, complicated sentences. Make your sentences clear, concise, and precise to encourage your readers to follow your ideas. Also, watch the length of your paragraphs. Long paragraphs discourage readers and tend to become incoherent.

- **Illustrations.** Do you need tables, photos, graphs, drawings, or videos to help your reader understand and remember key ideas? Illustrations combined with text often provide the best means of communicating with your readers.

- **Document design.** When you began drafting, if you used headings or names of report segments to help you organize your draft, you began at that point to design your document. *Document design* refers to the way you arrange information and display it on the page. The importance of how information looks on the page cannot be stressed enough. If you want your writing to be read, design the page or screen so that information is inviting and accessible.

Editing. Get into the habit of performing several "edits" for any document: one for mechanics—spelling, usage, punctuation, and sentence structure.

A second edit focuses on the document as a whole. How does it look? How does it sound? Is the important information easy to locate? Have you included all needed information in the document?

Another edit focuses on citing sources: check your documentation to be sure that you provide credit or sources for all information you have used. Be sure that when you use illustrations and ideas from other sources you give credit to the source.

In short, don't try to check for every possible error in one reading. Editing requires care, objectivity, patience, and diligence.

✎ Planning and Revision Checklist

Analyzing the Situation

☐ What is your subject or topic?

☐ What is the purpose of the document?

☐ Who are your readers, known and potential?

☐ Why are you writing? Why is this document required? What is the situation that led to the need for this document? Who cares?

Selecting Information

☐ What topics do you need to cover? What do your readers need to know? What do you want your readers to do?

☐ What structure do you plan to use? If you have required report sections, what are they?

☐ What information resources do you have available? What resources do you need to locate?

☐ What types of illustrations (e.g., tables, graphs, photos, diagrams) are you considering? How will they help to convey your message?

Arranging Information

☐ In what order should the information be placed? What does your reader need to know first?

☐ Have you sorted your material into specific groups?

☐ Can you include headings that announce each major section to your reader?

☐ Does all information have relevance to your purpose?

Drafting

☐ Have you inserted information under each of your headings?

☐ Have you recorded the sources of all information you will use so that you can develop correct citations after you have completed your draft?

☐ Have you noted where you will use illustrations? Have you noted the source of any illustration borrowed or adapted from another document, publication, or website?

Revising

☐ Have you stated clearly the purpose of your report?

☐ Does your information support your purpose?

☐ Is your tone appropriate?

☐ Will your readers be able to follow your logic?

☐ Have you included all required items—report sections and required information?

continues

☐ Have you checked all facts and numbers?

☐ Could any material be deleted?

☐ Is your document easy to read? Are your paragraphs well organized and of a reasonable length?

☐ Have you had someone read your draft and suggest improvements?

Editing

☐ Have you included all the formal elements that your report needs or is required to include?

☐ Is your system of documentation complete, consistent, and correct?

☐ Are your pages numbered?

☐ Are all illustrations placed in the appropriate locations within the text?

☐ Is the format consistent—font, size, placement of headings?

☐ Have you checked all points of the completed draft where your word processing application indicates that you might have errors in wording, sentence structure, mechanics, or spelling?

☐ Have you checked manually for misspellings and for mechanical errors such as misplaced commas, semicolons, colons, and quotation marks?

 EXERCISES

1. Phoenix Publishers has always provided free child care to all employees with children ages 3 months to 5 years. For its 50 years of operation, the company has taken great pride in being a family-friendly employer. Tough economic times for the book publishing industry and rising costs of operation for the child-care center, however, now require that the company begin charging parents $50 per week per child for the services of the child-care center. According to the president of the company, it was either that or freeze wages for all employees or lower the already slim dividend paid to the company's stockholders and risk a loss of investors. The president of the company directs you to write three letters regarding this important change: one to parents using the child-care center, one to all employees, and one to the stockholders. Note that parents will also receive the letter addressed to employees. Note also that some employees are also stockholders. The president recognizes the sensitivity of this policy change and thus will also expect from you a memo justifying the variations you made in the three letters.

2. Find a blog posting aimed at specialists in your field. Revise the posting for a specific group of readers outside your field who might need to know this information (e.g., business managers, city officials, school administrators). Attach the original to your revision, and submit both to your instructor.

3. Examine Figure 2–1 (addressed to parents) and Figure 2–2 (addressed to employers). What similarities and differences do you notice in the verbal and visual information included? How do the writers adapt information to each audience? What changes, if any, would make the two documents more suitable for their specific audiences?

4. Since Figure 2–2 addresses employers, how would you adapt this information for a message addressed to employees?

Handwashing: A Family Activity
Keeping Kids & Adults Healthy

Handwashing is an easy, inexpensive, and effective way to prevent the spread of germs and keep people healthy.

For kids, washing hands can be a fun and entertaining activity. It is simple enough for even very young children to understand. Handwashing gives children and adults a chance to take an active role in their own health. Once kids learn how to properly wash their hands, they can—and often do—show their parents and siblings and encourage them to wash hands, too.

Parents can help keep their families healthy by:

- Teaching them good handwashing technique
- Reminding their kids to wash their hands
- Washing their own hands with their kids

Improving Health

- Handwashing education in the community:
  » Reduces the number of people who get sick with diarrhea by 31%
  » Reduces diarrheal illness in people with weakened immune systems by 58%
  » Reduces respiratory illnesses, like colds, in the general population by 21%

Saving Time and Money

- Handwashing is one of the best ways to avoid getting sick and spreading illness to others.
- Reducing illness increases productivity due to:
  » Less time spent at the doctor's office
  » More time spent at work or school

Helping Families Thrive

Children who have been taught handwashing at school bring that knowledge home to parents and siblings. This can help family members get sick less often and miss less work and school.

Despite widespread knowledge of the importance of handwashing, there is still room for improvement. A recent study showed that only 31% of men and 65% of women washed their hands after using a public restroom.

For more details, visit www.cdc.gov/handwashing.

CDC Department of Health and Human Services
Centers for Disease Control and Prevention

CS234835-A

• **FIGURE 2–1** Document for Exercise 3

Source: US Department of Health and Human Services, Centers for Disease Control and Prevention. Handwashing: A Family Activity. Washington, DC: GPO, 2012. http://www.cdc.gov/healthywater/pdf/hygiene/hwfamily.pdf

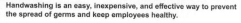

Handwashing: A Corporate Activity
Improving Health & Increasing Productivity

Handwashing is an easy, inexpensive, and effective way to prevent the spread of germs and keep employees healthy.

Handwashing gives people the opportunity to take an active role in their own health. Most handwashing studies have focused on child care or health care settings. The few that have looked at corporate settings show that promoting clean hands results in fewer employee sick days.

Improving Health

Germs can spread quickly. A healthier community means healthier employees. Handwashing education in the community:

• Reduces the number of people who get sick with diarrhea by 31%

• Reduces diarrheal illness in people with weakened immune systems by 58%

• Reduces respiratory illnesses, like colds, in the general population by 21%

Saving Time and Money

Handwashing is one of the best ways to avoid getting sick and spreading illness to others.

Sick employees are less productive even when they come to work. They may also spread illness to others at work. One recent study promoting clean hands in corporate environments showed:

• Fewer employee illnesses

• Less use of sick days

Helping Families and Workforces Thrive

Employees with healthy children spend less time away from work taking care of sick children, are more productive at work when not dealing with family illness, and get sick less often themselves.

Employers should promote employee handwashing and encourage them to also:

• Teach their children good handwashing technique

• Remind children to wash their hands

• Wash hands with their children

Despite widespread knowledge of the importance of handwashing, there is still room for improvement. A recent study showed that only 31% of men and 65% of women washed their hands after using a public restroom.

For more details, visit www.cdc.gov/handwashing.

 CDC
Department of Health and Human Services
Centers for Disease Control and Prevention

CS234835-B

● **FIGURE 2–2**   Document for Exercise 3

Source: US Department of Health and Human Services, Centers for Disease Control and Prevention. Handwashing: A Corporate Activity. Washington, DC: GPO, 2012. http://www.cdc.gov/healthywater/pdf/hygiene/hwcorporate.pdf

3

Writing Ethically

 Quick Tips

On the job, you won't always have a lot of time to analyze all the issues related to ethical dilemmas. You might have to decide quickly—in minutes or seconds.

If this is the situation you find yourself in, think of individuals in your company or in your profession you admire for their integrity and good judgment: it could be a favorite colleague, a supervisor, or a mentor. Ask yourself, "How would he or she manage this dilemma?" Allow your answer to this question to guide your actions.

Your Professional Obligations

None of us are isolated individuals, operating entirely separate from the traffic of human society. Your ethical obligations are several, often intersecting, and from time to time competing. Consider, for example, your duties to the following:

- **To yourself:** You will have to make decisions and take actions that allow you to support yourself financially while establishing (and maintaining) your reputation in your field. You can't quit (or lose) your job every time you object to a policy of your boss.

- **To your discipline and profession:** As a member of your profession, you have a responsibility to advance the knowledge and reputation of your field. You must share information with your colleagues that will improve the practices of your profession, clarify understanding, offer new insights, and promote better training of new students of your discipline. You must communicate in a manner that brings credit to your profession and inspires the next generation to want to study and join your profession.

- **To your academic institution:** You have a moral obligation to the institution that trained you for your profession. Your successes or failures will be indicative of the merits of that institution and its faculty. If you disgrace yourself by illegal or unethical actions, for example, investigating officials and the public might ask why you weren't taught better behavior.

- **To your employer:** Your responsibility as an employee is to serve the interests of your organization, to promote its products and services, to support its mission, and to shield confidential information and intellectual property, especially if doing so offers a competitive advantage.

- **To your colleagues:** You have a duty to your colleagues on the job to do your fair share of the work assigned and to do it with integrity, accuracy, and efficiency. You also have a responsibility to use no more than your fair share of the resources allotted and to take no more than your fair share of the credit (or blame) given.

- **To the public:** Your obligation to society is to promote the public good through greater safety, fuller liberty, and a better quality of life. Your decisions and actions on the job could allow communities to thrive in resilient and sustainable environments or to be poisoned by private greed and callous disregard for civic aspirations.

Also keep in mind that something could be legal but still unethical: for example, while it might be legal in your city to build a chemical storage facility adjacent to a hospital, the danger to immobilized patients from a chemical explosion—and their inability to evacuate quickly—might make this choice of location unethical. You would have to weigh the risk to the public versus the jobs that would be created in your city and the potential profits for your company.

In communicating on the job, you won't always have a clear choice of right versus wrong. Typically, none of your choices will be entirely satisfactory, and from time to time all your choices will be unsatisfactory. You will have to juggle your various obligations and determine which has priority. You can't simply do whatever the boss tells you to because you aren't merely an employee. You also have important responsibilities to yourself, your profession, your schools and teachers, your colleagues, and the public itself. You will have to make every effort to avoid being either submissive or self-righteous.

Codes of Conduct and Standards of Practice

Your professional association and your employing organization will each have codes of conduct that specify their expectations regarding appropriate behavior on the job. Many disciplines, such as engineering, accounting, medicine, and financial management, have specific regulations for professional conduct and for standards of practice. Continuing education and periodic testing on these rules may be a part of your certification. Violation of these rules can carry heavy penalties, particularly because these disciplines involve the health and financial resources of clients.

Examine the code of ethics for the National Society of Professional Engineers (Figure 3–1). Even if you are not an engineering major, you can see how the National Society of Professional Engineers dictates how engineers conduct their work.

National Society of Professional Engineers®

Code of Ethics for Engineers

Preamble

Engineering is an important and learned profession. As members of this profession, engineers are expected to exhibit the highest standards of honesty and integrity. Engineering has a direct and vital impact on the quality of life for all people. Accordingly, the services provided by engineers require honesty, impartiality, fairness, and equity, and must be dedicated to the protection of the public health, safety, and welfare. Engineers must perform under a standard of professional behavior that requires adherence to the highest principles of ethical conduct.

I. Fundamental Canons

Engineers, in the fulfillment of their professional duties, shall:
1. Hold paramount the safety, health, and welfare of the public.
2. Perform services only in areas of their competence.
3. Issue public statements only in an objective and truthful manner.
4. Act for each employer or client as faithful agents or trustees.
5. Avoid deceptive acts.
6. Conduct themselves honorably, responsibly, ethically, and lawfully so as to enhance the honor, reputation, and usefulness of the profession.

II. Rules of Practice

1. Engineers shall hold paramount the safety, health, and welfare of the public.
   a. If engineers' judgment is overruled under circumstances that endanger life or property, they shall notify their employer or client and such other authority as may be appropriate.
   b. Engineers shall approve only those engineering documents that are in conformity with applicable standards.
   c. Engineers shall not reveal facts, data, or information without the prior consent of the client or employer except as authorized or required by law or this Code.
   d. Engineers shall not permit the use of their name or associate in business ventures with any person or firm that they believe is engaged in fraudulent or dishonest enterprise.
   e. Engineers shall not aid or abet the unlawful practice of engineering by a person or firm.
   f. Engineers having knowledge of any alleged violation of this Code shall report thereon to appropriate professional bodies and, when relevant, also to public authorities, and cooperate with the proper authorities in furnishing such information or assistance as may be required.
2. Engineers shall perform services only in the areas of their competence.
   a. Engineers shall undertake assignments only when qualified by education or experience in the specific technical fields involved.
   b. Engineers shall not affix their signatures to any plans or documents dealing with subject matter in which they lack competence, nor to any plan or document not prepared under their direction and control.
   c. Engineers may accept assignments and assume responsibility for coordination of an entire project and sign and seal the engineering documents for the entire project, provided that each technical segment is signed and sealed only by the qualified engineers who prepared the segment.
3. Engineers shall issue public statements only in an objective and truthful manner.
   a. Engineers shall be objective and truthful in professional reports, statements, or testimony. They shall include all relevant and pertinent information in such reports, statements, or testimony, which should bear the date indicating when it was current.
   b. Engineers may express publicly technical opinions that are founded upon knowledge of the facts and competence in the subject matter.
   c. Engineers shall issue no statements, criticisms, or arguments on technical matters that are inspired or paid for by interested parties, unless they have prefaced their comments by explicitly identifying the interested parties on whose behalf they are speaking, and by revealing the existence of any interest the engineers may have in the matters.

4. Engineers shall act for each employer or client as faithful agents or trustees.
   a. Engineers shall disclose all known or potential conflicts of interest that could influence or appear to influence their judgment or the quality of their services.
   b. Engineers shall not accept compensation, financial or otherwise, from more than one party for services on the same project, or for services pertaining to the same project, unless the circumstances are fully disclosed and agreed to by all interested parties.
   c. Engineers shall not solicit or accept financial or other valuable consideration, directly or indirectly, from outside agents in connection with the work for which they are responsible.
   d. Engineers in public service as members, advisors, or employees of a governmental or quasi-governmental body or department shall not participate in decisions with respect to services solicited or provided by them or their organizations in private or public engineering practice.
   e. Engineers shall not solicit or accept a contract from a governmental body on which a principal or officer of their organization serves as a member.
5. Engineers shall avoid deceptive acts.
   a. Engineers shall not falsify their qualifications or permit misrepresentation of their or their associates' qualifications. They shall not misrepresent or exaggerate their responsibility in or for the subject matter of prior assignments. Brochures or other presentations incident to the solicitation of employment shall not misrepresent pertinent facts concerning employers, employees, associates, joint venturers, or past accomplishments.
   b. Engineers shall not offer, give, solicit, or receive, either directly or indirectly, any contribution to influence the award of a contract by public authority, or which may be reasonably construed by the public as having the effect or intent of influencing the awarding of a contract. They shall not offer any gift or other valuable consideration in order to secure work. They shall not pay a commission, percentage, or brokerage fee in order to secure work, except to a bona fide employee or bona fide established commercial or marketing agencies retained by them.

III. Professional Obligations

1. Engineers shall be guided in all their relations by the highest standards of honesty and integrity.
   a. Engineers shall acknowledge their errors and shall not distort or alter the facts.
   b. Engineers shall advise their clients or employers when they believe a project will not be successful.
   c. Engineers shall not accept outside employment to the detriment of their regular work or interest. Before accepting any outside engineering employment, they will notify their employers.
   d. Engineers shall not attempt to attract an engineer from another employer by false or misleading pretenses.
   e. Engineers shall not promote their own interest at the expense of the dignity and integrity of the profession.
2. Engineers shall at all times strive to serve the public interest.
   a. Engineers are encouraged to participate in civic affairs; career guidance for youths; and work for the advancement of the safety, health, and well-being of their community.
   b. Engineers shall not complete, sign, or seal plans and/or specifications that are not in conformity with applicable engineering standards. If the client or employer insists on such unprofessional conduct, they shall notify the proper authorities and withdraw from further service on the project.
   c. Engineers are encouraged to extend public knowledge and appreciation of engineering and its achievements.
   d. Engineers are encouraged to adhere to the principles of sustainable development[1] in order to protect the environment for future generations.

• **FIGURE 3–1** National Society of Professional Engineers' Code of Ethics

Source: http://www.nspe.org/Ethics/CodeofEthics/index.html

3. Engineers shall avoid all conduct or practice that deceives the public.
   a. Engineers shall avoid the use of statements containing a material misrepresentation of fact or omitting a material fact.
   b. Consistent with the foregoing, engineers may advertise for recruitment of personnel.
   c. Consistent with the foregoing, engineers may prepare articles for the lay or technical press, but such articles shall not imply credit to the author for work performed by others.
4. Engineers shall not disclose, without consent, confidential information concerning the business affairs or technical processes of any present or former client or employer, or public body on which they serve.
   a. Engineers shall not, without the consent of all interested parties, promote or arrange for new employment or practice in connection with a specific project for which the engineer has gained particular and specialized knowledge.
   b. Engineers shall not, without the consent of all interested parties, participate in or represent an adversary interest in connection with a specific project or proceeding in which the engineer has gained particular specialized knowledge on behalf of a former client or employer.
5. Engineers shall not be influenced in their professional duties by conflicting interests.
   a. Engineers shall not accept financial or other considerations, including free engineering designs, from material or equipment suppliers for specifying their product.
   b. Engineers shall not accept commissions or allowances, directly or indirectly, from contractors or other parties dealing with clients or employers of the engineer in connection with work for which the engineer is responsible.
6. Engineers shall not attempt to obtain employment or advancement or professional engagements by untruthfully criticizing other engineers, or by other improper or questionable methods.
   a. Engineers shall not request, propose, or accept a commission on a contingent basis under circumstances in which their judgment may be compromised.
   b. Engineers in salaried positions shall accept part-time engineering work only to the extent consistent with policies of the employer and in accordance with ethical considerations.
   c. Engineers shall not, without consent, use equipment, supplies, laboratory, or office facilities of an employer to carry on outside private practice.
7.   Engineers shall not attempt to injure, maliciously or falsely, directly or indirectly, the professional reputation, prospects, practice, or employment of other engineers. Engineers who believe others are guilty of unethical or illegal practice shall present such information to the proper authority for action.
   a. Engineers in private practice shall not review the work of another engineer for the same client, except with the knowledge of such engineer, or unless the connection of such engineer with the work has been terminated.
   b. Engineers in governmental, industrial, or educational employ are entitled to review and evaluate the work of other engineers when so required by their employment duties.
   c. Engineers in sales or industrial employ are entitled to make engineering comparisons of represented products with products of other suppliers.
8. Engineers shall accept personal responsibility for their professional activities, provided, however, that engineers may seek indemnification for services arising out of their practice for other than gross negligence, where the engineer's interests cannot otherwise be protected.
   a. Engineers shall conform with state registration laws in the practice of engineering.
   b. Engineers shall not use association with a nonengineer, a corporation, or partnership as a "cloak" for unethical acts.

9. Engineers shall give credit for engineering work to those to whom credit is due, and will recognize the proprietary interests of others.
   a. Engineers shall, whenever possible, name the person or persons who may be individually responsible for designs, inventions, writings, or other accomplishments.
   b. Engineers using designs supplied by a client recognize that the designs remain the property of the client and may not be duplicated by the engineer for others without express permission.
   c. Engineers, before undertaking work for others in connection with which the engineer may make improvements, plans, designs, inventions, or other records that may justify copyrights or patents, should enter into a positive agreement regarding ownership.
   d. Engineers' designs, data, records, and notes referring exclusively to an employer's work are the employer's property. The employer should indemnify the engineer for use of the information for any purpose other than the original purpose.
   e. Engineers shall continue their professional development throughout their careers and should keep current in their specialty fields by engaging in professional practice, participating in continuing education courses, reading in the technical literature, and attending professional meetings and seminars.

Footnote 1 "Sustainable development" is the challenge of meeting human needs for natural resources, industrial products, energy, food, transportation, shelter, and effective waste management while conserving and protecting environmental quality and the natural resource base essential for future development.

As Revised July 2007

"By order of the United States District Court for the District of Columbia, former Section 11(c) of the NSPE Code of Ethics prohibiting competitive bidding, and all policy statements, opinions, rulings or other guidelines interpreting its scope, have been rescinded as unlawfully interfering with the legal right of engineers, protected under the antitrust laws, to provide price information to prospective clients; accordingly, nothing contained in the NSPE Code of Ethics, policy statements, opinions, rulings or other guidelines prohibits the submission of price quotations or competitive bids for engineering services at any time or in any amount."

Statement by NSPE Executive Committee
In order to correct misunderstandings which have been indicated in some instances since the issuance of the Supreme Court decision and the entry of the Final Judgment, it is noted that in its decision of April 25, 1978, the Supreme Court of the United States declared: "The Sherman Act does not require competitive bidding."

It is further noted that as made clear in the Supreme Court decision:
1. Engineers and firms may individually refuse to bid for engineering services.
2. Clients are not required to seek bids for engineering services.
3. Federal, state, and local laws governing procedures to procure engineering services are not affected, and remain in full force and effect.
4. State societies and local chapters are free to actively and aggressively seek legislation for professional selection and negotiation procedures by public agencies.
5. State registration board rules of professional conduct, including rules prohibiting competitive bidding for engineering services, are not affected and remain in full force and effect. State registration boards with authority to adopt rules of professional conduct may adopt rules governing procedures to obtain engineering services.
6. As noted by the Supreme Court, "nothing in the judgment prevents NSPE and its members from attempting to influence governmental action . . ."

Note: In regard to the question of application of the Code to corporations vis-a-vis real persons, business form or type should not negate nor influence conformance of individuals to the Code. The Code deals with professional services, which services must be performed by real persons. Real persons in turn establish and implement policies within business structures. The Code is clearly written to apply to the Engineer, and it is incumbent on members of NSPE to endeavor to live up to its provisions. This applies to all pertinent sections of the Code.

National Society of Professional Engineers®

1420 King Street
Alexandria, Virginia 22314-2794
703/684-2800 • Fax:703/836-4875
www.nspe.org
Publication date as revised: July 2007 • Publication #1102

● **FIGURE 3–1**   Continued

ETHICAL PRINCIPLES

As technical communicators, we observe the following ethical principles in our professional activities.

LEGALITY

We observe the laws and regulations governing our profession. We meet the terms of contracts we undertake. We ensure that all terms are consistent with laws and regulations locally and globally, as applicable, and with STC ethical·principles.

HONESTY

We seek to promote the public good in our activities. To the best of our ability, we provide truthful and accurate communications. We also dedicate ourselves to conciseness, clarity, coherence, and creativity, striving to meet the needs of those who use our products and services. We alert our clients and employers when we believe that material is ambiguous. Before using another person's work, we obtain permission. We attribute authorship of material and ideas only to those who make an original and substantive contribution. We do not perform work outside our job scope during hours compensated by clients or employers, except with their permission; nor do we use their facilities, equipment, or supplies without their approval. When we advertise our services, we do so truthfully.

CONFIDENTIALITY

We respect the confidentiality of our clients, employers, and professional organizations. We disclose business-sensitive information only with their consent or when legally required to do so. We obtain releases from clients and employers before including any business-sensitive materials in our portfolios or commercial demonstrations or before using such materials for another client or employer.

QUALITY

We endeavor to produce excellence in our communication products. We negotiate realistic agreements with clients and employers on schedules, budgets, and deliverables during project planning. Then we strive to fulfill our obligations in a timely, responsible manner.

FAIRNESS

We respect cultural variety and other aspects of diversity in our clients, employers, development teams, and audiences. We serve the business interests of our clients and employers as long as they are consistent with the public good. Whenever possible, we avoid conflicts of interest in fulfilling our professional responsibilities and activities. If we discern a conflict of interest, we disclose it to those concerned and obtain their approval before proceeding.

PROFESSIONALISM

We evaluate communication products and services constructively and tactfully, and seek definitive assessments of our own professional performance. We advance technical communication through our integrity and excellence in performing each task we undertake. Additionally, we assist other persons in our profession through mentoring, networking, and instruction. We also pursue professional self-improvement, especially through courses and conferences.

Adopted by the STC Board of Directors
September 1998

• **FIGURE 3–2**   Society for Technical Communication's Ethical Principles for Technical Communicators

In contrast, the Society for Technical Communication, a leading international association for technical writers and graphic artists, has composed its guidelines as a list of six principles (Figure 3–2). Other professional associations publish highly detailed regulations to specify ethical practice in their field, such as the 400-page code of conduct of the American Institute of Certified Public Accountants or the 500-page code of medical ethics of the American Medical Association.

Your employer's code of conduct will identify the ideals of the organization and the behaviors expected of every employee. This could be a brief list of things to do and things to avoid doing or a comprehensive discussion of policies and procedures. Both codes of conduct and standards of practice might also be the access point to a wider array of ethics resources, including frequently asked questions, instructional videos, and helplines for reporting violations and soliciting advice.

Familiarize yourself with the codes of conduct or standards of practice that regulate ethical communication for your company and your profession: you might have to interpret and cite their guidelines to justify your decisions regarding ethical dilemmas on the job. Your knowledge of the available information materials about ethics will reinforce your credibility, support your judgment, and insulate you from intimidation.

Recognizing Unethical Communication

Writing clearly, accurately, and effectively is a crucial part of your job. And once a document leaves your control, it takes on a life of its own. It provides evidence of your work, the quality of your work, and the ethical standards by which you work. Every document you write is evidence of your integrity.

Essential to communicating ethically is to recognize the ways in which colleagues or supervisors on the job might violate or disregard standards of practice. Chief among the possibilities include plagiarizing, deliberately using imprecise or ambiguous language, manipulating statistics, using misleading visuals, promoting prejudice, failing to make information accessible, and distributing misinformation.

Plagiarism and theft of intellectual property. On the job, you may be responsible for the security of five kinds of intellectual property:

1. Copyrightable Material: A composition of original material fixed in a tangible medium, such as books, journals, software applications, computer programs, video or audio recordings, and illustrations. This includes materials available in digital files, e-mail messages, and website pages.

2. Trademark: A display of words or symbols communicated in text, illustrations, or sounds that identifies and distinguishes the goods and services of a manufacturer or supplier, such as the name or logo of a company.

3. Trade Secret: A design, formula, list, method, pattern, or process that offers a competitive advantage over parties who don't have the same information, such as a special recipe.

4. Invention: A new and unique design, device, method, or process that is subject to patent protection.

5. Tangible Research Property: Tangible items created during research related to copyrightable materials, trademarks, trade secrets, and inventions, such as databases, diagrams, drawings, notes, prototypes, samples, and associated equipment and supplies.

Copyrightable material is unique in that for certain purposes (e.g., criticism, news reporting, research, teaching) you have the right to borrow limited portions for presentation or publication without the explicit permission of the owner. If the borrowing is extensive, however, permission is necessary.

On the job, writers will often recycle the words and images from various documents of their company without identifying the original source. They will readily lift paragraphs from the corporate website, for example, to use in a business letter to a potential client, or they will borrow a table or a graph from the annual report to use again in a proposal to a potential funding agency. Such recycling of material—or *boilerplate*—is efficient, and it is entirely legal and ethical as long as the participating writers recognize and allow this sharing of effort. In such cases, the words and images that are being recycled are the intellectual property of the company itself. (If you have doubts about the propriety of such recycling within your company, ask the writer directly for his or her permission.)

If it isn't your company's materials that are being used, however, you must acknowledge the sources of borrowed words, images, and ideas. In the majority of documents you write (such as letters, e-mail messages, blog postings, and memos), the acknowledgment may be a brief and simple introduction to the borrowed material: for example, "As Dr. Shirley Olson discovered, it's possible to vaccinate mosquitoes to prevent their developing and passing the disease on to human beings."

In formal reports, however, a system of documentation appropriate to the field would be necessary to identify the source of the information and to give full credit to Olson. Your organization might develop a special style for such source citations or adopt a standard style guide such as *The Chicago Manual of Style* of the University of Chicago Press, the *Publication Manual of the American Psychological Association*, or the *MLA Style Manual* of the Modern Language Association.

To use the words, images, or ideas of others without attribution is *plagiarism*. It constitutes a theft of intellectual property and is highly unethical and potentially illegal. Your intentions are immaterial; that is, it would be plagiarism if it were deliberate or if it were entirely inadvertent. You must, therefore, be especially cautious to avoid plagiarism: your organization could find itself the subject of a criminal case or a civil suit, and you could lose your job and your reputation.

Note also that material is automatically copyrighted as soon as it is created: it need not carry a copyright notice, and the copyright need not be registered with the US Copyright Office (copyright.gov), though both actions will help to deter infringement.

If you quote a source (including material from e-mail messages, blog postings, or video clips), put the borrowed material inside quotation marks (or display it in a separate indented paragraph) and specify the source. If you paraphrase or summarize, you don't need quotation marks but you still must specify the source.

Make sure that even your acknowledged paraphrases and summaries differ from the wording and phrasing of the original passage so that you can't be accused of plagiarism.

Here, for example, is the original passage from a source:

> Ethylene oxide has a boiling point of 51°F. It is processed as a liquid through the application of pressure. As the temperature of the ethylene oxide increases, the pressure in the feed line will correspondingly increase. At the time of the explosion, the ambient temperature was around 93°F. The feed line was not insulated or cooled.

> (From United States Environmental Protection Agency, Office of Solid Waste and Emergency Response. 2000. EPA chemical accident investigation report, Accra Pac Group, Inc. North Plant, Elkhart, Indiana. EPA 550-R-00-001. Washington, DC.)

The following summary would constitute plagiarism even if the source were cited. Note how this passage changes words here and there but essentially duplicates the original passage:

> The boiling point of ethylene oxide is 51°F. Pressure is applied to process it as a liquid. As the temperature of the ethylene oxide rises, so does the pressure in the feed line. The temperature in the vicinity was roughly 93°F at the time of the explosion, and the feed line was neither insulated nor cooled.

The next summary would be considered ethically appropriate. Note how the order of the sentences as well as the words have been substantially changed. Nevertheless, the source of the information must still be identified:

> Ethylene oxide was pressurized as a liquid in a feed line that was neither insulated from external temperatures nor subjected to any kind of special cooling. Ethylene oxide boils at 51°, but the ambient temperature rose to approximately 93°F. Increased pressure from the boiling chemical inside the feed line caused the line to rupture, resulting in the explosion.

If the borrowing is a substantial portion of the original source (e.g., several paragraphs or a single image), you will have two ethical and legal duties:

1. Acknowledge the source.
2. Request permission from the owner of the intellectual property.

The extensive borrowing of copyrighted material will ordinarily be permitted but often with restrictions and often for a cost. You may contact the copyright owner directly or make your request through a service such as the Copyright Clearance Center (copyright.com). In this case, the borrowed material carries a note indicating both the source and the receipt of permission:

> From *Multiculturalism and Interculturalism: Debating the Dividing Lines,* by Nasar Meer, Tariq Modood, and Ricard Zapata-Barrero. Copyright 2016 by Oxford University Press. Reprinted by permission.

If you change the original material (e.g., summarizing passages, revising illustrations, pulling still slides from digital video), you would specify the adaptation of the borrowing:

From Lisk, Robin. (2016, August 10). Five Techniques to Relieve Anxiety [Video]. INN.com. http://www.inn.com/video/#/video/health/2016/08/10/hm.anxiety.inn. Adapted by permission.

Creative Commons (creativecommons.org) is a searchable source of materials (e.g., research reports, software, videos, music, photographs) that are specifically licensed by each creator for no-cost access and sharing. The only stipulations are that you must abide by the licensing conditions and you must cite the source of the material. Some materials are licensed to allow derivative creations or adaptations, while other materials may be modified only with the explicit permission of the creator.

For materials in the public domain (i.e., intellectual property for which copyright protection has already expired or material created by agencies of the government of the United States), no permission is necessary, but you must always cite the source of the information you are using.

Deliberately imprecise or ambiguous language. Unclear and ambiguous language is often just a result of a writer's negligence, but it could also indicate a deliberate effort to mislead or manipulate the reader by hiding or disguising information.

Writers can imply that things are better or worse than they really are through their choice of words. For example, a writer for a light bulb manufacturer could describe the company's newest product with the claim: "This light bulb was designed for nine years of operation." It could be that the bulb was designed with this objective in mind, but if it doesn't usually operate for this duration, the writer has created a deceit without telling a straight lie. And if this promised "nine years of operation" is contingent on the light operating only briefly each day, the deception is all the greater.

Negative assertions, in particular, will confuse and deceive readers. For example, consider a claim such as "No antivirus software does more or costs less than Infolator." With the words "does more" and "costs less" in the claim, you might be inclined to think that this software is truly better and lower in price. Notice, however, that this software doesn't really have to be better or lower in price to make its claim: it could have the same functions and be the same price as other antivirus software. The real claim here isn't that it's a better and cheaper product but that it's just as good and just as cheap.

Manipulation of numerical information. Manipulated statistics are a leading source of deception. For example, the writer of a recommendation report tries to give the impression that a controversial change in corporate policy on information privacy has wide support. She surveys only the twelve managers in the company;

seven complete the survey, and four of the seven approve of the change. In the recomtes, "A full 57 percent of those who completed the survey support the policy change." By not revealing that this "57 percent" is only four people and that it constitutes only a third of the company's managers, she magnifies the thin support for the controversial change. This isn't quite a lie, but it is certainly a deceptive distortion of the truth.

Use of misleading illustrations. Like words, illustrations have the capacity to misrepresent and mislead. For example, a company of one hundred employees has three who have physical disabilities. In the recruiting materials it carries to college campuses, the company displays photographs of a dozen of its employees doing different jobs, including all three of its people with physical disabilities. This portrayal constitutes unethical communication because it implies that people with physical disabilities constitute 25 percent of the employees—a gross distortion of the real situation. Prospective job candidates would be substantially deceived about the diversity of colleagues and the working environment this company offers.

Or consider a line graph, such as shown in Figure 3–3, that might be given to prospective clients of ABC Manufacturing. Here, we see a picture of volatile change year to year. If the increments on the line graph are changed from every year to every 4 years, however, ABC's disturbing record of erratic production almost vanishes entirely (see Figure 3–4). Viewers of this line graph would be given only the impression of relative stability. Prospective clients would be thoroughly deceived and would be quite surprised if ABC failed to meet its production goals in a subsequent year.

Promotion of prejudice. Writers also communicate unethically by voicing prejudice through their choice of words and illustrations. If you use titles for men but not for women (e.g., Mr. William Jones, advertising manager, and Harriet Smith, operations manager), you make women seem less credible and authoritative. If photographs in the company's annual report always show women sitting and staring at computer screens while men sit at big conference tables listening to reports and making decisions, the pictures imply that men occupy (or deserve) important positions as executives while women are suited to clerical positions.

As a communicator, make sure that you don't reinforce or inspire prejudice and bigotry. Your ethical obligation requires you to offer only valid and reliable findings, fair and unbiased analyses, and logically justified conclusions.

Failing to make information accessible. Your job as a communicator is to make pertinent information readily and equally available to your audience regardless of their physical abilities. This obligation applies to text and images as well as audio and video materials. Make your text easy for assistive technologies to navigate by keeping the display as simple as possible. Support your illustrations with textual descriptions of the images. Compose closed captioning for audio materials, and

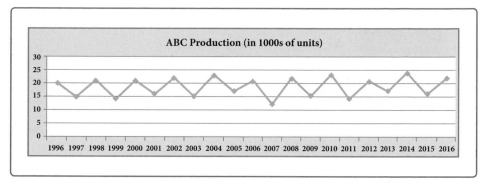

• **FIGURE 3–3** ABC Production Reported in 1-Year Intervals

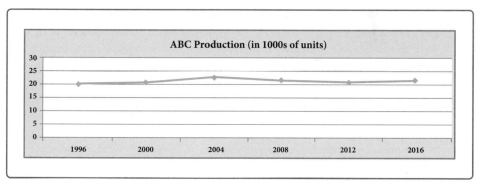

• **FIGURE 3–4** ABC Production Reported in 4-Year Intervals

supplement video materials with closed captioning (for people of limited hearing) or audio descriptions (for people of limited vision). Failure to make information accessible injures the individuals thus excluded and indicates your disregard for their merits and dignity. Your insensitivity also damages your reputation and the reputation of your organization.

Uncritical use of information. On the Internet especially, you will find all kinds of erroneous information, manipulated images, distorted depictions, and dubious claims. If you use the material you find without judging its accuracy, assessing the credibility of your sources, or verifying validity across multiple sources, you could easily be distributing dangerous and damaging misinformation. Your negligence here makes you look both unethical and inept. To build a reputation for integrity, you must exercise caution and accept responsibility for the accuracy of the information you communicate.

| CASE 3–1 | **Accuracy in Reporting** |

You currently work as an environmental engineer for Ventus Power, which operates wind energy facilities in five states: Arizona, Nevada, New Mexico, Texas, and Wyoming. Included in your responsibilities is monitoring of wildlife impact at the Sweetwater facility and the periodic preparation of reports to be submitted to the US Department of Energy and the US Fish and Wildlife Service (USFWS).

You have just prepared a report indicating that fatal bird collisions with the wind turbines at Sweetwater are at a new high for the third consecutive quarter. You know this evidence of a persistent and growing hazard to migratory birds will trigger a visit from regulators and a potential interruption of operations. And mitigation efforts will likely cost the company at least $500,000.

Your supervisor, Garrett Driscoll, says the higher level could be regarded as a "temporary blip" and asks you to "adjust" the data so that the facility appears to be running safely. He explains, "We can't afford the expensive repair. It might also cost us jobs. And the publicity will be terrible—antagonizing the environmentalists and discouraging investors."

You decide to write a memo to Driscoll telling him that, as a professional engineer, the National Society of Professional Engineers (NSPE) code of ethics does not allow you to misrepresent information. You must report this information as you know it. You recognize that Driscoll will disagree with your decision and will likely share the letter with his bosses. In your message to Driscoll, you explain why you must report this information according to your profession's standards of practice. Before you compose your letter, you again review the NSPE ethics guidelines (Figure 3–1).

Question: Why should you write to Driscoll instead of just telling him?

CASE DOCUMENT 3–1

DATE: October 24, 2016
TO: Garrett Driscoll
SUBJECT: **High Levels of Fatal Bird Collisions at Sweetwater**

Garrett:

I understand your concern regarding the cost of solving our facility's problem with fatal bird collisions. The $500,000 reparation expenditure will indeed affect our company financially. However, the high mortality rate cannot be viewed as a mere technicality. The facility's unintentional bird kills are again at a new high, and the number will persist or grow if we do not take measures to mitigate. Excessive fatal collisions put us in legal jeopardy as well as do callous injury to bird populations. Adjusting the data so that the facility appears to be operating without incident would be an explicit violation of an engineer's ethical standards.

As an environmental engineer, I adhere to the Code of Ethics instituted by the National Society of Professional Engineers. I will report this data to the USFWS and the Department of Energy as is. As soon as we have the okay from the corporate office for expenditures, I will assess the situation and determine a suitable fix.

I believe that accurately reporting the statistics will positively contribute to our reputation and the integrity of this company.

Questions. Do you believe this response will be satisfactory? How could you improve it?

CASE 3–2   Teaching Ethics by Example

You complete your report and submit it to the vice president of operations for signatures and submission to the USFWS. You also keep a copy on your computer. You are satisfied that you have followed the actions required by the NSPE code of ethics. Driscoll is annoyed but quiet, and you assume that he has resigned himself to your decision. Soon thereafter you leave Sweetwater for a supervisory position at the Albuquerque facility.

A year later, you read in the newspaper that the US Department of Justice is investigating Ventus Power for violating the Migratory Bird Treaty Act. It appears that scores of fatal collisions of migratory birds, including eagles, at the Sweetwater Wind Facility were never reported to the Fish and Wildlife Service and no mitigation efforts were instituted.

Almost immediately you find yourself interrogated by investigators from the Department of Justice and the USFWS as well as corporate officials at Ventus Power. Wisely, you still have a copy of your report on your computer as well as a copy of your message to Driscoll.

After both legal entities clear you of any responsibility for the failure to report, you decide to write an internal memo to the engineers who answer to you in your new job about the importance of effective communication and of following engineering ethics in all situations.

CASE DOCUMENT 3–2

Date: Monday, December 12, 2017
To: Ventus Engineering Corps
Subject: Proper Reporting Procedures for Ethics Violations and Other Issues

Purpose

A recent incident with our corporation has prompted me to write this open memo to the engineering corps of our company to explain proper reporting procedures. In case you were unaware, Ventus Power has been cited by the US Department of Justice for violations of the Migratory Bird Treaty Act in failing to report fatal collisions of migratory birds (including eagles) with the turbines at the Sweetwater Wind Power Facility. An investigation by the company identified the persons at fault, and they have been appropriately reprimanded. As some of you may know, I served at one point as the engineer in charge of wildlife impact monitoring at the Sweetwater facility and, as such, found myself one of the first to be questioned about bird collisions. I had to report unacceptable mortality levels and advised mitigation efforts while in my position. Because I followed proper reporting procedure, the company will not bring any litigation or punishments against me. The following guidelines make sure that necessary information makes its way to the right people and that we deal with problems properly before they can harm the people or environments we operate around. These guidelines can also safeguard you.

Introduction

At some point in your career you may have to report an issue you are experiencing, especially issues of ethical violations. Remember that we are bound at all times to our code of ethics. These guidelines keep us, our customers, and our company safe. Operations or procedures done in an unethical manner may lead to people being hurt or killed, can bring litigation against our company, and will lead to the termination of any employees found in violation. For this reason, it is extremely important that everyone follow proper procedures for communications and reporting so that problems can be dealt with as quickly as possible and so that you can be sure you will not be considered a responsible party.

Everyday Communication

These communications include project reports and other regular incremental reports. They need to be written as simply as possible with an emphasis on any items or issues that need immediate attention from your supervisor. Supervisors use their direct subordinates as their eyes and ears on projects, looking for any problems or issues that will require their action. Remember, if something needs to be fixed, it should be mentioned. These simple guidelines should cover nearly 99% of reports you will find yourself writing, but there are special cases that require a more involved approach.

Special Cases

These special cases occur whenever a serious problem exists. This should be obvious, such as anything that could lead to lawsuits against our company, any time product quality is not at an acceptable level, and especially any time you feel you are being asked to violate the code of ethics. These situations require your full attention. For these reports, tell events exactly as they appear to you, including background information that clearly defines the problem you are facing, any efforts you have made to correct the problem, and any assistance you need to solve it. Send copies of your letter to whoever needs to know this information. Normally this will be your supervisor and anyone else in your department who works with the affected system. If your supervisor asks you to cover up or ignore this problem, it is your responsibility to report the problem up the chain of command to your supervisor's supervisor since that is what they were supposed to do. You can decide how to handle your supervisor's response, although if you feel that he or she has truly violated the code of ethics, it is your ethical responsibility to report it.

Conclusion

In conclusion, I appreciate the hard work every one of you puts into your job and I hope that this level of dedication continues. With proper reporting procedures, we can make sure that all problems are dealt with properly and that everyone remains safe. And as a warning to those who think that sweeping an issue under the carpet or agreeing to a supervisor's request to do so is acceptable, it is possible that nothing will happen to you in the short run, but problems always find their way to the surface, and once they do it's too late to protect yourself. So be smart and keep our customers, our company, and your career safe by always reporting issues. Thank you for your time and continued efforts to make this company the best that it can be.

Questions: How effective is this memo at explaining the importance of following a code of conduct, of acting ethically, and of communicating clearly? What changes would make it more persuasive or inspiring for the intended readers?

Managing Unethical Situations

Always keep in mind that unethical behavior has potentially serious consequences. It could damage your reputation as well as cost you your job and your ability to find a new job. You could forfeit the right to practice your profession. You could also be subject to civil suits for damages caused by your unethical behavior. If your actions were also illegal, you could be fined or jailed.

If you find yourself asked by a boss, colleague, or client to do something that you consider unethical, therefore, don't hesitate to ask polite questions. Don't make accusations or immediately declare the request unethical. Instead, explain your dilemma carefully. Ask the individual to clarify the request. He or she might be unaware of the difficult moral position you're being put in by such a request or might immediately modify the request.

Note also that people of different cultures might have different ethical perspectives. For example, the dominant ethical perspective in your culture might assume that to be fair you must consider all people as equal and give privilege to none, but in your colleague's culture this idea might be considered unfair because family, friends, and neighbors are ordinarily thought to deserve special consideration. In this situation, you would be wise to acknowledge each side's intention to be ethical as you try to find a solution that is satisfactory to both. Here again it might be a case of competing right answers instead of a question of right or wrong.

If explanations and negotiations don't satisfy you and if time allows, visit with a supervisor or mentor for guidance. Review the code of conduct of your profession or company for passages that might support or challenge your position.

If your investigation and deliberation fail to quiet your moral doubts, explain (in writing, if possible) that you don't feel comfortable doing X but you could do Y. In other words, identify both the thing you can't do (the unethical action solicited) and the thing that you could do (the ethical alternative).

If asked to justify your decision, cite the appropriate passages of your profession's or company's code of conduct. Make it clear that it isn't only you who is rejecting the request but that it is your profession or your company that proscribes this behavior. Again, don't be impolite and don't accuse.

If you see something occurring on the job that you think is unethical, discuss the situation with your supervisor or mentor (in writing, if possible). If your investigation fails to satisfy you that the activity is justified, consider your several ethical obligations, including to yourself, to your profession, and to the public. Always make the decision that you could live with if your decision were made public—the decision that would make your parents and teachers proud of you.

As a final piece of advice, keep in mind that you bring two important credentials to a job: a knowledge of your field and a reputation for integrity. If you don't have a reputation for integrity, your knowledge of your field can't be trusted.

✎ Ethics Decision Checklist

☐ What is the nature of the ethical dilemma?

☐ What are the specific aspects of this dilemma that make you uncomfortable?

☐ What are your competing obligations in this dilemma?

☐ What advice does a trusted supervisor or mentor offer?

☐ Does your company's code of conduct address this issue?

☐ Does your professional association's code of conduct address this issue?

☐ What are you unwilling to do? What are you willing to do?

☐ How will you explain or justify your decision?

 EXERCISES

1. Visit the website of two major employers in your field. Locate the code of conduct for each. What are the similarities and differences in the two codes of conduct? What are the values that each employer espouses? What does each employer expect from its employees? What does each employer expect from its executives? What additional ethics resources does each employer make

available to employees? What can you tell about each employer from the code of conduct? Based on your review of the two codes of conduct, which employer would you prefer to work for? Why? Summarize your findings in a memo, and share this memo with your colleagues.

2. The Ethisphere Institute (ethisphere.com) every year publishes a list of the world's most ethical companies. Check this year's list, and choose one of the companies to investigate. What makes it one of the world's most ethical companies? Do you agree or disagree with this assessment? Report your findings in a slide presentation addressed to majors in your field.

3. You have been assigned as the project manager for a new bridge that your company hopes to build in a foreign city. You visit that city with your company's vice president of international operations, scouting the site for the bridge and meeting with city officials in preparation for making a bid on the project. In a private meeting with the minister of public projects, the vice president pulls out a platinum wristwatch in a box from Tiffany's and gives it to the minister. "This is for you from the president of my company!" announces the vice president. The minister seems quite pleased with the expensive gift and assures the vice president that the bid from your company would be "positively winning" or something to that effect. You can't swear to it because the minister's English is inexact, but it does seem that your company is being promised the bridge project.

   You know that your company has a policy barring gifts to foreign officials, but you don't know that it necessarily applies in this case. In the taxi on the way to the building site, you ask the vice president about the gift. He advises you that policies are "nice words" but that "this is how we have to do business here" and mentions that the minister and the president of the company are personal friends. He smiles and changes the subject of conversation, but you are still worried that the gift giving was unethical and might also be illegal.

   Managing this project would give a real boost to your career. It could lead to more international opportunities and higher visibility in your profession as well as executive positions at your company and possibly job offers from prestigious competitors. It could also terminate your career if it were determined that you were involved in bribing a foreign official.

   What should be your next steps in addressing this dilemma? What sources of information could you consult—about applicable laws, about ethics, about intercultural communication? Who could you talk to? What documents or messages, if any, should you write?

4. The US Centers for Disease Control publishes a report with a one-page infographic summarizing key medical advice regarding women and the drinking of alcohol during pregnancy (see Figure 3–5). Aimed at preventing fetal alcohol syndrome, the infographic urges women of childbearing age to avoid drinking alcohol if engaging in sex without using birth control. A firestorm erupts

almost immediately on social media with accusations that the CDC is depicting women through words and images in ways that are prejudicial—as responsible for their injuries if victims of violence, as exclusively responsible if infected with a sexually transmitted disease, as existing only to be carriers of babies, and as sexually active only if slim and trim.

If you were a technical communicator at the CDC responsible for monitoring social media, how would you address the comments of the critics?

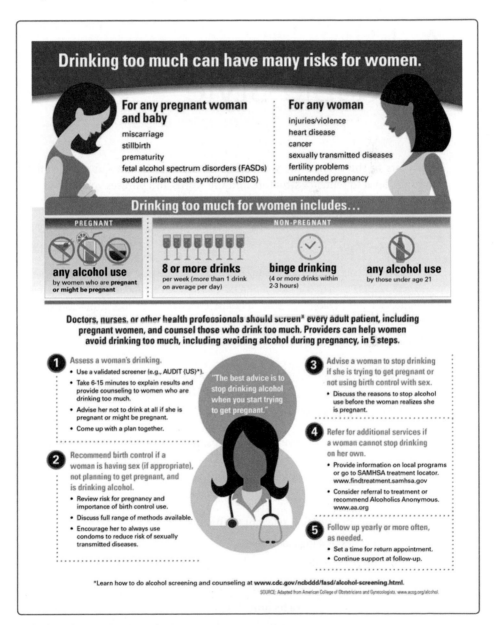

• **FIGURE 3–5** Document for Exercise 4

4

Achieving a Readable Style

Style refers to the overall way you express your ideas in a document, from the paragraph level to sentence structure and the words and phrases you choose.

 Quick Tips

If you want your report to be read, use a style that your readers can follow easily. If your readers can't understand your report as they read it and must constantly reread sentences and paragraphs, they may just disregard it or toss it out. Unreadable documents usually result from ineffective style.

The Paragraph

We define a *paragraph* as a group of sentences that begins with a statement of their central idea. The supporting sentences build on the idea stated in this topic sentence and should occur in a logical order. In short:

- Begin each paragraph with a topic sentence that summarizes the ideas to come.
- Include only information relevant to the topic sentence.
- Place sentences in a logical order.
- Avoid long paragraphs.

Examples for study. Effective report segments result from effective paragraphs. Examine the introduction to a technical report shown in Figure 4–1. Because each paragraph begins with a topic sentence (easy to locate and retrieve), you can read

Introduction

In 1996, the U.S. Department of Energy (DOE), Office of Fossil Energy, asked Argonne National Laboratory (ANL) to conduct a preliminary technical and legal evaluation of disposing of nonhazardous oil field wastes (NOW) into salt caverns. The conclusions of that study, based on preliminary research, were that disposal of oil field wastes into salt caverns is feasible and legal. If caverns are sited and designed well, operated carefully, closed properly, and monitored routinely, they can be a suitable means for disposing of oil field waste (Veil et al. 1996). Considering these findings and the increased U.S. interest in using salt caverns for nonhazardous oil field waste disposal, the Office of Fossil Energy asked ANL to conduct a preliminary identification and investigation of the risks associated with such disposal.

Report Purpose

As Chapter 8 discusses, report introductions must have purpose statements to tell readers what the report will do. Having the purpose statement at the beginning of a paragraph helps readers find the purpose statement.

The purpose of this report is to evaluate the possibility that adverse human health effects (carcinogenic and noncarcinogenic) could result from exposure to contaminants released from the caverns in domal salt formations used for nonhazardous oil field waste disposal. The evaluation assumes normal operations but considers the possibility of leaks in cavern seals and cavern walls during the post-closure phase of operation. It does not consider the risks associated with emissions from surface equipment operating at the site, nor does it consider the risks associated with surface oil leaks or other equipment-related spills or accidents.

The study focuses on possible long-term risks to human health. It does not address potential ecological effects, although such effects could result. Also, risks associated with naturally occurring radioactive materials (NORM) are not addressed. This preliminary assessment estimates risks associated with disposal in a single generic cavern only. No attempt has been made to address the possibility or likelihood that several caverns may be located in relatively close proximity and that more than one cavern could be a source of contamination to a given receptor. Also, no attempt has been made to evaluate the possible impacts of synergistic effects of multiple contaminants on a single receptor.

Because the history of salt cavern use for solid waste disposal is very limited, no readily available data could be accessed for this study. As a result, data from similar operations and professional judgment were used to develop the possible release mechanisms assumed in this hypothetical, generic analysis. The validity of the results would be enhanced if real data could be used. As data are generated on the use and post-closure operations of salt caverns used for solid waste disposal, they should be incorporated to update this study.

Report Development Process

Moves from broad to specific in supporting information

In this assessment, several steps were followed to identify possible human health risks. At the broadest level, these steps include identifying a reasonable set of contaminants of possible concern, identifying how humans could be exposed to these contaminants, assessing the toxicities of these contaminants, estimating their intakes, and characterizing their associated human health risks. The risk assessment methodology and

• FIGURE 4–1 Effective Introduction to a Technical Report

techniques used in this report are based in large part on two documents. The first document is a training manual that was developed for a risk assessment workshop sponsored by DOE (DOE 1996). The second is the Risk Assessment Guidance for Superfund (U.S. Environmental Protection Agency [EPA] 1989).

Report Plan

The remainder of this report consists of nine sections. Section 2 provides background on the development, use, and closure of salt caverns that may be used for disposal of nonhazardous oil field wastes and possible cavern release scenarios. Section 3 identifies contaminants of potential concern that could cause harm to human health. Sections 4, 5, and 6 provide information for assessing potential exposure pathways that the contaminants of concern could take to reach human populations. Specifically, Section 4 describes fate and transport mechanisms of the contaminants of concern; Section 5 describes specific hydrogeologic conditions of locations where salt caverns are most likely to be used for oil field disposal (Gulf Coast, Texas, and New Mexico); and Section 6 describes potential release modes that could cause contaminants to leak from the cavern and be transported to areas where human populations may be exposed. Section 6 also estimates possible concentrations of the contaminants to which humans could be exposed under various release scenarios. Section 7 describes the toxicity of those contaminants that could come in contact with humans, given the fate and transport mechanisms identified in Section 5 combined with the potential exposure pathways described in Section 6. Section 8 estimates the potential intakes of those contaminants by humans and characterizes the risks to which those humans may be subjected on the basis of the intake of the contaminants (the potential for harm), their toxicities, and the release assumptions. Section 9 addresses the sensitivity of the estimated risks to operating procedures and potential regulatory structures, and Section 10 summarizes the results of the analyses.

> Topic sentence helps readers anticipate presentation of the nine sections. Naming each section allows the writer to show relationships among the sections.

• **FIGURE 4–1**   (Continues)

just the topic sentences and have a clear sense of the introduction. While the final paragraph is the longest of this introduction, its structure allows you to follow easily the development of the paragraph.

The next example paragraph (Figure 4–2) uses a list to draw the reader's eyes to the central idea presented in the paragraph, the criteria for approval. The topic sentence introduces the paragraph and the list.

The listing strategy emphasizes the six items that take front and center in the Ashton Lake project. In this situation, listing highlights the rhetorical immediacy of the problem. A traditional paragraph with linear text would obscure the concerns (Figure 4–3).

Avoid excessive use of any writing technique—too many short paragraphs, too much enumeration (first, second, third, etc.), as well as too many lists. Concise paragraphs that begin with topic sentences and well-structured sentences of moderate length create clear, readable documents. Compare Figures 4–4 and 4–5. Which do you find easier to read? Why?

The City of Ashtonville Economic Development Council presented six main concerns to the City Parks and Recreation Department about the proposed Ashton Lake development proposal:

1. Financial feasibility of the project
2. Ability to raise the necessary funds
3. Project maintenance, including long-term
4. Protection/Enhancement of the shoreline
5. Effect of project on changing lake water levels
6. Erosion of project's materials caused by water/sand

The EDC expressed a major concern that the project may not be as family-friendly as needed because it would require extensive funding to develop the recommended recreation facilities. In addition, the concerns stated suggest that the project, if approved, will have to be funded strictly by private donations rather than city or county parks money.

• **FIGURE 4–2** Effective Use of a List Within a Paragraph

The City of Ashtonville Economic Development Council presented six main concerns to the City Parks and Recreation Department about the proposed Ashton Lake development proposal: financial feasibility of the project; ability to raise the necessary funds; project maintenance, including long-term; protection/enhancement of the shoreline; effect of project on changing lake water levels; and erosion of project's materials caused by water/sand. The EDC expressed a major concern that the project may not be as family-friendly as needed because it would require extensive funding to develop the recommended recreation facilities. In addition, the concerns stated suggest that the project, if approved, will have to be funded strictly by private donations rather than city or county parks money.

• **FIGURE 4–3** Ineffective Run-in List

Basic Principles of Effective Style

Effective writers adjust their style to the needs of their readers: (1) readers' knowledge of the subject, (2) readers' expectations about style based on the specific kind of writing, (3) readers' probable reading level based on the context in which they will read the document, and (4) the writer's relationship to readers—the professional roles of both writer and readers.

Determine your readers' knowledge of the subject. The reader's familiarity with the subject will determine how many specialized terms you can use. If the reader has a thorough knowledge of the subject, you can use acronyms, specialized nomenclature, and jargon that readers in a specific discipline regularly read and use. If the reader has little knowledge of the subject, limit the use of specialized vocabulary or perhaps define the terms. Another possibility is to substitute phrases or words that clearly express your meaning.

How an MRI Works

Brain perfusion refers to the blood circulation around the brain while exchanging oxygen and nutrients between the blood and the brain tissue. The effectiveness of brain perfusion depends on "blood pressure, blood velocity [...], and diffusion rates of oxygen and nutrients" [1]. An MRI scanner can monitor the perfusion of the brain to determine if a neurological disease is present.

An MRI scan uses a magnetic field to generate enhanced images that show certain parts of the body. The hydrogen atom provides a crucial component in the generation of an MRI scan. The hydrogen atom contains a single proton that interacts with the magnetic field generated by the MRI. The hydrogen atoms in a specific area of the human body will behave differently due to the presence of a magnetic field. The MRI scanner will detect the behavior of these hydrogen atoms to generate an enhanced image of that area.

In many MRI procedures, a contrast agent is used to further enhance the image of the body by increasing the brightness of the tissue being examined. The injection of contrast agents into the body will help the MRI scan retrieve information such as "blood flow [...] and related physiological parameters" [2]. These contrast agents work by altering the local magnetic field in the tissue being examined to generate an illuminated image of that tissue on the MRI scan.

Unfortunately, contrast agents have limited functions when physicians attempt to study the human brain. The brain contains a "membrane structure" known as the blood brain barrier, which protects the brain from foreign chemicals in the blood, while still allowing essential metabolic function [3]. The blood brain barrier filters out any unwanted chemicals that may travel through the brain. Since the blood brain barrier is very effective in its function, the chemicals in the contrast agents will be filtered out by this *membrane structure*. If a neurological disease is present in the brain, it can weaken the blood brain barrier and allow these unwanted chemicals to pass through [3]. The contrast agent would be useful for studying the disease in this situation.

Many methods have been created to penetrate the blood brain barrier. My proposed method will strengthen the interaction between hydrogen atoms with the magnetic field via electrical signals. During an MRI scan, an electrical signal will be emitted around the neck area. This signal will mark the hydrogen atoms that are in the vicinity of the emitted signal. Because no external chemicals are introduced, the blood brain barrier will not detect any foreign chemicals and will let the marked hydrogen atoms into the brain. The hydrogen atoms will travel through the bloodstream around the brain acting as beacons. They will then interact with the magnetic field of the MRI, allowing the MRI to enhance the image of the brain, much like contrast agents [4].

● **FIGURE 4–4**   Original Introduction to a Research Proposal

In a work context, the better you know the people who will read your writing, the better you can design your reports, memos, and e-mails. You can also ask people who know your readers to aid you in finding out preferences in organization, style, and length. When you begin a job assignment, ask your supervisor for his or her preferences in these three areas.

Determine whether a particular style will be expected. Use the company style sheet and templates. But remember that you still must adapt what you say to your intended readers.

Adjust the style to the readers, the purpose, and the context. Most business and technical writing should strive for as much conciseness as possible because of the large quantity of information that most readers confront. E-mail messages should have concise paragraphs and concise sentences. Even in complex, highly technical reports, readers value conciseness: the longer the report, the less likely that anyone will read all of it.

Introduction

Importance of Brain Perfusion

Brain perfusion refers to the blood circulation around the brain while the circulatory system exchanges oxygen and nutrients between the blood and the brain tissue. Brain perfusion depends on "blood pressure, blood velocity [...], and diffusion (dispersal) rates of oxygen and nutrients" [1]. An MRI scanner can monitor the perfusion of the brain to determine if a neurological disease is present.

How an MRI Works

A magnetic resonance imaging (MRI) scan uses a magnetic field (an attractive force) to show enhanced images of certain parts of the body. A dynamic MRI scan displays a fuller and more detailed result, unlike a conventional MRI. The hydrogen atoms in the human body are very important in an MRI procedure. One hydrogen atom contains a single proton that behaves differently when the MRI emits a magnetic field around that area. The MRI scanner will detect the behavior of these hydrogen atoms to generate an enhanced image of that area.

In many MRI procedures, a contrast agent (a compound fluid) is used to further enhance the image of the body by increasing the brightness of the tissue being examined. The injection of contrast agents into the body will help the MRI scan retrieve information such as blood flow and other related physical characteristics [2].

Limits of the MRI

Unfortunately, contrast agents have limited functions when physicians attempt to study the human brain. The brain contains a "membrane structure" known as the blood brain barrier, which protects the brain from foreign chemicals in the blood, while still allowing essential metabolic function [3]. The blood brain barrier filters out any unwanted chemicals that may travel through the brain. Since the blood brain barrier is very effective in its function, the chemicals in the contrast agents will be filtered out by this *membrane structure*. If a neurological disease is present in the brain, it can weaken the blood brain barrier and allow these unwanted chemicals to pass through [3]. The contrast agent would be useful for studying the disease in this situation.

Value of My Method in Improving the MRI

Many methods have been created to penetrate the blood brain barrier. My proposed method will strengthen the interaction between hydrogen atoms with the magnetic field via electrical signals. During a dynamic MRI scan, an electrical signal will be emitted around the neck area. This signal will mark the hydrogen atoms that are in the vicinity of the emitted signal. Because no external chemicals are introduced, the blood brain barrier will not detect any foreign chemicals and will let the marked hydrogen atoms into the brain. The hydrogen atoms will travel through the bloodstream around the brain acting as beacons. They will then interact with the magnetic field of the MRI, allowing the MRI to enhance the image of the brain, much like contrast agents [4].

Subject and Purpose

This research will focus on developing a new method of performing a dynamic MRI scan on the human brain. This new method involves emitting an electrical signal into the body to interact with the hydrogen atoms and performing a dynamic MRI scan. To determine the effectiveness of this new method, I will analyze the MRI scan to determine the rate of perfusion in the brain.

• **FIGURE 4–5**   Revised Introduction to a Research Proposal

 Note: Conciseness does not equal brevity. When you write concisely, you include all that you need to say without extra words and phrases that contribute little to the main idea. Brevity means that you aim only for economy, rather than completeness of thought.

Keys to Building Effective Sentences

English sentences work more effectively—they become easier to read, understand, and remember—if they follow five basic guidelines.

1. Watch sentence length.
2. Keep subjects and verbs close together.
3. Avoid pompous language; write to express, not to impress.
4. Avoid excessive use of *is/are* verb forms.
5. Use active voice for clarity.

Watch sentence length. Documents composed of consistently long sentences can become difficult to read. Sentence length should vary, but consider revising sentences longer than fifteen to twenty words. Even legal documents can benefit from shorter sentences and have improved as a result of Plain Language laws that now govern insurance policies and many other legal documents. Many government entities want their public documents written in concise, easily understood sentences. For example,

Before:
This Appendix contains a brief discussion of certain economic and demographic characteristics of the Area in which the County is located and does not constitute a part of this Official Statement. Information in this Appendix has been obtained from the sources noted, which are believed to be reliable, although no investigation has been made to verify the accuracy of such information.

Characteristics of Bad and Good Writing

Bad Writing	Good Writing
Few verbs/number of words per sentence	Many verbs/number of words per sentence
Excessive *is/are* verb forms	Concrete verbs
Abstract nouns	Concrete nouns
Many prepositional phrases	Few prepositional phrases
Few clauses	Linked clauses
Passive voice	Active voice
Separation of key words: subject-verb, actor-action	Clear actor–action relationship
Long, rambling sentences	Specific, precise sentences
Main idea of sentences difficult to process	Meaning of sentences is easy to find and follow
Sentences must be read several times	Meaning is clear after one reading

After:
This Appendix contains a brief discussion of certain economic and demographic characteristics of the Area in which the County is located. The Appendix does not constitute part of this Official Statement. Information in the Appendix has been obtained from the sources noted. They are believed to be reliable. However, the accuracy of the information has not been verified.

Keep subjects and verbs close together. A recipe for sentence clarity: keep the subject of the sentence and the verb close together, and emphasize verbs. The more verbs in a sentence, the sharper and more direct the sentence. We call this characteristic *verb/word ratio*. For example,

John <u>loves</u> Mary because she <u>inherited</u> money. (verb/word ratio = 2/7)

versus

Mary's inheritance of money <u>was</u> one of the reasons for John's interest in Mary. (verb/word ratio = 1/12)

In this simple example, you can see the point: the more verbs, the sharper the sentence.

Let's take this method a step further: lengthy sentences become less distracting to the reader if the writer structures them to enhance clarity and readability. To achieve clarity, build sentences with clauses and as many verbs and verbals (to + verb; verb ending in "ing") as possible. For example,

When they plan investment portfolios, financial planners recommend a variety of investments because they resist rapid economic changes. (18 words)

This sentence includes three clauses:

When they <u>plan</u> investment portfolios
financial planners <u>recommend</u> a variety of investments
because they <u>resist</u> economic change.

Note that the sentence follows the three guidelines:

Interlocking clauses (three in this sentence)
Specific action verbs: *plan, recommend*, and *resist*
Subject next to the verb in each clause:

they plan
planners recommend
they resist

The sentence has a verb/word ratio of 3/17.

If the writer did not follow the guidelines and avoided verbs, the sentence might read like this:

In plans for investment portfolios, a variety of investments is recommended by financial planners because of their resistance to economic changes.

In this sentence the verb/word ratio is 1/21. The sentence lacks directness and conciseness. Compare the two versions. Can you see the difference? In English sentences, the more verbs and verbals a writer uses, the easier it is for the reader to understand the sentence(s). In addition, make the actor the subject and follow it with a concrete verb that identifies what the actor does.

For most writing, use specific concrete subjects and verbs.

Instead of:
There <u>is</u> now no effective existing mechanism for introducing into the beginning initiation and development stages requirements on how <u>to guide</u> employees on how <u>to minimize</u> errors in product development efforts.

(verb/word ratio = 3/31; one *is* verb and two verbals)

Use:
The company <u>has</u> no way <u>to guide</u> employees on how <u>to minimize</u> product development errors during the early development stages.

(Note that the sentence begins with the actor in the subject position and two verbals. verb/word ratio = 3/20)

When a sentence lacks a clear subject/agent doing the action (verb), writers can often drift into the phrases "there is," "there are," "there was," and "there were," which have no meaning and deter conciseness and directness.

Instead of:
Our lack of pertinent data <u>prevented</u> determination of committee action effectiveness in funding target areas that <u>needed</u> assistance the most.

(Note that the sentence incorporates two clauses and two verbs. verb/word ratio = 2/21)

Or, even worse:
There was a lack of pertinent data determination of committee action effectiveness in funding target areas that needed assistance the most.

Use, assuming we know who did what action:
Because we <u>lacked</u> pertinent data, our committee <u>could not determine</u> whether we <u>had targeted</u> funds to areas that <u>needed</u> assistance the most.

(Note the revision of the sentence has four interlocking clauses and four verbs; placing clear subjects close to their verbs makes clear who does what action. verb/word ratio = 4/22)

Or use:
We <u>didn't have</u> enough data: we <u>could not decide</u> if the committee <u>had sent</u> funds to areas that <u>needed</u> them most.

When we break the sentence into two sentences, we still have four verbs. In addition, the short sentence followed by the longer, explanatory sentence also clarifies the meaning.

Note that in this revision we place the actor as the subject of the sentence/clause and then follow the subject with the verb—a concrete, descriptive verb that explains what the actor/subject will do.

Avoid pompous language; write to express, not to impress. The concept of simplicity relates to the concept of naturalness. Writers often believe they must sound learned, aloof, and sophisticated to impress readers. The idea that direct writing lacks sophistication frequently derives from writing done in secondary school. Teachers encourage high school students to expand their vocabularies. Academic writing in college reinforces the importance of using jargon-laden language to convince the professor that the student knows the subject and the terminology of the discipline. Instructors may reward students for using bookish language in research papers and may view long and complicated sentences as evidence of a student's skill with grammar and mechanics. On the job, however, verbose writing may be ignored or misread by readers who are interested in gleaning information relevant to their job needs as quickly as possible.

Remember how direct/indirect words affect tone:

(A) We encourage you to anticipate the amount of correspondence you accumulate and suggest you endeavor to answer it promptly.

(B) Please expect large amounts of e-mail and try to answer it quickly.

Note that you can read (B) more easily and quickly than (A). The tone of (A) sounds pompous.

Remember that writing exists for human beings, and few of us enjoy writing that seems harder to read than it needs to be. What constitutes "difficult" writing depends on the reader, the topic, and the purpose of the document. But direct, concise writing that uses a conversational style will usually be appreciated by your readers. Using shorter rather than longer sentences also helps readers follow your thoughts:

Please give immediate attention to insure that the pages of all documents prepared for distribution are numbered sequentially and in a place of optimum visibility. This is needed to facilitate our ability to refer to items during meetings.

versus

Please correctly number the pages of all documents. Place numbers in the upper right-hand corner. Sequential numbering helps us locate material during meetings.

or

Please number all pages sequentially.

Three additional examples:

1. It has recently been brought to my attention that only a small percentage of the employees in our division are contributors to the citizens' health research fund supported

by this firm. This fund is a major source of money for the encouragement of significant discoveries and innovations made in behalf of research relevant to community health.

versus

I have discovered that only a small percentage of employees in our division contribute to the citizens' health research fund. Our firm supports this research because the products of this research improve community health.

2. As a result of their expertise, the consulting team is provided with the opportunity to make a reasonable determination of the appropriate direction to proceed regarding their selection of information systems.

versus

The consulting team has the expertise to select the best information systems.

3. It is our contention that the necessary modifications should be made to make the system operational because its complete replacement is economically prohibitive.

versus

We believe that the system should be modified to make it operational. Complete replacement costs too much.

Avoid excessive use of _is/are_ verb forms. Choosing specific, concrete verbs for clarity means avoiding forms of the "be" verb, if possible. As the following sentences illustrate, excessive use of "be" verbs often obscures action verbs. Many times, a "be" verb is the best choice (as this sentence exemplifies). However, you can lessen the tendency to use "be" verbs by doing the following:

- Avoid beginning sentences with _there is_ or _there are, there was_ or _there were._
- Avoid beginning sentences with phrases such as _it is clear that, it is evident that,_ and _it should be noted that._
- Choose a specific verb rather than _is, are, was,_ and _were_ verb forms.

"Be" verbs often create a longer, less direct sentence:

Delegation is a means of lessening the manager's workload.

versus

Managers who delegate <u>reduce</u> their workload.

Examine the following three examples:

1. Our appraisal system is broken: it is inefficient, it is unfair, and it is costly.

versus

Our appraisal system lacks efficiency, fairness, and thrift.

2. My decision <u>is based on the assumption</u> that he will soon be retiring.

versus

I <u>assumed</u> he will soon retire.

3. Our office <u>has been provided with the authority to be instrumental in the selection</u> of a new computing system.

versus

Our office will select the new computing system.

As these examples and the ones that follow show, the clearest sentences focus on the agent and the action (the verb):

1. There are two systems presently available for testing job candidates.

versus

Two available systems can test job candidates.

2. There are several national and global organizations dedicated to promoting environmental sustainability for healthcare facilities.

versus

Several national and global organizations promote environmental sustainability for healthcare facilities.

Use active voice for clarity. The structure of a sentence—the arrangement of words—affects the clarity of the sentence. In active voice, the agent that does the action occurs next to the verb, the agent and the action both appear in the sentence, and the agent appears as the subject of the sentence.

      agent   *verb*
The department teaches the course every spring term.

      agent   *verb*
Our office submits all travel vouchers within 24 hours of their completion.

The result? Clear, concise, direct sentences.

Before:
(A) Attempts were made by the division staff to assess the project.

After:
                agent   *verb*
(B) The division staff attempted to assess the project.

Sentence (A) uses passive voice. Sentence (B) uses active voice: the agent (staff) occurs as the subject and appears next to the verb (attempted).

Research to determine the most readable sentence structures indicates that active-voice sentences may be more readable than passive-voice sentences. Readers often need the agent (the actor) placed near the action (the verb) to determine the sentence meaning. The agent and the action contain the essence of the sentence. The following examples illustrate this concept.

The door is to be locked at 6:00 p.m.

This sentence, which does not specify the agent, could mean either of the following:

The guard (or some designated person) will lock the door at 6:00 p.m.
The last person leaving the building at 6:00 p.m. must lock the door.

As both revisions illustrate, to understand a sentence, readers must know the agent and the action carried out by the agent. When you write, be sure your sentences indicate who or what performs the action.

Passive-voice sentences often intentionally do not include the agent doing the action to hide responsibility. The result may produce a sentence more verbose and less accurate than an active-voice version. Passive-voice sentences often use "there is" and "there are" constructions. Even in engineering writing, such as articles for academic journals, many editors want active-voice sentences because of their increased clarity. As in the examples below, the use of active subjects will usually make an explanation easier to read and easier to understand:

Before:
With the growing request of high-quality multimedia service, especially in portable systems, efficient algorithms for audio and/or video data processing have been developed. These algorithms have the characteristics of high-complexity data-intensive computation. For these applications, there exist two extreme implementations. One is software implementation running on a general-purpose processor and the other is hardware implementation in the form of application-specific integrated circuit (ASIC). In the first case, it is flexible enough to support various applications but may not yield sufficient performance to cope with the complexity of application. In the second case, optimization is better in respect of both power and performance but only for a specific application. A coarse-grained reconfigurable architecture fills the gap between the two approaches, providing higher performance than software implementation and wider applicability than hardware implementation.

After
(Note: We separate sentences in the original passage here so that you can follow the changes in sentence structure.):
To respond to growing requests for high-quality multimedia services, especially in portable systems, engineers have developed efficient algorithms for audio and/or video data processing.

These algorithms exemplify high-complexity, data-intensive computation.

In addition, these applications use two extreme implementations: (1) a *software implementation* running on a general-purpose processor and (2) a *hardware implementation*, an application-specific integrated circuit (ASIC). The first has flexibility needed to support various applications, but it may not yield sufficient performance to cope with complex applications. In the second, we can optimize power and performance but only for a specific application. A coarse-grained reconfigurable architecture fills the gap between the two approaches. This architecture provides higher performance than software implementation and wider applicability than hardware implementation.

Breaking long sentences into shorter ones and creating short paragraphs can also produce clarity:

Before:
To ensure dimensional quality of manufactured products, a crucial step is to take coordinate measurements of the geometric features to reconstruct product surface and then to check their compliance with tolerance specifications: my research develops a method to integrate the coordinate measurements from measuring devices of different resolutions for a better reconstruction of the product surface.

After:
To ensure dimensional quality of manufactured products, researchers must take coordinate[d] measurements of the geometric features. Their goal is to reconstruct product surface and then check surface compliance with tolerance specifications.

My research develops a method to integrate the coordinate[d] measurements from measuring devices of different resolutions to better reconstruct the product surface.

Word Choice

To write concise sentences, use clear, concise words and phrases. Avoid using longer words when shorter ones will do just as well. (Write to express, not to impress.)

Instead of	Write	Instead of	Write
accordingly	so	hence	so
accumulate	gather	implement	carry out
acquaint	tell	initiate	begin
acquire	get	maximum	most
activate	begin	modification	change
aggregate	total	nevertheless	but, however
assist	help	objective	aim
communicate	write, talk, tell	optimum	best
compensation	pay	personnel	people, staff
consequently	so	procure	get
continue	keep up	purchase	buy
demonstrate	show	terminate	end
discontinue	stop	transmit	send
endeavor	try	utilize	use
facilitate	ease, simplify		

Eliminate dead phrases—words that add nothing to the meaning of the sentence.

to the extent that	in view of
with your permission	inasmuch as
hence	as a matter of fact
with reference to	for the purpose of
in connection with	in order
with respect to	
as already stated	

Avoid words that sound knowledgeable without being specific. Many are technical words that have been overused and poorly adapted to nontechnical situations.

parameters	warrants further investigation
logistical interface	broad-based
contact	dynamics
impact	infrastructure
input/output	longitudinal study
conceptualize	matrix
formalize	meaningful
multifaceted	monolithic
systematized	paradigm
prioritize	participatory involvement
time frame	resource utilization
hard date	viability
in-depth study	

Avoid redundant phrases.

absolutely complete	human volunteer
absolutely essential	insist and demand
agreeable and satisfactory	my personal opinion
anxious and eager	necessary essentially
basic fundamentals	past memories
complete absence	point in time
consensus of opinion	right and proper

each and every

exactly identical

example to illustrate

few in number

first and foremost

general consensus

green in color

sincere and earnest

small in size

summarize briefly

thought and consideration

true facts

very unique

Avoid business jargon.

Instead of	Write
consideration was given	I considered
prior to the	before
at the present writing	now
effect an improvement	improve
in the neighborhood of	about
beg to advise	tell
cognizant of	know
thanking you in advance	I would appreciate
endeavor	try
viable alternative	possibility
in regard/reference to	about
send under separate cover	send separately
return same to the above	return to us
needless to say	[omit]
it goes without saying	[omit]
in the normal course of procedure	normally
in this day and age	today
in my opinion	I believe
it is our opinion	we think
on a daily basis	daily
on the grounds that	because
pursuant to our agreement	as we agreed
we are not in a position to	we cannot
without further delay	now
please be advised that	[omit]

Squeaky Clean Prose

Your goal as a writer is to achieve a message as clean as possible, with nothing unclear in the wording or phrasing to call attention to itself or distract the reader from your ideas.

The following excerpt from *DNA: The Secret of Life* addresses readers interested in science and possessing a basic understanding of genetics. Note the structure of each sentence, the use of topic sentences, and the development of each paragraph.

The great size of DNA molecules posed a big problem in the early days of molecular biology. To come to grips with a particular gene—a particular stretch of DNA— we would have to devise some way of isolating it from all the rest of the DNA that sprawled around it in either direction. But it was not only a matter of isolating the gene; we also needed some way of "amplifying" it: obtaining a large enough sample of it to work with. In essence we needed a molecular editing system: a pair of molecular scissors that could cut the DNA text into manageable sections; a kind of molecular glue pot that would allow us to manipulate those pieces; and finally a molecular duplicating machine to amplify the pieces that we had cut out and isolated. We wanted to do the equivalent of what a word processor can now achieve: to cut, paste, and copy DNA.

Developing the basic tools to perform these procedures seemed a tall order even after we cracked the genetic code. A number of discoveries made in the late sixties and early seventies, however, serendipitously came together in 1973 to give us so-called "recombinant DNA" technology—the capacity to edit DNA. This was no ordinary advance in lab techniques. Scientists were suddenly able to tailor DNA molecules, creating ones that had never before been seen in nature. We could "play God" with the molecular underpinning of all of life. This was an unsettling idea to many people. Jeremy Rifk in, an alarmist for whom every new genetic technology has about it the whiff of Dr. Frankenstein's monster, had it right when he remarked that recombinant DNA "rivaled the importance of the discovery of fire itself."

Source: Watson, James. *DNA: The Secret of Life*, pp. 87–88. Knopf, 2003. Used by permission.

This excerpt uses a variety of sentences of moderate length, close subject–verb patterns, familiar words, and a description of recombinant DNA in words easily understood by the nonscientific reader: the passage concisely and vividly expresses the meaning of recombinant DNA.

CASE 4-1

Readable style can make a major difference between your document being read or not read. Case Document 4–1A is the original version of a proposal for a research award. Case Document 4–1B is the revised proposal, which received the award. What differences do you see in the two versions?

CASE DOCUMENT 4–1A

Statement of Grant Purpose
Charles Johnson, New York, Mechanical Engineering
Development of a Comprehensive Model of a Lead-cooled Fast-spectrum Nuclear Reactor

This research proposes to work in conjunction with the FAST Reactors Group (Fast-spectrum Advanced Systems for power production and resource managemenT) at the Paul Scherrer Institute (http://fast.web.psi.ch) to develop and apply a computer-based model of a next-generation nuclear reactor, named the European Lead-cooled System (ELSY) [1]. This reactor design is one of several ideas being pursued by the European Commission, the executive branch of the European Union, in order to prepare the way for future nuclear technology development. The FAST Reactors Group has been at the forefront of this research, developing a unified code system capable of analyzing the dynamic behavior of new design ideas. The proposed research will seek to develop and apply the system specifically to the ELSY design.

The importance of ELSY and other next-generation reactors lies in their commitment to sustainability, safety, and security. While climate change has emphasized the need for non-carbon emitting energy sources, and even Greenpeace founder Patrick Moore now believes that nuclear power must play a predominate role among these sources [2], many people are still not comfortable with the thought of building new reactors. Three of the biggest issues contributing to this are radioactive waste storage, safety, and the threat of nuclear proliferation. Current technology and policies are capable of addressing these issues; but future technology could both produce more proliferation-resistant reactors of enhanced safety and minimize the amount of waste generated by them. Thirteen countries, including the United States and Switzerland, are working toward these goals under a program called the Generation-IV International Forum (GIF). GIF has chosen six promising reactor designs on which to focus, and one of them is a Lead-cooled Fast Reactor. ELSY is the European Commission's version of this design.

In 2006, the European Commission funded a 36-month effort to develop the initial design of ELSY [3]. From this came a mid-sized reactor rated at 600MWe, perfectly suited for the interconnected European electrical grids (GIF 2008). Its main advantages include its fuel efficiency, its high safety level, its unattractiveness as a tool for proliferation, and its reduced production of high-level radioactive waste. However, the corrosive nature of lead must be taken into account. Furthermore, advanced designs intend to operate at 550°C, which will introduce additional challenges for the reactor.

The FAST Reactors Group has been successful at developing computer models that account for the challenges presented by the new designs chosen by GIF. Specifically, the group has developed a unified code system capable of expert analysis in the areas of coupled neutronics, thermal hydraulics, and fuel behavior. The FAST code system incorporated well-established codes from each of the separate areas, linking them together and adapting them where necessary for simulation of fast reactor features [4]. The codes include ERANOS, TRACE, FRED, and PARCS, which simulate coupled 3D static neutronics, transient system thermal hydraulics, thermal mechanics, and 3D reactor kinetics respectively. The diagram of the FAST code system is shown in Fig. 1.

The overall goal of the research is to apply the FAST code system to and update it for the ELSY design. To carry out this research, I would arrive in Switzerland in September 2016 and then pursue the following course of action:

1. Become familiar with the FAST code system (September 2016)
2. Learn the ELSY fuel, core, and system parameters (October 2016)
3. Analyze and test the available computer models: ERANOS, TRACE, FRED (November, December 2016, January 2017)

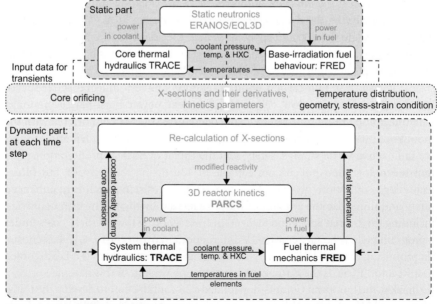

Figure 1. Diagram of the FAST code system.

4. Develop the PARCS core kinetic model and interfaces between different parts of the FAST code system (February, March 2017)
5. Apply model to steady-state simulation and a limited number of transients (April, May 2017)

I would return to America in June 2017.

The proposed research would not only benefit me as I develop toward my ultimate goal of designing more sustainable nuclear reactors but would also benefit the FAST Reactors Group as my work would be integrated into the new EU LEADER project to be started in 2017. Moreover, as the world moves toward more environmentally-friendly power sources, it will be vital to make the case for nuclear power to be part of the energy portfolio and to keep striving for excellence in the nuclear industry, both with current and future technology.

L. Cinotti, et al. "The ELSY Project," Proceeding of the International Conference on the Physics of Reactors (PHYSOR), Interlaken, Switzerland, 14–19 September, 2008.

P. Moore. "Going Nuclear: A Green Makes the Case," Washington Post 2006.

"European Lead-cooled System," Studiecentrum voor Kernenergie 2007.

K. Mikityuk. "FAST Code System: Review of Recent Developments and Near Future Plans," Proceedings of ICONE17, Belgium, 2009.

CASE DOCUMENT 4–1B

STATEMENT OF GRANT PURPOSE
Charles Johnson, New York, Mechanical Engineering
Development of a Comprehensive Model of a Lead-cooled Nuclear Reactor

Summary

I propose to work with the FAST Reactors Group [1] at the Paul Scherrer Institute to develop a computer model of a next-generation nuclear reactor. To prepare for future nuclear technology development, the executive branch of the European Union, the Executive Commission is pursuing several ideas, including the design on which my research would focus. The FAST Reactors Group has been at the forefront of this research and has developed a computer code capable of analyzing the behavior of the European Commission's new design ideas.

Background and Rationale for This Work

While climate change has emphasized the need for non-carbon emitting energy sources, many people are still not comfortable with building new reactors. Radioactive waste storage, potential safety issues, and the threat of nuclear proliferation prevent the public from enthusiastically supporting nuclear energy. Some countries view the technology more favorably than others, but these three issues require the foremost attention of the entire nuclear industry. Consequently, current technology and policies have made the nuclear industry safer than even the average office workplace.

However, future technology could do better. Thirteen countries, including the United States and Switzerland, are working to produce safer, more proliferation-resistant reactors under a program called the Generation-IV International Forum. These new reactors will also generate less radioactive waste. The forum has chosen to focus on six promising reactor designs, and it hopes to be able to make them available for commercial production by 2030 [2].

During my time at the Paul Scherrer Institute, I would develop a model for a design that falls under one of the forum's six categories. This category focuses on lead-cooled fast reactors. "Lead-cooled" means that liquid lead would absorb the heat produced by the reactor, thus "cooling" it. "Fast" refers to the energy of the neutrons flying around the reactor. In 2006, the European Commission funded a 36-month effort to develop a design for this type of reactor, called the European Lead-cooled System [3]. Its main advantages include its fuel efficiency, its high safety level, its unattractiveness as a tool for proliferation, and its reduced production of highly radioactive waste. However, the corrosive nature of lead must be taken into account. Furthermore, advanced designs intend to operate at 550°C, which will introduce additional challenges for the reactor.

Previous Work of the FAST Reactors Group

The FAST Reactors Group has successfully developed a computer model that accounts for the challenges presented by the new reactor designs. Before the group's work, several separate computer codes were able to analyze particular features of today's technology. However, these well-established codes did not work together. The group has been able to link them together to produce a unified computer code capable of analyzing the reactor as a whole. Moreover, the group updated the code to analyze features of fast reactors, creating the ability to simulate the new reactor designs [4].

Methods and Schedule

My overall goal is to develop the group's computer code to work well when analyzing the European Lead-cooled System. To carry out this research, I would arrive in Switzerland in September 2016 and then pursue the following course of action:

	09/2016	10/16	12/16	01/2017	02/17	03/17	04/17	05/17
Learn group's code	■							
Learn about design		■						
Analyze and develop code			■	■	■	■		
Apply code to European Lead-cooled System							■	■

I would return to America in June 2017.

Conclusion
The proposed research would benefit not only me as I develop toward my ultimate goal of designing more sustainable nuclear reactors but it would also benefit the FAST Reactors Group. My work would be integrated into one of the group's projects to be launched in 2017. Moreover, as the world must move toward more environmentally-friendly power sources, it will be vital to make the case for nuclear power to be part of the energy portfolio and to keep striving for excellence in the nuclear industry, both with current and future technology.

FAST stands for Fast-spectrum Advanced Systems for power production and resource management. The group's website can be found at http://fast.web.psi.ch.
"Preparing Today for Tomorrow's Energy Needs." Generation-IV International Forum, http://www.gen-4.org/
L. Cinotti, et al. "The ELSY Project," Proceeding of the International Conference on the Physics of Reactors (PHYSOR), Interlaken, Switzerland, 14–19 September, 2008.
K. Mikityuk. "FAST Code System: Review of Recent Developments and Near Future Plans," Proceedings of ICONE17, Belgium, 2009.

✎ STYLE CHECKLIST

Planning

☐ How will you adjust your writing style to accommodate your readers' knowledge of the subject?

☐ How will you meet your readers' expectations about style for the specific kind of document you are writing?

☐ Can you determine the appropriate reading level for the context in which the document will be read?

☐ How will you adjust your style so that it is appropriate for the professional relationship you have with your readers?

Revision

☐ Do paragraphs begin with a topic sentence? Do the subsequent sentences in the paragraph build on the idea in the topic sentence?

☐ Are most sentences twenty words or shorter? Could you make any longer sentences shorter?

☐ Are subjects and verbs close together in your sentences?

☐ Have you used specific nouns and concrete verbs?

☐ Have you avoided ponderous and impersonal language?

☐ Have you avoided *is/are* verb forms whenever possible?

☐ Are most of your sentences written in active voice? Could you change any sentences with passive voice to active voice?

☐ Have you defined everything that might require defining?

☐ Could you write any of your sentences with equal clarity but with fewer words?

⊘ EXERCISES

1. Which recommendations in this chapter operate in the revisions of the following sentences? Why do these principles result in clearer, more concise revisions?

   i. A stroke is an increasingly common problem associated with the brain. It is caused when a vein or artery in the brain is obstructed. This results in loss of consciousness from the loss of oxygen in the brain.

      A stroke, an obstruction in a vein or artery in the brain, results in loss of consciousness as the brain does not receive oxygen.

   ii. There is another method that is more invasive to the patient but proves to be successful enough to treat stroke.

      Another more invasive method proves successful in treating stroke.

iii. There are many proposed solutions for ridding the United States of oil dependency.

> Many proposed solutions target reduction of US oil dependency.

iv. A significant amount of new research has come out recently about a bodily phenomenon called brown fat. Brown fat is regarded with interest because of its potential use as a new therapy for obese patients. Brown fat is nothing new—it's been known for centuries as "baby fat," the stuff that makes newborn infants so adorably chubby.

> Recent research on human "brown fat" shows that it may help obese patients. Brown fat, known for centuries as "baby fat," makes newborn infants chubby.

v. People in this field work in hospitals and dedicate their time to the mental well-being of children in hospitals as well as to helping the parents be able to support their children and cope with the trauma.

> People in this field work in hospitals. These professionals focus on the mental well-being of hospitalized children and help the parents support their children as they cope with trauma.

vi. Asperger's syndrome, sometimes called high functioning autism disorder, is a disability in young people that is frequently overlooked by educators and medical personnel.

> Asperger's syndrome, a high functioning autistic disorder in young people, is frequently overlooked by educators and medical personnel.

or

> Educators and medical personnel frequently overlook Asperger's syndrome, a high functioning autism disorder occurring in young people.

vii. Even though Hurricane Gustav left its mark, there was a minimal amount of damage to the town, and finally there was a sigh of relief regarding energy prices.

> Even though Hurricane Gustav left its mark, the town suffered minimal damage and gave a sigh of relief regarding energy prices.

viii. A shoemaking company has signed a deal to share military technology to make their shoes waterproof, at the same time making the fabric breathable.

> A shoemaking company signed a deal to share military technology, making their shoes waterproof and breathable.

ix. There are many issues facing New Orleans related directly to hurricanes and tropical storms. Recently most of the problems caused from these storms have been traced back to the loss of wetlands of New Orleans during the past century. Despite recent pushes by community and political leaders in New Orleans to help restore these very little has been done, leaving much of the city vulnerable to hurricanes and allowing sea level to continue to rise.

New Orleans faces many issues related to hurricanes and tropical storms. Most problems caused by these storms have occurred because of loss of wetlands around New Orleans during the past century. [a. Despite efforts by community and political leaders in New Orleans to restore wetlands, little has been done.] Much of the city becomes increasingly vulnerable to hurricanes as the sea level continues to rise. [b. Little has been done, despite efforts by community and political leaders in New Orleans to restore wetlands.]

x. Efforts were made on the part of the director's committee for completion of an evaluation of the recommendations of the report.

The director's committee tried to complete their evaluation of the report recommendations.

2. Revise the following:

   i. Though there have been numerous economic anthropology studies in Mexico, they have largely been in three locations.
   ii. The intervening steps taken in the site are with the intention to make the site more attractive. This is done by establishing services, heading offices, cultural centers, and commercial firms in the downtown district.
   iii. For adequate housing in such circumstances, this paper here presents a model. The model is a step-by-step development process which is aimed at empowering the poor users. This process includes:
   - Innovative financing
   - Creating employment opportunity
   - Make the users aware of sustainable systems for drinking water, sewage disposal, other utilities, and shelter improvement
   - Demonstrate the knowledge of the mentioned systems through a construction of a public facility for them.
   - Transfer the knowledge by employing the users in the construction and maintenance of the facility.
   iv. At the same time, these fuels must be easy to burn cleanly. This is typically characteristic of liquid and gaseous fuels. Lastly, it is best that the fuel be a liquid or a gas because of the ability to pump both. This means that the existing infrastructure can be used to distribute the fuel.
   v. Lots of money around the world is poured into cancer research.
   vi. There has been success using lymphotropic paramagnetic nanoparticles for imaging prostate cancer.
   vii. There is plenty of money and room for big companies to get their foot in and discover new technologies to fight cancer.
   viii. Recently there has been a surge in research and development for a material called grapheme. One area grapheme is being applied to is solar cells.
   ix. Sustainable development is the process of moving human activities into a pattern that can be perpetually sustained. It is an approach that seeks to meet the needs of the present while protecting the resources that will be needed in the future.

x. The results of the survey will be reported following a discussion of the methods that were used by the researchers to solicit participants for the survey and to encourage complete answers to each of the ten questions.

3. Examine Figure 4–6. Revise this memo to eliminate as many of the thirteen "be" verbs as possible. What impact does this change have on the clarity and length of the sentences?

 USDA | United States Department of Agriculture | Research, Education, and Economics | National Institute of Food and Agriculture | 1400 Independence Avenue SW Washington, DC 20250

May 6, 2016

TO: State Agricultural Experiment Station Directors
Cooperative Extension Directors
1890 Research Deans-Directors
1890 Extension Directors

FROM: Sonny Ramaswamy
Director – NIFA

SUBJECT: Clarification of NIFA Policy for Integrated Research and Extension Reporting

This correspondence is to clarify the intent of a memo I sent to all 1862 and 1890 Deans and Directors on November 8, 2012, which described a new policy that all Plans of Work (POW) and Annual Reports be submitted to NIFA at the "state level." This policy, which was approved by the Office of Management and Budget via the Federal Register process, requires that each 1862 and 1890 Research and Extension organization within a state work together to develop and submit a combined Plan of Work and Annual Report.

The intent of this policy is to provide NIFA a mechanism for efficiently identifying the *programmatic* collaboration and integration between Research and Extension (and multiple Land-Grant Universities in some states). NIFA must be adept at responding to stakeholders and decision-makers at all levels of government when data is needed to demonstrate the greater return on investment when Research and Extension are strategically informing their respective programs. When NIFA program leaders review POWs and Annual Reports, they look for evidence that there are open lines of communication and feedback loops in place within organizations to ensure Research and Extension are working together to plan integrated work as appropriate. Having a single Plan of Work and Annual Report for each institution or state and insular area helps us do just that.

NIFA's need to collect data about integrated programs via a single Plan of Work does not imply any intent or requirement to force administrative restructuring at Land-Grant institutions. Our intent is simply to see evidence in a single report that Research and Extension are collaborating at the program level to strengthen the impact of said programs. How an institution(s) chooses to structure their administrative units to achieve this is at their discretion.

I have asked my staff in the Planning, Accountability, and Reporting office to include this topic as an agenda item in upcoming webinars and newsletters related to Plan of Work reporting so that we can ensure any further clarification you may need is provided.

• **FIGURE 4–6** Document for Exercise 3

4. Examine Figure 4–7. If you were the author of this document and were given a little extra time to make it more readable, what changes would you make?

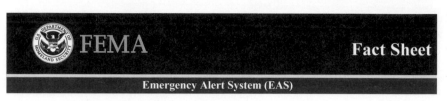

Emergency Alert System (EAS)

What is the Emergency Alert System (EAS)?

The Emergency Alert System (EAS) is a national public warning system that requires broadcasters, cable television systems, wireless cable systems, satellite digital audio radio service (SDARS) providers and direct broadcast satellite (DBS) providers to provide the President with communications capability to address the American people within 10 minutes during a national emergency. The system also may be used by state and local authorities, in cooperation with the broadcast community, to deliver important emergency information, such as weather information, AMBER alerts and local incident information targeted to specific areas.

FEMA, in partnership with the Federal Communications Commission (FCC) and National Oceanic and Atmospheric Administration (NOAA), is responsible for implementation, maintenance and operations of the EAS at the federal level. The President has sole responsibility for determining when the national level EAS will be activated. FEMA is responsible for national-level EAS, tests, and exercises.

EAS Modernization and Primary Entry Point (PEP) Stations

The modernization of the EAS begins with the FEMA adoption of a new digital standard for the distribution of alert messages to the broadcast community. The Integrated Public Alert and Warning System (IPAWS) uses the Common Alerting Protocol (CAP) standard and new distribution methods to make the EAS more resilient and to provide enhanced alerting capabilities.

Primary Entry Point (PEP) stations are broadcast stations located throughout the country with a direct connection to FEMA and resilient transmission capabilities. These stations provide the initial broadcast of a Presidential EAS message. FEMA increased the number of PEP facilities to provide direct coverage to over 90 percent of the American people.

History of the Emergency Alert System (EAS)

In 1951, the CONtrol of ELectromagnetic RADiation, originally called the "Key Station System" or CONELRAD, initiated a special sequence and procedure on participating stations tuned to 640 & 1240 kHz AM which was designed to warn citizens. In 1963, the Emergency Broadcast System (EBS) was initiated to address the nation through audible alerts. It did not allow for targeted messaging. EBS upgraded in 1976 to provide for better and more accurate handling of alert receptions. During this time, EBS was expanded for use during peacetime at state and local levels. In 1997, the Emergency Alert System (EAS) was designed for the President to speak to the American people within 10 minutes of a national emergency. EAS messages are composed of a digitally encoded header, attention signal, audio announcement and digitally encoded end-of-message marker.

The EAS Remains a Critical Component of IPAWS

In 2006, President Bush signed Executive Order 13407 directing the Department of Homeland Security (DHS) to create a comprehensive public alert and warning system for the United States. FEMA was directed to lead the effort and adopted a set of standards and protocols which support IPAWS. IPAWS is a modernization and integration of the nation's existing and future alert and warning systems, technologies and infrastructure. Federal, state, territorial, tribal and local government alert and warning systems able to integrate with the national alert and warning infrastructure providing a broader range of message options and communications pathways for the delivery of alert and warning information to the American people before, during, and after a disaster by providing one message over more media to more people for the preservation of life and property.

For more information: http://www.fema.gov/ipaws
To contact the IPAWS Program Management Office: ipaws@fema.dhs.gov

1/1/2016

• **FIGURE 4–7** Document for Exercise 4

5

Designing Documents

Effective writing is more than just choosing the right words. And it's more than correct sentences organized in logical paragraphs. To be effective, your document must also look like it will be easy to read and easy to understand, and it must be accessible to all your readers regardless of their physical abilities.

You have many choices about how a page or screen will appear to your audience. For example, you could integrate illustrations, include audio and video materials, incorporate links to files or sites with more information, invite comments or encourage the sharing of information through e-mail or social media. It is also easy to overwhelm and confuse your audience with these elements. This chapter will help you make the careful choices that allow for easy reading and navigation of your pages and screens.

 Quick Tip

On the job, you will probably write and receive more e-mail messages than any other kind of document. And ordinarily you will write your messages and your readers will receive your messages on mobile devices. If you design your e-mail messages for quick and easy reading, your recipients will be able to decipher your intended meaning efficiently and respond readily and appropriately to your instructions, questions, and requests regardless of their location. Here are six simple guidelines to follow:

1. **Keep your messages brief.** E-mail is especially effective for brief messages that the recipient will read and reply to quickly. Long, scrolling messages with extensive detail are often better relegated to attachments that could be accessed and studied later.

2. **Use short paragraphs.** Short paragraphs separated by blank space encourage quick reading and make it easy for your recipient to perceive and retrieve the chief points of your message.

3. **Use the subject line to specify your message.** A clear and specific subject line will preview your message for your recipients, making their reading easier and aiding their understanding. You will also be helping recipients to sort and

find your messages later, especially if you write a separate message for each topic you address. If your message includes a sizable attachment that might be slow to access from a wireless mobile device, identify it here in brackets—[BIG DOC], [BIG PIC], [AUDIO], or [VIDEO]—so that your recipients know whether to look at it immediately or wait for a more convenient time.

4. **Use headings to identify the sections of your message.** Headings make it easier for your recipient to skim your message for its chief points and assist in later retrieval of specific information in your message.

5. **Use hyperlinks selectively and strategically.** Hyperlinks allow you to quickly direct your readers to related information about your subject (e.g., news articles, research reports, videos, blogs). Every hyperlink, however, offers your reader the opportunity to exit your message. If the reader jumps immediately to the linked site and finds it especially interesting or engaging, the remainder of your message could easily go unread. Avoid inserting excessive hyperlinks in your message: one per paragraph is quite enough. If possible, list the hyperlinks to pertinent sources of information in the closing paragraph of your message.

6. **Ask simple yes/no questions.** Make it easy for your recipient to reply to your message with the briefest possible answer. For example, ask, "Should I . . . ?" instead of "How would you like me to handle this situation?" If a series of yes/no questions are necessary, number the questions.

Understanding the Basics of Document Design

Readers judge a document by how it looks as much as by what it contains, and first impressions are especially important. A dense page of long paragraphs will often discourage or annoy readers even before they begin reading. A page designed to help readers locate important information, however, may add to the persuasiveness of your position or convince your readers to put a little more effort into finding what they need and understanding what they find.

Effective document design is built on principles of visual perception—on how human beings perceive and interpret visual information. For example,

- *Contrast*: Different items must be visibly different, especially more important and less important items.
- *Alignment*: Related items must be aligned with each other, and every item must be aligned with some other item (or risk looking misaligned).
- *Proximity*: Related items must be positioned close together.
- *Size*: Greater size implies greater importance.
- *Repetition*: Repetition of design creates unity and builds familiarity.

The following six guidelines will help you adapt the principles of perception to plan your document's visual design:

- Determine which decisions are yours to make.
- Choose a design that fits your situation.
- Plan your design from the beginning.
- Make your design accessible.
- Reveal your design to your readers.
- Keep your design consistent.

Know what decisions are yours to make. Many organizations have standard formats or templates for reports, letters, proposals, e-mail messages, blogs, wikis, social media postings, and websites. Before you develop your document, determine the pertinent design requirements. Don't change the format arbitrarily just to be different. If you think the template you are assigned to use isn't appropriate for your audience and your message, find out who makes decisions on such design issues and make a case for the changes that you consider necessary.

Choose a design that fits your situation. Don't make your document any more complicated than the audience and situation requires. For example, you don't typically need a table of contents or a glossary for reports that are under five pages. Appendices are worthwhile additions to a document, but only if you believe your readers will actually use this extra level of detail.

You'll impress readers most by providing just the information they need in a way that makes it easy for them to find it and understand it. Most people read technical and business documents selectively. They scan the document, looking for sections that are relevant to their needs. They try to grasp the main points quickly because they have other responsibilities, limited time, and many more documents to read. Remember that your readers don't get paid to read documents: they get paid to make decisions and take actions. The more time needed to read your document, the less productive is their paid time on the job. And always keep in mind that they could be reading your document on their smartphones, tablets, or laptops while traveling to and from the job site, while sitting in meetings, and while multitasking in their offices.

For example, users working with a software application are unlikely to read the entire user's manual. They resort to the manual or to online help when they have a specific problem or need instructions for a specific task and can't figure it out by trial and error. They want to get to the right page or screen immediately. They want the instructions to stand out on the page or screen. Look at Figures 5–1 and 5–2. Notice how the numbered steps make for quick reading and easy understanding relative to the long and confusing paragraph of instructions.

Plan your design from the beginning. Before you start writing, consider how you will organize and display your information. Ask yourself the following questions:

- How will your readers use the document? Will they read it from beginning to end? Will they want to skim it and grab the main points without reading more? Will they want to jump to a specific topic? Even if they read the document through once, will they want to come back later and find a specific point quickly?

- How familiar are your readers with the subject of the document? How much support will they need in understanding and navigating the information you offer?

- How important is the subject of your document for your readers? Will they be highly interested and motivated to read or indifferent and inattentive?

- How familiar are your readers with the kind of document you are writing? Do they come to the document with certain expectations about how the information will be organized and exhibited?

- Will your readers view this document on paper or on a screen?

- Will your readers be stationary or mobile?

- Will your readers be focused or distracted in their attention to your document?

If, for example, readers are likely to skim your document, adding a table of contents, a subject index, and headings on every page will help them find information quickly. (The rest of this chapter includes techniques for developing effective designs to help people find what they need.)

If readers don't know much about the topic of the document, a glossary of key-words and abbreviations might be necessary. Illustrations or videos could also be important because such readers won't have a store of pictures in their minds on which to call for support. Links to pertinent interactive resources or social media sites might keep otherwise unmotivated readers attentive and engaged.

If readers are unfamiliar with a particular kind of document, they might benefit from a simple and explicit design that avoids distracting variations or unnecessary decorative elements. Experienced readers, however, might have rigid design expectations: for example, hiring managers assume that letters of application will adopt a standard format of business correspondence (with return address, salutation, signature, etc.) and are usually annoyed with applications that violate their expectations.

If the document is going to be read on a screen, you may have both more constraints and more choices than if the document were printed on paper. Illustrations and color will be easier and less expensive to include in a digital document than in a paper document. You can also incorporate audio or video in your document as well as hyperlinks to related materials and sources of information. Readers, however, may adjust the size of text on a screen to their level of comfort, which would change your intended spatial relationship of words and images on the screen. If readers are mobile, your document could be competing for their attention with all the distractions in their changing environment. In this case, headings, lists, illustrations, and blank space will help readers to keep track of their location in your document and to recognize the essential items of information.

All managers must report on the achievement of sales targets by their full-time and part-time salespeople, specifically the number of salespeople meeting weekly, monthly, and quarterly sales targets. These figures are based on the official sales targets issued by the vice president of Merchandising and the number of full-time and part-time salespeople identified by Human Resources as assigned to each department at the beginning of the reporting period.

To submit reports, start by logging in to your digital profile in Corporate Measures (http://www.corporatemeasures.com/login/managers). Then, in the Manager section, select Weekly, Monthly, or Quarterly Sales Report, as appropriate. Next, in the Department section, locate each department managed during the reporting period and choose the pencil icon to edit the information for that department. Then, in the list of information for each department, scroll to Sales Targets. Then enter the appropriate figures for both full-time and part-time salespeople. Finally, in the adjacent text box, explain anything new or different during this reporting period that might be related to the achievement of sales targets (e.g., special promotions, changes in aisle displays, suggestive selling techniques).

• **FIGURE 5–1** Instructions in Ineffective Paragraph Style

You don't know just by looking at this passage that it includes a set of instructions, and you can't readily determine the number of steps involved.

All managers must report on the achievement of sales targets by their full-time and part-time salespeople, specifically the number of salespeople meeting weekly, monthly, and quarterly sales targets.

These figures are based on

a. official sales targets issued by the vice president of Merchandising
b. the number of full-time and part-time salespeople identified by Human Resources as assigned to each department at the beginning of the reporting period

To submit reports,

1. Log in to your digital profile in CorporateMeasures (http://www.corporatemeasures.com/login/managers).

2. In the Manager section, select Weekly, Monthly, or Quarterly Sales Report as appropriate.

3. In the Department section, locate each department you managed during the reporting period and choose the pencil icon to edit the information for that department.

4. In the list of information for each department, scroll to Sales Targets.

5. Enter the appropriate figures for both full-time and part-time salespeople.

6. In the adjacent text box, explain anything new or different during this reporting period that might be related to the achievement of sales targets (e.g., special promotions, changes in aisle displays, suggestive selling techniques).

• **FIGURE 5–2** Instructions in Effective List Style

This passage makes it easy for you to recognize it as a set of instructions, and you know immediately that it's a six-step procedure.

Make your design accessible. Your entire intended audience deserves quick and equal access to your information. Your job as a communicator is thus to make sure that text and images as well as audio and video materials are readily accessible regardless of the audience's physical abilities. Make text accessible by simplifying the display of paragraphs, headings, and lists and avoiding unnecessary variations in the size, color, style, and design of the type. Screen readers and similar assistive technologies are typically more efficient and less subject to error if the information display is relatively plain and simple (e.g., black text on a white background, in a standard size, in a single column). In addition, writing in plain language makes information more accessible to individuals with learning or reading disabilities (e.g., dyslexia).

Make your illustrations accessible by using alternative text, captions, or titles to describe the images in graphs, charts, diagrams, drawings, and photographs. Your objective here is to indicate the nature and significance of the image (e.g., "Figure 1. Line graph of annual sales showing consistent growth from $9M USD in 2001 to $57M USD in 2017.").

If you are using audio materials, make the information accessible to people of limited hearing by incorporating closed captions or by including a complete script. For video materials, closed captions or a script might also be appropriate; if you are addressing individuals with limited vision, however, you would offer a paired audio that describes the images and actions in the video.

Keep in mind that if you design information to be accessible, you will also make it more usable for more people in more situations: that is, the more accessible your design, the more universal your design.

Reveal your design to your readers. Research on how people read and process information shows that readers must see how information is organized in order to make sense of it. That is, as you read, you try to do two things at the same time: you try to make sense of the passage you're reading and you try to make sense of how this passage fits with previous passages and what it contributes to the entire document. The more difficult it is to do one or the other, the more difficult the document is to read.

Tables of contents and headings reveal the organization, scope, and direction of your document and give readers a clear overview or map with which to proceed. Using headings (at least one on every page or screen) in a memo, e-mail message, or brief report will show the structure and logic of the discussion and help readers recognize, remember, and retrieve your major points. Longer documents definitely need headings and a table of contents that lists the headings. In digital documents, a table of contents typically links to the pages listed.

Keep your design consistent. Consistency in design makes for easy reading. After you consider your audiences, the information you have to deliver, and the ways that people will read and use your document, you must develop a page or screen design that will work for your situation. After you decide on the appropriate design, don't change it arbitrarily. You want your readers to proceed confidently and comfortably through the document. You want them to know immediately when they are beginning a new section or when they are in another part of the same section because they recognize the differences in the design of the headings at each level.

To achieve this consistency, identify the different types of information in your document and use the styles function of your word processing application to standardize the design. This involves three steps:

1. Identify all the types of information you will need to display, such as

   - paragraphs
   - quotations
   - lists
   - examples
   - equations and formulas
   - levels of headings

2. Create a design that always shows the same type of information in the same way throughout your document (e.g., size of type for headings versus paragraphs versus quotations, style of numbering for instructions versus regulations).

3. Use the style function of your word processing program to label and fix the design of each type of information.

Designing Effective Pages and Screens

Visually effective pages and screens are designed on a grid so that readers know where to look for information. Using space inside the text, around the illustrations, and in the margins will keep pages uncluttered and information easy to locate. The right line length and margins will also help people read easily.

The following suggestions will help you develop effective pages and screens:

- Use blank space to frame and group information.
- Choose a type design that is legible.
- Space the lines of text for easy reading.
- Use a medium line length.
- Use a ragged right margin.
- Position words and illustrations in a complementary relationship.

Use blank space to frame and group information.   Don't think of blank space as wasted or empty space. Space is a critical element in design for both paper and screens because it makes information easier to find and read. Look at Figures 5–3 and 5–4. Which do you think is easier to read?

From: Lydia Schlosser (lydia.schlosser@crsu.edu)
Date: Tuesday, January 23, 2018, at 8:15 a.m.
To: All Department Heads (departments@crsu.edu)
Subject: New Media Request Procedures

A recent study of our media center request procedures indicates that we are not fulfilling media requests as efficiently as possible. A number of problems surfaced in the study. First, many requests, and particularly special requests (e.g., poster printing, laminating, close captioning of audio/videocasts, audio/video editing) are submitted in the evening after the media center has closed for the day or in the early morning before it has opened. As a result, the media center has an enormous backlog of orders to fill before it can begin servicing the orders submitted after 8:30 A.M., when it officially opens. This backlog may throw the center two or three hours behind schedule. Media requests submitted throughout the day cannot be completed promptly. By 2:00 P.M., requests submitted may not be filled that same day. If special orders arrive unexpectedly, even a routine copy request may take two business days to complete.

To remedy the situation, we will change to the following media request procedure beginning Monday, February 5. The media center will close at 3:00 every afternoon. Two employees will work at the center from 3:00 until 5:00 to complete all routine orders by 5:00. If you submit requests by 3:00, the center will have them ready by 5:00. In short, all requests will be filled the day they are submitted. However, do not leave requests after 3:00, as these will not be processed until the following day. However, we guarantee that if you submit your request 3 between 8:00 and 3:00, you will have it fulfilled that day.

Special requests—including over 100 copies of one item, single/multiple copies of any document over 50 pages, front/back copying of one item up to 50 copies, poster (large-scale) printing, laminating, jobs involving trimming or folding, close captioning of audio/video, and audio/ video editing—will require that notice be given to the media center at least one day in advance (more time for unusual projects such as audio/video in excess of five minutes or close captioning in multiple languages). That way, the center can prepare for your request and be sure to have it ready for you.

A copy of the special request form is attached to this message. Please complete this form and e-mail it to me at the media center so that I can schedule all special jobs. If you submit a special request without having completed the form, your project will be completed after all other special requests are completed.

Allow plenty of time for routine jobs—at least two hours, and three if possible. Beginning February 5, give all media requests to the receptionist in the media center. Be sure you attach complete instructions. Give your name, employee number, e-mail address, and department. Describe the project in detail, such as the number of copies required, paper color, collating and binding for multipage copies, color or black and white printing.

Pick-up procedures also change February 5. All jobs, after they are complete, will be placed in each department's mail box. No jobs will be left outside the media center after closing time. No jobs will be left with the receptionist. Large orders that will not fit mail boxes will be delivered to your office. Electronic projects will be delivered to the e-mail address listed on the request form.

If you have questions about this new procedure, please contact me (by voice or text) at 742-2500.

• **FIGURE 5–3** E-mail Message That Violates Design Guidelines

Notice how important instructions are buried in the middle of paragraphs, making it impossible to skim this page-long memo or easily retrieve key pieces of information.

From: Lydia Schlosser (lydia.schlosser@crsu.edu)
Date: Tuesday, January 23, 2018, at 8:15 a.m.
To: All Department Heads (departments@crsu.edu)
Subject: New Procedures for Ordering Jobs from the Media Center

EFFECTIVE DATE: MONDAY, FEBRUARY 5, 2018

To handle orders more quickly and efficiently, the Media Center is changing its procedures. Please notify all staff in your department and ask them to follow these new procedures.

Special Orders versus Routine Requests
Decide if you have a special order or a routine request. A special order is
- over 100 copies of one item
- single/multiple copies of any document over 50 pages
- front/back copying of one item up to 50 copies
- poster (large-scale) printing
- laminating
- jobs involving trimming or folding
- close captioning of audio/video
- audio/video editing

Procedure for a Routine Request
1. Attach complete instructions to your request. Include
   - your name, employee number, e-mail address, and department
   - project details such as number of copies
   - special instructions such as collating and binding for multipage copies
2. Give all routine requests to the receptionist in the Media Center.
3. Allow 2 hours for your order to be filled.

NOTE: Routine requests submitted between 8:00 A.M. and 3:00 P.M. will be processed by 5:00 P.M. on the same day.

Procedure for a Special Order
1. Fill out the attached Special Order form.
2. E-mail the completed form to mediarequest@crsu.edu at least one day in advance of the day you need the job completed (two days in advance for unusual projects such as audio/videocasts in excess of five minutes or closed captioning in multiple languages).

NOTE: This early notice will allow us to schedule your job with the appropriate staff and equipment and complete your job promptly.

The Copy Center will close at 3:00 P.M. Orders submitted after 3 P.M. will be processed the next business day.

Copy Pick-Up Procedures
Jobs will be delivered to your department's mailbox. If the order is too large for the mailbox, it will be delivered to your office. Electronic jobs will be delivered to your e-mail address.

If you have questions, call or text 742-2500.

• **FIGURE 5–4** A Revision of Figure 5–3.
Notice how the headings help you to recognize the categories of information covered in the memo, while the bullets and numbers allow you to skim for key points.

You can incorporate blank space into documents in several ways. A critical location for blank space is at the margins. Here, blank space encloses and contains the information, keeping the page or screen from looking crowded and chaotic. Clear and generous margins make your information look organized and coherent.

If your document will be read on paper, also think about how it will be bound. If you are putting your work in a ring binder, for example, you will have to make sure that holes don't punch through the text. For unbound copies set all your margins at 1 inch, but for bound copies set the left margin at 1½ inches and the remaining margins at 1 inch.

If you are putting text on both the front and back of a page, put the extra ½ inch of space in the left margin of odd-numbered pages and in the right margin of even-numbered pages. In your word processing application, choose mirror margins to alternate for right-hand (odd-numbered) pages and left-hand (even-numbered) pages. If you cannot set alternating margins, set both the right and the left margins at 1½ inches.

The space in the margins is important, but you must do more. Graphic designers call margins *passive space* because margins only define the block of the page or screen for readers. Graphic designers know that *active space*—the space inside the text—makes the real difference in designing effective pages or screens. Blank space helps readers to find information quickly, keep track of the location of information, process the information in identifiable chunks, and retrieve the information later.

Here are three techniques to bring active space to your pages and screens:

- Use headings frequently (at least once per page or screen). Put each heading above its related text with an extra line of space before the heading to separate it visually from the preceding text.

- Use bulleted lists for three or more parallel points. Use numbered lists for steps in instructions. Lists are often indented inside the text, and each item may be separated from the others by blank space.

- Use an extra line of blank space to separate paragraphs or indent the first line of each paragraph. In digital documents, make your paragraphs even shorter than you would in paper documents so that there is notable space even on the small screens of mobile devices. On a smartphone or tablet, one instruction or one short sentence may make an appropriate paragraph.

Choose a type design that is legible. Type comes in two major categories: serif and sans serif (i.e., with or without serifs, the extra decorative flourishes on the tips of the individual letters). Historically, serif types (e.g., Century or Times) were advised for paper documents and sans serif types (e.g., Arial or Helvetica) for electronic documents. Technological advances in digital publishing, however, have made serif and sans serif designs equally legible in a wide variety of sizes.

Type style (regular, bold, italic) also influences legibility. Regular type is easiest to read and will usually be your default choice. Bold type is typically reserved for headings and isolated words or phrases that merit emphasis. Italic type is challenging to decipher (for both human beings and machines), especially in long passages: reserve it for special words, book and film titles, and social invitations.

Black text on a white background (or dark text on a light background) offers the easiest reading display and is the standard for page and screen. Other colors for text or background could easily prove distracting or disorienting. If you can't justify a variation—or if a variation is only decorative—avoid it: you will never go wrong using the colors that readers expect to find.

The issue for you is to make sure that the specific type you choose is legible in the size, color, style and design that it will likely appear on the page or screen that you are composing.

Space the lines of text for easy reading. Single spacing of documents is standard, usually with an extra line of blank space inserted between paragraphs.

Letters and memos of one or two short paragraphs will often be double-spaced. Drafts of documents submitted for review and editing are also typically double-spaced to give writers and editors more room in which to write corrections and comments. If you use double spacing, either indent the first line of each paragraph or insert an extra line of blank space between paragraphs.

Adjust the line length to the size of the page or screen. The number of words that fit on a line depends in part on the size and style of type that you are using. Long lines of text fatigue readers and make them lose their place in moving from the right margin back to the left margin of the next line. Short lines are also a challenge because readers finish reading each line rapidly and are thus shifting almost continuously from the right margin to the left margin of the next line. Figure 5–5 illustrates the problems with both long and short lines of text.

Long lines of type often prove challenging for people to read. The problem is that readers find it difficult to get back to the correct place at the left margin. They may find themselves inadvertently skipping a line or rereading the same line. If they make this mistake more than once or twice, they may get tired or annoyed and stop reading altogether. The smaller the type, the harder it is for most people to read long lines of type. If readers feel fatigued, they will slow their reading, interrupt their reading, or stop reading altogether.

Very short lines of type can look choppy
on a page and make comprehension
difficult because readers must work
harder to put the words together in
coherent groups.

• **FIGURE 5–5** Line Length
Very long lines and very short lines are hard to read.

Use a ragged right margin. Although text is almost always lined up on the left margin, it is sometimes also aligned on the right margin, creating a tidy rectangle of text. The text of this book, for example, aligns on both the left and the right margins. Most of the examples in the figures in this chapter, however, align on the left but not on the right. The technique of making all the text align exactly on both the left and the right margin is called *justifying the text*. If the text aligns on the left but not on the right, it is *left-aligned* or has a *ragged right margin*.

Justified text gives a document a formal appearance, whereas ragged right text gives a document a friendly, informal feeling. Justified text is often more difficult for readers because every line is the same length, thus eliminating a visual signal that helps readers both to keep track of each line and to locate the next line of text. Digital documents ordinarily have ragged right margins in order to make reading on screens of various sizes as easy as possible. Screen readers and similar assistive technologies also typically do a better job with ragged right margins.

Position words and illustrations in a complementary relationship. The visual and verbal information in your document must reinforce each other, working together to communicate your message. Position each illustration as close as possible to the text that relates or refers to it. Keep in mind also that readers will typically examine the illustrations on a page or screen before giving their attention to the words in captions, headings, and paragraphs. Make sure the text adjacent to each illustration supports the message of the illustration and doesn't lead to misinterpretation. Notice, for example, how the position of the diagram in Figure 5–6 could cause readers to think it displays the kind of extension cord that ought to be discarded instead of the kind with desired safety features. A quick fix here would be to change the neighboring heading to "Look for Key Safety Features" so that the words truly complement the diagram.

Helping Readers Locate Information

To help your readers find what they need and make sense of what they find, you must plan a useful structure for your document and make that structure evident to your readers. In the previous sections of this chapter, we showed you how to use page and type design to make your document clear and easy to use. In this section, we explain how to give your readers signs of your document's overall structure.

On the job, readers of technical and business documents rarely examine every page; instead, they read selectively. They may glance over the table of contents to see what the document is about and then pick and choose the sections to read by finding headings that match their needs and interests. They may go straight to the search function for digital documents (or the index of paper documents) to locate a specific topic. They may skim through the document, stopping only when a heading or example or illustration strikes them as important. They may go back to the document later to retrieve or verify specific facts. Long reports or proposals may be divided

CPSC Safety Alert

Household Extension Cords Can Cause Fires

Choose the right extension cord for the job, and use it as specified. Extension cords can overheat and cause fires when used improperly. Overheating is usually caused by overloading or connecting appliances that consume more watts than the cord can handle. Damaged extension cords can also cause fires. Extension cords should only be used temporarily. Protect young children by keeping them away from extension cords and unplugging the cords when not in use.

Overheating or Damage
- Overheating can occur at the plug, at the socket, or over the entire length of the cord. Hot plugs and sockets are often caused by deteriorated connections to the cord's wires.
- Look for visible signs of excessive wear or damage to the plug, sockets or insulation. Replace damaged extension cords.

Prevent Overloading
- If any part of the extension cord is hot while in use, it is a warning sign that it may be overloaded. Check if the extension cord is properly rated for the products that are plugged into it. Also, inspect the cord along its entire length to ensure it has not been damaged.
- Do not overload your extension cord by using it to power appliances beyond its capacity. You can check its capacity, or rating, by looking at the tag on the cord or its packaging.

Protect Extension Cords from Damage
- Do not run extension cords under carpets, through doorways or under furniture.
- Only use an extension cord outdoors if it is marked for outdoor use.
- Never alter a cord to change its length or perform inadequate repairs such as taping up damaged insulation. Do not trim, cut or alter the plug blades in any way.
- Unplug an extension cord when it is not in use. The cord is energized when it is plugged in and can overheat if shorted.

Discard Older Extension Cords
- Discard cords that are old and/or are missing important safety features, including safety closures, polarized blades and a large plug face that covers the outlet's slots and is easy to grasp to unplug.

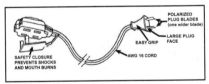

- Extension cords should be at least 16 AWG, unless they are 18 AWG with fuse protection. AWG refers to the size of the wires in the cord. The wire size is imprinted on the cord's surface.

Check cords to make sure they have been listed by a recognized national testing laboratory, such as Underwriters Laboratories (UL), Intertek (ETL) or CSA-International (CSA).

Publication 5032 0098010 42012

U.S. CONSUMER PRODUCT SAFETY COMMISSION • (800) 638-2772 • www.cpsc.gov • www.SaferProducts.gov

• **FIGURE 5–6** Ifneffective Word–Illustration Relationship
The diagram and adjacent heading communicate opposite messages instead of reinforcing each other.

among several readers, with each reviewing the section pertinent only to his or her field (e.g., accountants checking the budget, engineers examining the schematics, managers assessing the timeline) before coming together to discuss the subject of the document or sharing comments and questions through a wiki or e-mail list.

Following are four ways to help your readers find information easily:

- Use frequent headings.
- Write descriptive headings.
- Design distinctive headings.
- Use page numbers and headers or footers.

Use frequent headings. Frequent headings help readers know where they are in a document at all times. In a report, for example, you want a heading for every section and subsection, with at least one heading on every page or screen. You want to keep each topic short (one to three paragraphs) and give each topic a heading.

Write descriptive headings. *Headings* are the short titles that you use to label each section and subsection of your document. Even brief documents, such as letters and e-mail messages, can benefit from headings. Compare Figures 5–7 and 5–8 to see how useful headings are in a brief message.

Headings are the roadmap to your document, identifying the key topics and revealing the direction of thought. These five suggestions will help you write useful headings:

- Use concrete language.
- Use questions, verb phrases, and sentences instead of nouns alone.
- Use standard keywords if readers expect them.
- Make the headings at a given level parallel.
- Make sure the headings match any list or table of contents in the document.

Use concrete language. Generic headings such as "Introduction" or "Conclusion" give no indication of the topic you are discussing. Make your headings specific to your document. Make your headings reveal the subject and claims of your document. Readers should be able to read only your headings, without any of the accompanying text, to get a clear sense of your overall message.

Use questions, verb phrases, and sentences instead of nouns alone. The best way to write headings is to put yourself in your readers' place. Will readers come to your document with questions? If so, then questions will make good headings. Will they come wanting instructions for doing tasks? If so, then verb phrases that match the actions they need to take will make good headings. Will they come seeking knowledge about a situation? If so, then statements of fact about that situation will make good headings.

In addition, avoid headings that are individual nouns or strings of nouns: such headings are often perceived as ambiguous. For example, a heading such as "Usability Questionnaire Completion" makes it impossible to predict the kind of information that this section will offer. Much clearer would be headings such as "How Do I Complete the Usability Questionnaire?" or "What Is the Deadline for Completing the Usability Questionnaires?" or "Who Must Complete the Usability

UNITED STATES OF AMERICA
FEDERAL TRADE COMMISSION
WASHINGTON, D.C. 20580

Bureau of Consumer Protection
Division of Enforcement

Julia Solomon Ensor
Attorney

Email: jensor@ftc.gov
Direct Dial: (202) 326-2377

February 9, 2016

FEDERAL EXPRESS

Hector Penton, COO
Origin PC Corporation
12400 SW 134 Court, Suite #1
Miami, FL 33186

Dear Mr. Penton:

We received your submissions on behalf of Origin PC Corporation ("Origin"). During our review, we raised concerns that certain marketing materials may have overstated the extent to which Origin computers are made or "built" in the United States.

As we discussed, unqualified "Made in the USA" or "Built in the USA" claims on marketing materials likely suggest to consumers that all products advertised in those materials are "all or virtually all" made in the United States. The Commission may analyze a number of different factors to determine whether a product is "all or virtually all" made in the United States, including the proportion of the product's total manufacturing costs attributable to U.S. parts and processing, how far removed any foreign content is from the finished product, and the importance of the foreign content or processing to the overall function of the product.

In this case, you explained that Origin completes final assembly of computers in the United States using significant imported parts. Accordingly, to avoid deceiving consumers, Origin removed all "Built in the USA" claims from its marketing materials, and replaced them with the claim, "24/7 US Based Support."

Based on your actions and other factors, the staff has decided not to pursue this investigation any further. This action should not be construed as a determination that there was no violation of Section 5 of the Federal Trade Commission Act, 15 U.S.C. § 45. The Commission reserves the right to take such further action as the public interest may require. If you have any questions, you can reach me at (202) 326-2377.

Sincerely,

Julia Solomon Ensor
Staff Attorney

● **FIGURE 5–7** Letter Without Headings

The design of this letter makes it impossible to determine the kinds of information it covers. You can't skim it for a quick overview or easily locate the key points.

UNITED STATES OF AMERICA
FEDERAL TRADE COMMISSION
WASHINGTON, D.C. 20580

Bureau of Consumer Protection
Division of Enforcement

Julia Solomon Ensor
Attorney

Email: jensor@ftc.gov
Direct Dial: (202) 326-2377

February 9, 2016

FEDERAL EXPRESS

Hector Penton, COO
Origin PC Corporation
12400 SW 134 Court, Suite #1
Miami, FL 33186

Dear Mr. Penton:

We received your submissions on behalf of Origin PC Corporation ("Origin"). During our review, we raised concerns that certain marketing materials may have overstated the extent to which Origin computers are made or "built" in the United States.

FTC Questions "Built in USA" Claim

As we discussed, unqualified "Made in the USA" or "Built in the USA" claims on marketing materials likely suggest to consumers that all products advertised in those materials are "all or virtually all" made in the United States. The Commission may analyze a number of different factors to determine whether a product is "all or virtually all" made in the United States, including the proportion of the product's total manufacturing costs attributable to U.S. parts and processing, how far removed any foreign content is from the finished product, and the importance of the foreign content or processing to the over all function of the product.

Origin Agrees to Correct Claim

In this case, you explained that Origin completes final assembly of computers in the United States using significant imported parts. Accordingly, to avoid deceiving consumers, Origin removed all "Built in the USA" claims from its marketing materials, and replaced them with the claim, "24/7 US Based Support."

FTC Could Take Further Action

Based on your actions and other factors, the staff has decided not to pursue this investigation any further. This action should not be construed as a determination that there was no violation of Section 5 of the Federal Trade Commission Act, 15 U.S.C. § 45. The Commission reserves the right to take such further action as the public interest may require. If you have any questions, you can reach me at (202) 326-2377.

Sincerely,

Julia Solomon Ensor
Staff Attorney

• **FIGURE 5–8** Revision of Figure 5–7
The design of this letter emphasizes the key pieces of information. The headings offer a quick overview of the letter and make it easy to skim.

Questions are useful as headings in a brochure.
    What does the gypsy moth look like?
    How can we protect trees from gypsy moths?
    How often must we spray?
    What is the chemical composition of the pesticide?

Verb phrases are useful in instruction manuals.
    Verb phrases can be gerunds, like these:
        Choosing the right kind of graph
        Importing the data
        Preparing the image
        Adding a title and caption
    Verb phrases can be imperatives, like these:
        Make your attendance policy clear.
        Explain your grading scale.
        Announce your office hours.
        Identify titles and authors of required textbooks.
        Review all assignments and due dates.

Short sentences are useful in memos and reports.
    We doubled sales this past year.
    We served 40% more clients.
    We logged 500 hours of overtime this year.
    We need three more staff positions.

• **FIGURE 5–9** Different Structures You Can Use for Effective Headings

Questionnaire?" Figure 5–9 shows how effective it can be to use questions, verb phrases, and statements as headings.

Use standard keywords if readers expect them. You may be working on a document for which readers expect to see a certain set of headings in a certain order, as in a standard proposal format. In that case, organize your material in the order and with the headings that your readers expect. Figure 5–10 shows the headings you might use in a standard proposal format.

Make the headings at a given level parallel. Like the items in a list, headings at any given level in a document should be parallel. Parallelism is a very powerful tool in writing because it helps readers to recognize the similarity among the listed items. See for yourself the difference parallelism makes by comparing the two sets of headings in Figure 5–11.

Project Summary
Project Description
Facilities and Equipment
Personnel
Rationale and Significance
Budget Plan of Work

• **FIGURE 5–10** Keywords as Headings in a Proposal

Nonparallel Headings	Parallel Headings
Graph Modifications	Modifying a graph
Make data changes	Changing the data
To insert or delete columns	Inserting or deleting columns
How to adjust color or gradients	Adjusting the color or gradients
Titles and legends can be added	Adding titles and legends

• **FIGURE 5–11**   Nonparallel and Parallel Headings

Headings that use the same sentence structure—parallel headings—are easier for users to follow.

Make sure the headings match any list or table of contents in the document. To check how well your headings tell your story and how well you've maintained parallel structure in headings, create an outline view or a table of contents for your draft document. Readers can then use the table of contents to locate a particular section. They know they're in the right place if the heading for that section matches the wording in the table of contents. This function is especially important in digital documents, where readers will navigate by jumping directly from a heading in the table of contents to a screen of information. If the heading on the screen they come to doesn't match the heading that they clicked on in the table of contents, they may be unsure of their location. Their confusion will quickly lead to irritation and a loss of trust in your attention to accuracy.

Design distinctive headings. Headings do more than outline your document. They also help readers find specific parts quickly, and they show the relationship among the parts. To help readers, headings must be easily distinguished from the text, and each level of heading must be easily distinguished from all other levels. Figure 5–12 is a good example of a document with four levels of headings. You can see how the writer uses color to distinguish all headings from the text and then uses type size, capitalization, and position on the page to distinguish each level of heading from the other levels.

These seven suggestions will help you design distinctive headings:

- Limit the number of heading levels.
- Create a pattern for the headings and stick to it.
- Match size to importance.
- Put more space above a heading than below it.
- Keep each heading with the section it covers.
- Consider using numbers with your headings.

Limit the number of heading levels. Don't make the hierarchy of levels more complicated than it needs to be. Headings are supposed to be an aid to reading, not an obstacle. The more levels of headings you use, the more that readers must do to keep the hierarchy straight in their minds. Paper documents don't need more than four levels of headings. If you have more than four levels, consider dividing the material into two major sections or two separate documents. Digital documents don't need more than two levels of headings: readers scrolling through documents displayed on various sizes of screens will easily lose track of different levels of headings.

Create a pattern for the headings and stick to it. Although your choices depend in part on the technology you are using, you almost certainly have several options for showing levels of headings. Figure 5–12 demonstrates a variety of ways to show different levels of headings. You can combine these to create the pattern for your headings. For example, you can change size, position, and capitalization to show the different levels of headings.

Match size to importance. Changing the type size is one way to indicate levels of headings. If you use different type sizes, make sure that you match the size to the level of importance. If the headings are different sizes, readers expect first-level headings to be larger than second-level headings, second-level headings to be larger than third-level headings, and so on, as shown in Figure 5–12. The lower-level headings can be the same size as the text, but no level of heading should be smaller than the text because this would violate readers' expectations. If you use different type sizes for different heading levels, don't make the differences in type sizes excessive.

Put more space above a heading than below it. Headings announce the topic that is coming next in your document. Therefore, you want the heading to lead the reader's eye down the page or screen into the text that follows. One way to do that is to have more space, on the page or screen, above the heading rather than below it. In this way, the heading and its accompanying text constitute a visible chunk of information.

Controlling Soil-Borne Pathogens

Types of Soil-Borne Pathogens
Simply stated, the effects of soil-borne pathogens . . .

Soil-borne fungi
At one time, it was thought that all soil-borne fungi were damaging to plant growth. Fungi are the most common . . .

Basiodiomycetes. The Basiodiomycetes are a class of 25,000 fungi with species that include mushrooms and . . .

Phycomycetes. The class of Phycomycetes is a highly diversified type of fungus. It is the . . .

Plant parasitic nematodes
Nematodes are small, unsegmented . . .

Treatments for Soil-Borne Pathogens

• **FIGURE 5–12**   Four Levels of Headings in a Report

If you are going to use a rule with the heading, consider putting it *above* the heading rather than below it. A rule above the heading creates a "chunk" that includes both the heading and the text that it covers. A rule above the heading also draws the reader's eye down into the text that follows instead of up and away from that text.

Keep each heading with the section it covers. Don't leave a heading at the bottom of a page when the text appears on the next page. Make sure you have at least two lines of the first paragraph on the page with the heading. In some cases, you may want each topic to be on a separate page so that the heading and all the text of a topic appear together. Most word processing programs have functions that help you keep headings from being stranded at the bottom of a page and that allow you to set up your document so that all headings of a certain level start on a new page.

Consider using numbers with your headings. In many companies and agencies, the standard for organizing reports and manuals is to use a numbering system with headings. Figure 5–13 shows two such numbering systems: the alphanumeric system and the decimal system. If you use a numbering system with your headings, include the numbers before the entries in your table of contents.

```
TITLE
I.  FIRST-LEVEL HEADING
      A. Second-Level Heading              alphanumeric system
         1. Third-level heading
         2. Third-level heading
      B. Second-Level Heading
II. FIRST-LEVEL HEADING
      A. Second-Level Heading
         1. Third-level heading
            a. Fourth-level heading
            b. Fourth-level heading
         2. Third-level heading
      B. Second-Level Heading


TITLE
1 FIRST-LEVEL HEADING
      1.1 Second-Level Heading             decimal system
         1.1.1 Third-level heading
         1.1.2 Third-level heading
      1.2 Second-Level Heading
2 FIRST-LEVEL HEADING
      2.1 Second-Level Heading
         2.1.1 Third-level heading
            2.1.1.1. Fourth-level heading
            2.1.1.2. Fourth-level heading
         2.1.2 Third-level heading
      2.2 Second-Level Heading
```

● **FIGURE 5–13** Two Types of Numbering Systems

A numbering system allows you to refer precisely and concisely to a section of the report by the number of its heading (e.g., Section II.A.3.b or Section 4.3.7). This function could be especially important if the document is subject to review by multiple readers from different divisions or remote locations.

Numbering systems also aid accessibility: that is, screen-reading assistive technologies easily decipher numerically identified headings but routinely misinterpret headings and levels of headings characterized by typographical variations (e.g., bold, italics).

With numbering systems, if you add or delete a section later, you must take care to revise the numbering of the report. Note also that a lot of readers find numbering systems confusing and, for example, would fail to recognize immediately that a section numbered 6.8.23 would come after a section numbered 6.8.2.3: therefore, keep the numbering and the levels of numbering as simple as possible.

Use page numbers and headers or footers. In addition to clearly worded and visually accessible headings, page numbers and running headers and footers are important aids to efficient reading.

Number the pages. Page numbers help readers keep track of where they are and provide easy reference points for talking about a document. Always number the pages of drafts and final documents that people are going to read on paper.

If the document is going to be read on screens, inserting page numbers could be unnecessary. On websites, for example, readers jump from topic to topic and page to page instead of proceeding in a specified order from start to finish. In addition, word processing applications keep track automatically of the number of pages and typically display this information in the bottom margin of the document window as a reference point for writers and readers. If readers are likely to print your digital document, however, they will certainly appreciate your inclusion of page numbers. In slides prepared for oral presentations, you will help your audience to track your progress and to stay attentive if you identify the slides with a numbering notation like this:

Slide 1 of 10

Short manuscripts and reports that have little prefatory material almost always use Arabic numerals (1, 2, 3). The common convention is to center the page number below the text at the bottom of the page or to put it in the upper outside corner (upper left corner for left-hand pages, upper right corner for right-hand pages). Always leave at least two lines of space between the text and the page number. Put the page number in the same place on each page. Page numbers at the bottom of the page often have a hyphen on each side, like this:

-17-

As reports grow longer and more complicated, the page-numbering system also may need to be more complex. If you have a preface or other material that comes before the main part of the report, it is customary to use lowercase Roman numerals (i, ii, iii) for that material and then to change to Arabic numerals for the body of the report.

In addition, the title page doesn't show the number but is counted as the first page. The page following the title page is number 2 or ii.

You also must know whether the document will carry text on one side of the page or both. If both sides will carry text, you may have to number several otherwise blank pages in word processing files. New chapters or major sections usually start on a right-hand page. The right-hand page always has an odd number. If the last page of your first chapter is page 9, for example, and your document will be double-sided, you must include a blank page 10 so that the first page of your second chapter will be a right-hand page 11 when the document is printed, copied, and bound.

The body of a report is usually paginated continuously, from page 1 to the last page. For the appendices, you may continue the same series of numbers, or you may change to a letter-plus-number system. In that system, the pages in Appendix A are numbered A-1, A-2, and so on. The pages in Appendix B are numbered B-1, B-2, and so forth. If your report is part of a series or if your company has a standard report format, you will need to make your page numbering match that of the series or standard format.

Numbering appendices with the letter-plus-number system has several advantages:

- It separates the appendices from the body. Readers can tell how long the body of the report is and how long each appendix is.

- It indicates that a page is part of an appendix and identifies which appendix. It makes pages in the appendices easier to locate.

- It allows the appendices to be printed separately from the body of the report. Sometimes the appendices are ready before the body of the report has been completed, and being able to print the appendices first may save time and help you meet a deadline.

- It allows the pagination of either an appendix or the body to be changed without requiring changes in the other parts.

Include headers or footers. In long documents, it helps readers if you give identifying information at the top or bottom of each page. Information at the top of the page is a *header*; information at the bottom of the page is a *footer*. Organizations often have standard practices regarding the information to be displayed in headers and footers. A typical header for a report might show the author's name, the title of the report, and the date. It might look like this:

Lisa Miller   Feasibility Study   November 2017

In this case, the page numbers would likely appear in the footer.

A typical header for a letter might show the name of the person receiving the letter, the page number, and the date. It might look like this:

Dr. Emilia Rodriguez   -2-   November 16, 2017

or

Dr. Emilia Rodriguez   Page 2   November 16, 2017

Note that headers and footers rarely appear on the first pages of documents because first pages already carry identifying information like the title, author, recipient, and date. Word processing applications allow you to designate the first page as different and start headers and footers on the second page.

Testing Your Design

After you have considered all the issues of information design and before you distribute the final copy of your document, take time to test your design on a group of representative readers to determine the usability of your document. The more important your document and the longer your document, the more users you will want to include in your test. Your objective is to make sure that readers will be able to locate and retrieve the information they need to take actions and make decisions.

To conduct your usability test, give each of your representative readers a copy of the document and ask him or her to find the answers to several questions about the subject and about the document itself. For example, in a report about bicycle accidents, you might ask readers to find the answers to the following questions:

Are the number of bicycle accidents increasing or decreasing?

What is the leading cause of bicycle accidents?

Which section of the report identifies the medical costs related to bicycle accidents?

What is the major recommendation of the report?

You would observe each representative reader as he or she navigates the document and assess the time it takes to answer each question as well as the accuracy of the answers.

You would thereafter interview the representative readers, either individually or jointly, about their experience with your document: What did they like? What did they dislike? What helped them find the answers to the questions? What delayed or disrupted their finding the answers? What would make the document easier to use?

The results of this usability test will help you identify how to revise the design of your document to serve the needs, interests, and abilities of your audience.

Document Design Checklist

☐ Is your document clear, organized, attractive, and accessible? Is your text easy to read?

☐ If your document is supposed to conform to a standard template, does it? Have you left adequate margins? (If necessary, have you left extra room for binding?)

☐ Is the spacing between the lines and paragraphs consistent and appropriate?

☐ Can the reader tell easily where sections and paragraphs begin? Are words and illustrations positioned appropriately?

☐ Are the headings informative? Unambiguous? Consistent? Parallel?

☐ Will readers get an overall picture of the document by reading the headings?

☐ Is the hierarchy of the headings obvious?

☐ Can readers easily determine what is heading and what is text?

☐ If readers want to find a particular section quickly, will the size and position of the heading help them?

☐ If you are using a numbering system, is it consistent and correct?

☐ Have you checked the page breaks to be sure that you do not have a heading by itself at the bottom of a page?

☐ Are the pages of a paper document numbered? Are there appropriate headers or footers?

☐ Did you test representative readers for their ability to locate information easily?

 EXERCISES

1. Visit the websites of two government agencies in your city or state that might employ majors in your field (e.g., the Department of Agriculture, the Department of Transportation). What kinds of documents do you find at each site? Do the agencies seem to have a standard template for the design of their documents? From looking at a number of the documents at each site, how would you describe their design? Which agency does a better job of designing documents for its audience and purpose? How would you make the design of each agency's documents more effective?

2. Roger Line is a claims adjuster for Framers Insurance. His job is to review insurance claims, accept justified charges, reject unsupported charges, solicit additional evidence as necessary, and issue payments to property owners.

Hello,

I have completed the review of the mitigation invoice for your homeowner's claim and am issuing a payment for the undisputed amount of $10,411.01 with the following deductions applied:

--Foyer - changed extraction to account for hard surface (-$14.52) --Master Bedroom - The sketch is not accounting for door openings - using field adjusters measurement for drywall removed of 57.59 sf (-$11.31) Adjuster antimicrobial application to account for door opening (-$1.92) Field adjusters photos show no furniture blocked or padded in this room. Framers agrees to pay for 2 hours contents manipulation (-$177.93) --Master bath - Floor measurement appear to not account for the tub/shower and vanity areas. Using field adjusters measurement of 87 sf for floor measurements (-$87.95) --Stairway - the field adjusters measurement for the drywall removed in this area is 392.17 sf vs. 606.39 as invoiced (-$128.53) Changed application of antimicrobial to match (-$34.28) Please provide photo to support trim in stairway, this is not a common charge and field adjusters estimate does not support (-$3.78) --Bathroom - Flooring measurement does not account for tub or vanity area. Using field adjusters measurement of 25 sf (-$14.70) --Living room / Dining room - Field adjusters photos do not support blocking & padding of furniture. Framers agrees to pay for 4 hours contents manipulation based on photos provided. (-$355.86) --Wet Bar - field adjusters measurement and photos support the vanity was 1.5 sf vs. 4 as invoiced (-$11.85) --Downstairs Bath - floor measurement does not account for vanity or tub area. Using field adjusters measurement of 29.5 sf (-$35.04) Field adjuster paid for toilet detach - deducted charge to avoid duplicate payment (-$15.78) Field adjusters photos show vanity, sink & faucet were not detached. Please provide photo support (-$37.86) Field adjusters measurement for drywall removed in this room is 51.46 sf vs. 118 as invoiced (-$39.92) Changed application of antimicrobial measurement to match affected area measurement per previous changes (-$42.73)

If you have additional information to help support any of the deducted charges, please send to me for supplemental consideration.

Sincerely,
Roger Line, AIS, AINS
Property Restoration Field Coordinator
Framers Insurance Company, Inc.

• **FIGURE 5–14** Original E-mail Message for Exercise 2

Figure 5–14 is the original version of Roger's e-mail message to a property owner. After reading through his explanation carefully, he revises the design to make it more usable for his reader. Figure 5–15 is Roger's revised version. Compare and contrast the two versions of Roger's message. What specific changes has he made to the original version? How do these changes improve the readability of his message? Which of the changes do you think has the greatest impact? What additional changes do you think are necessary?

Hello,

I have completed the review of the mitigation invoice for your homeowner's claim and am issuing a payment for the undisputed amount of $10,411.01 with the following deductions applied:

- Foyer: Changed extraction to account for hard surface (-$14.52).

- Master Bedroom: The sketch does not account for door openings. Using field adjuster's measurement for drywall removed of 57.59 sf (-$11.31). Adjuster anti-microbial application to account for door opening (-$1.92). Field adjuster's photos show no furniture blocked or padded in this room. Framers agrees to pay for 2 hours contents manipulation (-$177.93).

- Master bath: Floor measurement appears to not account for the tub/shower and vanity areas. Using field adjuster's measurement of 87 sf for floor measurements (-$87.95).

- Stairway: Field adjuster's measurement for the drywall removed in this area is 392.17 sf vs. 606.39 sf as invoiced (-$128.53). Changed application of antimicrobial to match (-$34.28).

  NEEDED: Please provide photo to support trim in stairway as this is not a common charge and field adjuster's estimate does not support (-$3.78).

- Bathroom: Flooring measurement does not account for tub or vanity area. Using field adjuster's measurement of 25 sf (-$14.70).

- Living room / Dining room: Field adjuster's photos do not support blocking & padding of furniture. Framers agrees to pay for 4 hours contents manipulation based on photos provided. (-$355.86).

- Wet Bar: Field adjuster's measurement and photos support the vanity was 1.5 sf vs. 4 sf as invoice (-$11.85).

- Downstairs Bath: Floor measurement does not account for vanity or tub area. Using field adjuster's measurement of 29.5 sf (-$35.04). Field adjuster paid for toilet detach; deducted charge to avoid duplicate payment (-$15.78). Field adjuster's measurement for drywall removed in this room is 51.46 sf vs. 118 sf as invoiced (-$39.92); changed application of antimicrobial measurement to match affected area measurement per previous changes (-$42.73).

  NEEDED: Field adjuster's photos show vanity,sink & faucet were not detached. Please provide photo support (-$37.86).

If you have additional information to help support any of the deducted charges, please send to me for supplemental consideration.

Sincerely,
Richard Line, AIS, AINS
Property Restoration Field Coordinator
Framers Insurance Company, Inc.

● **FIGURE 5–15**   Revised E-mail Message for Exercise 2

RANSOMWARE
What It Is and What To Do About It

WHAT IS RANSOMWARE?
Ransomware is a type of malicious software cyber actors use to deny access to systems or data. The malicious cyber actor holds systems or data hostage until the ransom is paid. After the initial infection, the ransomware attempts to spread to shared storage drives and other accessible systems. If the demands are not met, the system or encrypted data remains unavailable, or data may be deleted.

HOW DO I PROTECT MY NETWORKS?
A commitment to cyber hygiene and best practices is critical to protecting your networks. Here are some questions you may want to ask of your organization to help prevent ransomware attacks:

1. **Backups**: Do we backup all critical information? Are the backups stored offline? Have we tested our ability to revert to backups during an incident?
2. **Risk Analysis**: Have we conducted a cybersecurity risk analysis of the organization?
3. **Staff Training**: Have we trained staff on cybersecurity best practices?
4. **Vulnerability Patching**: Have we implemented appropriate patching of known system vulnerabilities?
5. **Application Whitelisting**: Do we allow only approved programs to run on our networks?
6. **Incident Response**: Do we have an incident response plan and have we exercised it?
7. **Business Continuity**: Are we able to sustain business operations without access to certain systems? For how long? Have we tested this?
8. **Penetration Testing**: Have we attempted to hack into our own systems to test the security of our systems and our ability to defend against attacks?

HOW DO I RESPOND TO RANSOMWARE?
Implement your security incident response and business continuity plan. It may take time for your organization's IT professionals to isolate and remove the ransomware threat to your systems and restore data and normal operations. In the meantime, you should take steps to maintain your organization's essential functions according to your business continuity plan. Organizations should maintain and regularly test backup plans, disaster recovery plans, and business continuity procedures.

Contact law enforcement immediately. We encourage you to contact a local **FBI**[1] or **USSS**[2] field office immediately to report a ransomware event and request assistance.

There are serious risks to consider before paying the ransom. We do not encourage paying a ransom. We understand that when businesses are faced with an inability to function, executives will evaluate all options to protect their shareholders, employees, and customers. As you contemplate this choice, consider the following risks:

- Paying a ransom does not guarantee an organization will regain access to their data; in fact, some individuals or organizations were never provided with decryption keys after having paid a ransom.
- Some victims who paid the demand have reported being targeted again by cyber actors.
- After paying the originally demanded ransom, some victims have been asked to pay more to get the promised decryption key.
- Paying could inadvertently encourage this criminal business model.

---

[1] https://www.fbi.gov/contact-us/field/listing_by_state
[2] http://www.secretservice.gov/contact/

• **FIGURE 5–16** Document for Exercise 3

3. Examine Figure 5–16. How would you revise the design of this document to make it more usable for its likely readers? Try changing the three major headings from questions to phrases. Which is more engaging? Which is better suited to the purpose of this document? What other changes will make it easier for readers to recognize and retrieve key information?

6

Designing Illustrations

In communicating technical information, you must often use illustrations either in addition to words or instead of words to convey your message. How do you determine whether illustrations are necessary? Ask yourself two questions:

- What do you want the reader to do or think after reading your document?
- How will illustrations help you to achieve your objective?

 Quick Tips

Illustrations come in two categories: tables and figures. Tables display numbers and words in columns and rows. Every other kind of illustration is a figure.

If you want to summarize information to make it easier to remember or retrieve, use a table.

If you want to emphasize information to clarify or reinforce its meaning, use a figure.

Creating Illustrations

In deciding when and how to use illustrations, remember the following guidelines:

- **Simplify your illustrations.** Keep your illustrations as simple as possible so that your reader has no difficulty understanding your message. Avoid distracting your reader with unnecessary details or decorative flourishes.
- **Use computer applications critically.** Graphics applications and clip art allow you to include a wide variety of illustrations in your document. To make sure your tables and figures are effective, however, you must evaluate the choices available. Graphics applications, for example, might create artistic but misleading graphs, and clip art might exhibit a pictorial style that isn't quite serious enough or detailed enough to do justice to your subject.

You must choose illustrations that display your information clearly and correctly and reinforce your message instead of detracting and distracting from it.

- **Consider size and cost.** Calculate the impact of illustrations on the expected length of your document. Illustrations will often increase the size of a document and, if printed, add to the cost of production and distribution.
- **Make your illustrations accessible.** To make your illustrations accessible to people with limited vision, include captions and alternative text descriptions that indicate the meaning of the image (e.g., This aerial-view diagram of the building identifies the location of the electrical fire and resulting smoke damage to adjacent areas.). Note also that a portion of people experience color vision deficiency, usually red versus green or blue versus yellow: avoid designing illustrations in which color coding is the only access to meaning.
- **Consider the cultural implications of color.** The colors in your illustrations might have implicit meanings for your readers (e.g., the color red implies danger in some cultures but happiness in others). If communicating across multiple cultures, use light desaturated colors to neutralize unintended associations.
- **Title your illustrations.** Give each table and figure a title that clearly indicates the message you wish it to convey (e.g., The Rise in Single-Family Housing Prices for 100 Major U.S. Cities, 2000–2020).
- **Number your illustrations.** If you use several illustrations in your report, assign each one a number to make it easy to refer to them. Number the tables and figures separately (e.g., Table 1, Table 2, Figure 1, Figure 2).
- **Alert your readers.** Always alert your readers to illustrations by referring to them in the text. Every time you refer to the illustration, use the table or figure number (e.g., see Table 1). Announce the illustration—what it is or shows— then add any verbal explanation your reader will need to fully understand it. Don't lead readers through a complicated explanation and only later refer to the illustration. Point them to the illustration immediately so that they can shift back and forth between the explanation and the illustration as necessary.
- **Position your illustrations strategically.** Place each illustration as close to the passage it explains as possible.
- **Identify your sources.** If you borrow or adapt a table or figure from another source, identify that source (and, if necessary, permission to use it) below the illustration (e.g., Source: *A Practitioner's Guide to Statistical Illustrations*, p. 157, by Xiling Xiao. Copyright 2017 by Oxford University Press. Reprinted by permission.).

Tables.

- Note that the purpose of a table is to summarize information in specific categories to assist the viewer in accessing and retrieving this information.
- Every column in a table should have a heading that identifies the information below it. In a table of numbers, include the unit of measurement, such as "miles per hour." For large numbers, add a designation such as "in thousands" or "in millions" to the column heading (and delete the corresponding zeros from the numeric data). Headings should be brief. If headings need more explanation,

include this information in a footnote below the table. Use lowercase letters, numbers, or symbols (e.g., a, *, or 1) to indicate footnoted material.

- If possible, box your table to separate it from surrounding paragraphs.
- Keep tables as simple as possible. Include only data relevant to your purpose.
- Consider omitting lines between rows and columns to avoid giving your table a crowded appearance. If possible, use white space to separate rows and columns.

Figures 6–1 through 6–3 show three ways of presenting tabular data.

Table 4.

Refugee Arrivals by Age, Sex, and Marital Status: Fiscal Years 2012 to 2014

Characteristic	2014		2013		2012	
	Number	Percent	Number	Percent	Number	Percent
AGE						
Total	69,975	100.0	69,909	100.0	58,179	100.0
0 to 17 years	24,647	35.2	23,647	33.8	18,876	32.4
18 to 24 years	10,143	14.5	10,399	14.9	9,700	16.7
25 to 34 years	14,608	20.9	15,328	21.9	13,491	23.2
35 to 44 years	9,223	13.2	9,543	13.7	7,446	12.8
45 to 54 years	5,612	8.0	5,504	7.9	4,409	7.6
55 to 64 years	3,192	4.6	3,098	4.4	2,441	4.2
65 years and over	2,550	3.6	2,390	3.4	1,816	3.1
SEX						
Total	69,975	100.0	69,909	100.0	58,179	100.0
Female	33,208	47.5	32,117	45.9	26,799	46.1
Male	36,767	52.5	37,792	54.1	31,380	53.9
MARITAL STATUS						
Total	69,975	100.0	69,909	100.0	58,179	100.0
Married	26,168	37.4	26,789	38.3	22,322	38.4
Single	39,739	56.8	39,392	56.3	32,608	56.0
Other*	4,068	5.8	3,728	5.3	3,249	5.6

*Includes persons who were divorced, separated, widowed, or of unknown marital status.

Source: U.S. Department of State.

• **FIGURE 6–1** Table This table is effectively designed using thin vertical lines and generous white space to separate the columns and thin horizontal lines to separate the three categories of information, making the entire table look easy to navigate. Note that the row labels are aligned on the left for easy reading, while the numerical data in the rows are aligned on the right (or on the decimal point) to allow the easiest possible comparison of information across the rows and columns. The column headings are centered, while the subheadings ("Number" and "Percent") are aligned on the right with their corresponding rows of numerical data. The column headings and subheadings are displayed in bold, making the use of color here nice but unnecessary. The use of all capital letters for the headings of the three major categories is also unnecessary but consistent and unlikely to disrupt reading. Note that the title of this table is "Fiscal Years 2012 to 2014" but the columns are arranged left to right from 2014 to 2012, creating a possibility for misinterpretation. Changing the subtitle to "Fiscal Years 2014, 2013, and 2012" or putting the columns in chronological order from left to right would avoid this potential confusion.

Source: US Department of Homeland Security, Office of Immigration Statistics. *Refugees and Asylees:2014.* Washington, DC: GPO, 2016. p. 4.

Table 8. Derailed cars for train Q40927 (hazardous materials highlighted in yellow).

Position	Reporting Marks	Load	Commodity	Car Type	Fire Damage
1	CSXT 135017	Empty	Scrap or waste paper	A332 box	No
2	TTPX 82350	Empty	Plate, iron, or steel	F253 flat	No
3	EEC 1095	Empty	Malt liquors	A402 box	No
4	CSXT 130476	Empty	Pulpboard or fiberboard	A302 box	No
5	CSXT 141318	Empty	Pulpboard or fiberboard	A302 box	Yes
6	TTPX 811227	Empty	Beams, iron, or steel	F453 flat	No
7	UNPX 128076	218,279 pounds	Sodium chlorate class 5.1, UN1495	C712 covered hopper	Yes
8	GATX 61416	Residue	Fluorosilicic acid, class 8, UN1778	DOT-111A100W5	Yes
9	ACFX 67747	198,680 pounds	Terephthalic acid	C414 covered hopper	Yes
10	ACFX 66451	199,257 pounds	Terephthalic acid	C414 covered hopper	Yes
11	ACFX 68042	199,099 pounds	Terephthalic acid	C214 covered hopper	Yes
12	ACFX 67966	198,460 pounds	Terephthalic acid	C414 covered hopper	No
13	NOKL725103	Empty	Plate, iron, or steel	F343 flat	No
14	AEX 12778	Empty	Monoammonium phosphate	C113 covered hopper	No
15	HOKX 111773	199,000 pounds	Sodium hydroxide solution, class 8, UN 1824	DOT-111A100W1	No

• **FIGURE 6–2** Table This table is easy to read with effective spacing to separate the columns and rows of information. The headings for each column are centered and displayed in distinctive bold type. Rows designated for special attention are effectively highlighted in bright yellow but dominate the light coloring of the row of headings. Here is a case where white type on a black background would be appropriate to differentiate the row of headings. Nevertheless, this is a table that viewers could quickly navigate to compare and contrast information or to retrieve specific pieces of evidence.

Source: National Transportation Safety Board. Highway–Railroad Grade Crossing Collision, Rosedale, Maryland, May 28, 2013. Highway Accident Report NTSB/HAR-14/02. Washington, DC: GPO, 2014. p. 40.

Table 4
Pedalcyclists Killed and Injured and Fatality and Injury Rates by Age and Gender, 2013

Age (Years)	Male Killed	Male Population (thousands)	Male Fatality Rate*	Female Killed	Female Population (thousands)	Female Fatality Rate*	Total Killed	Total Population (thousands)	Total Fatality Rate*
<5	3	10,152	0.30	0	9,716	0.00	3	19,868	0.15
5–9	15	10,509	1.43	2	10,062	0.20	17	20,571	1.83
10–14	30	10,553	2.84	2	10,098	0.20	32	20,650	1.55
Children (≤14)	48	31,214	1.54	4	29,875	0.13	52	61,089	0.85
15–19	49	10,846	4.52	8	10,313	0.78	57	21,159	2.69
20–24	37	11,679	3.17	17	11,116	1.53	54	22,795	2.37
25–29	35	10,960	3.19	7	10,620	0.66	42	21,580	1.95
30–34	26	10,962	2.43	4	10,583	0.38	30	21,264	1.41
35–39	25	9,785	2.55	4	9,819	0.93	29	19,604	1.48
40–44	40	10,360	3.86	4	10,489	1.04	44	20,849	2.11
45–49	66	10,498	6.29	10	10,710	1.10	76	21,208	3.58
50–54	79	11,071	7.14	12	11,488	0.74	91	22,559	4.03
55–59	91	10,282	8.85	12	10,912	0.52	103	21,194	4.86
60–64	52	8,674	5.99	7	9,448	0.35	59	18,122	3.26
65–69	41	6,913	5.93	4	7,696	0.23	45	14,609	3.08
70–74	22	4,884	4.50	2	5,724	0.14	24	10,608	2.26
75–79	12	3,390	3.54	1	4,288	0.32	13	7,678	1.69
80+	17	4,412	3.85	1	7,397	0.60	18	11,809	1.52
Seniors (≥65)	92	19,600	4.69	8	25,104	0.32	100	44,704	2.24
Total‡	645	155,652	4.14	97	160,477	0.60	742	316,129	2.35

Age (Years)	Male Injured	Male Population (thousands)	Male Injury Rate*	Female Injured	Female Population (thousands)	Female Injury Rate*	Total Injured	Total Population (thousands)	Total Injury Rate*
<5	**	10,152	**	**	9,716	**	**	19,868	**
5–9	1,000	10,509	73	**	10,062	**	1,000	20,571	50
10–14	3,000	10,553	325	1,000	10,098	53	4,000	20,650	192
Children (≤14)	4,000	31,214	136	1,000	29,875	29	5,000	61,089	83
15–19	5,000	10,846	486	1,000	10,313	130	7,000	21,159	313
20–24	7,000	11,679	623	1,000	11,116	117	9,000	22,795	376
25–29	5,000	10,960	428	1,000	10,620	81	6,000	21,580	257
30–34	4,000	10,962	331	1,000	10,583	61	4,000	21,264	197
35–39	2,000	9,785	164	**	9,819	**	2,000	19,604	101
40–44	3,000	10,360	327	**	10,489	**	4,000	20,849	176
45–49	3,000	10,498	282	1,000	10,710	89	4,000	21,208	184
50–54	3,000	11,071	245	**	11,488	**	3,000	22,559	139
55–59	1,000	10,282	142	**	10,912	**	2,000	21,194	82
60–64	1,000	8,674	142	**	9,448	**	1,000	18,122	76
65–69	1,000	6,913	88	**	7,696	**	1,000	14,609	70
70–74	1,000	4,884	117	**	5,724	**	1,000	10,608	61
75–79	**	3,390	**	**	4,288	**	**	7,678	**
80+	**	4,412	**	**	7,397	**	**	11,809	**
Seniors (≥65)	2000	19,600	88	1,000	25,104	22	2,000	44,704	51
Total	40,000	155,652	258	8,000	160,477	50	48,000	316,129	152

Source: FARS 2012 Final File, 2013 ARF. NASS GES 2013. Bureau of the Census projections.
* Rate per million population.
** Less than 500 injured, injury rate not shown.
‡ Total includes 5 males fatality of unknown age. One pedalcyclist of unknown gender is not included.
Note: Injured totals may not equal sum of components due to independent rounding.

• **FIGURE 6–3 Ineffective Table** This table tries to offer a lot of information in a single location but makes reading and retrieval difficult as a consequence. Notice that bold, centered type displayed in horizontal blue bars is used for both first-level and second-level headings. In addition, the headings occur both at the top of the table (which is to be expected) but again in the middle of the table (which would be unexpected unless you realize this display is really two groupings of information divided by a thin line of white space). In each grouping, special attention is given to two rows highlighted in blue—rows that inexplicably compile statistics on children 14 and under and seniors 65 and older even though neither grouping has the highest or lowest rate of fatalities or injuries. Unnecessary vertical grid lines and cells with double asterisks only exacerbate the visual confusion. At least two separate tables would make for easier reading: one for numbers and one for rates or one for fatalities and one for injuries.

Source: US Department of Transportation, National Highway Traffic Safety Administration. *Bicyclists and Other Cyclists*. Traffic Safety Facts DOT HS 812 151. Washington, DC: GPO, 2015. p. 3.

Bar and column graphs.

- The purpose of a bar graph is to compare and contrast two or more subjects at the same point in time, while the purpose of a column graph is to reveal change in a subject at regular intervals of time.

- Avoid putting excessive information in a bar or column graph and thereby complicating the reader's ability to decipher it. Consider using a separate graph to communicate each point.

- Be sure to label the horizontal, or *x*, axis and the vertical, or *y*, axis—what each measures and the units in which each is calibrated. Readers can't understand your graph if they don't know what you are measuring or how it is measured.

- For bar graphs, start the *x* axis at zero and equally space the intervals on the *x* axis to avoid distorting the length of the bars. For column graphs, start the *y* axis at zero and equally space the intervals on the *y* axis to avoid distorting the height of the columns. In a bar graph, neither axis is a measure of time; in a column graph, the *x* axis is ordinarily a measure of time.

- Color can enhance the effect of a graph, but excessive color can reduce comprehension and distort information. Use the same color for all bars or columns that represent the same categories of information. Avoid using color as simple decoration.

- Make the graph accurate. Typical graphics applications allow you to add a range of special effects. However, artistic graphs are not always either effective or accurate. Three-dimensional bars and columns are often deceptive because readers have difficulty visually comparing the relative lengths of the bars and the relative heights of the columns. If you choose three-dimensional bar or column graphs, watch for possible distortion.

- Limit the number of divisions. The more a bar or column is divided, the more difficult it is to understand. Bars or columns with more than four or five divisions are especially challenging for readers and could lead to confusion or misinterpretation.

- For divided bar and column graphs, use color or solid shading to distinguish divisions instead of textured patterns that may create distracting optical illusions.

- Put labels on or near the bars or columns. Keep in mind that explanatory legends (or keys) will slow comprehension because the reader is required to look back and forth between the bars and the legend. Placing labels on bars can be difficult, however, especially if you use colored bars. Even black text will often be difficult to read on any colors but very light ones. In short, use a legend only if you can't use labels. And if you do use a legend, locate it as close to the bars or columns as possible.

- Avoid crowding the bars or columns within a graph. Such visual clutter makes a graph look difficult to interpret. Using three-dimensional bars or columns will also reduce the number that will fit in a given space. Effective and inviting graphs leave generous space between the bars or columns.

See Figures 6–4 through 6–7 for examples of bar and column graphs.

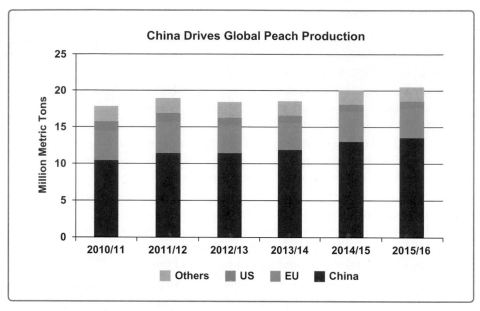

● **FIGURE 6–4** Column Graph This divided column graph is clean and concise and makes its key point explicit in the title. The columns are labeled clearly for each period, arranged in chronological order from left to right, and spaced generously. The colors used for the sections of each column are bright and distinctive from each other unless you have a color vision deficiency. Numerical labeling of each section and a total for each column would offer more specific information, but the horizontal grid lines here are sufficient to support the claim in the title. Instead of the inefficient use of a separate legend to identify the meaning of each color, the sections of the first or last column could be labeled "Other" "US," "EU," and "China" for quicker and easier access to the same information.

Source: US Department of Agriculture, Foreign Agricultural Service. *Fresh Peaches and Cherries: World Markets and Trade*. Washington, DC: GPO, 2015. p. 1.

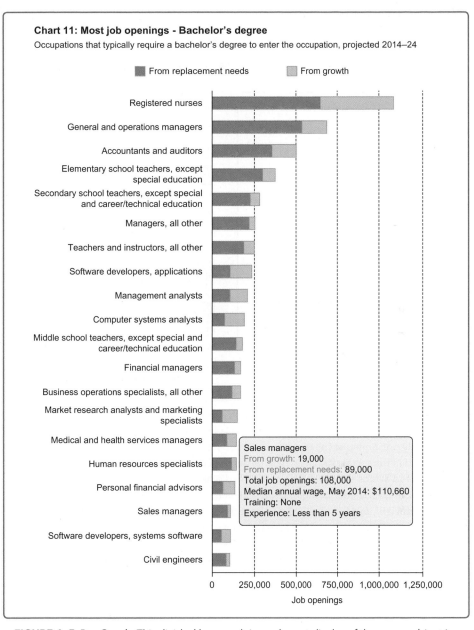

Chart 11: Most job openings - Bachelor's degree

Occupations that typically require a bachelor's degree to enter the occupation, projected 2014–24

■ From replacement needs   ■ From growth

- Registered nurses
- General and operations managers
- Accountants and auditors
- Elementary school teachers, except special education
- Secondary school teachers, except special and career/technical education
- Managers, all other
- Teachers and instructors, all other
- Software developers, applications
- Management analysts
- Computer systems analysts
- Middle school teachers, except special and career/technical education
- Financial managers
- Business operations specialists, all other
- Market research analysts and marketing specialists
- Medical and health services managers
- Human resources specialists
- Personal financial advisors
- Sales managers
- Software developers, systems software
- Civil engineers

Sales managers
From growth: 19,000
From replacement needs: 89,000
Total job openings: 108,000
Median annual wage, May 2014: $110,660
Training: None
Experience: Less than 5 years

0   250,000   500,000   750,000   1,000,000   1,250,000

Job openings

• **FIGURE 6–5** Bar Graph This divided bar graph is a coherent display of the same subject in different fields during the same period of time. The ordering of the bars from largest to smallest assists the viewer in navigating the 20-item list and making rapid sense of it. The two-item legend at the top of the display is effectively positioned, but the same distinction could be more efficiently conveyed by a simple labeling of the top divided bar. Sufficient white space separates each bar from its neighbors, and the vertical gridlines are clear and unobtrusive. Because this illustration is online, it allows explanatory pop-up windows with more details about projected job openings in each of the fields displayed (e.g., sales managers).

Source: US Department of Labor, Bureau of Labor Statistics. *Career Outlook*. Projections of Occupational Employment 2014-2024. Washington, DC: GPO, 2015. http://www.bls.gov/careeroutlook/2015/article/projections-occupation.htm.

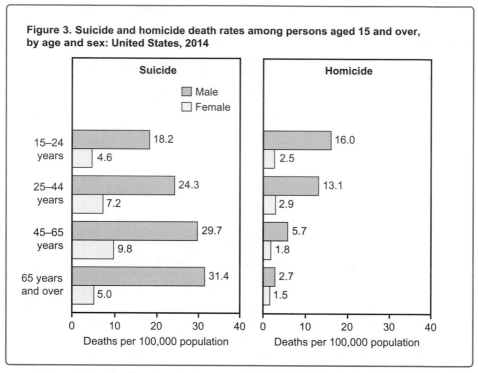

Figure 3. Suicide and homicide death rates among persons aged 15 and over, by age and sex: United States, 2014

• **FIGURE 6–6** Bar Graph This pair of bar graphs makes several clear and vivid points regarding the frequency of suicides and homicides for men and women. The legend is appropriately positioned at the top of the first set of bars, but might as easily have been incorporated in the top set of bars. The blue and gold coloring of the bars creates sufficient distinction (and avoids the stereotype of blue and pink). White space is generous, and each bar is numerically labeled (making grid lines unnecessary). The resulting illustration is crisp, clean, and efficient, making it easy to note that the suicide rate for men increases with age, that the homicide rate for men decreases with age, and that men are more likely than women to be victims of suicide or homicide at every age.

Source: US Department of Health and Human Services, National Center for Health Statistics. *Health, United States, 2015: With Special Features on Racial and Ethnic Health Disparities.* DHHS Publication No. 2016-1232. Washington, DC: GPO, 2016. p. 11.

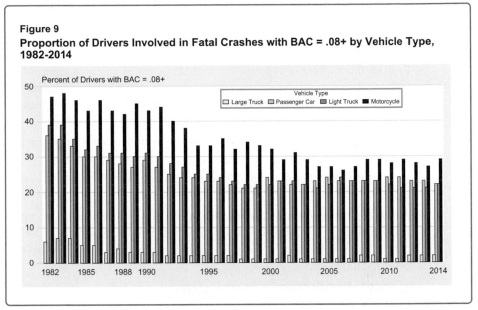

Figure 9
Proportion of Drivers Involved in Fatal Crashes with BAC = .08+ by Vehicle Type, 1982-2014

• **FIGURE 6–7** Ineffective Column Graph This column graph is difficult to decipher because it displays too many columns at too many points in time in too narrow a space. The distracting visual effect makes the data look crowded and cramped instead of readily accessible. Inconsistent labeling of the horizontal axis (with intervals of 3 years, 2 years, 5 years, 4 years) only multiplies the potential confusion. The graph would be more effective, for example, if only the odd-numbered years were displayed. In addition, the color of the columns for the large truck category is almost indistinguishable from the white background of the graph, making these particularly short columns easy to overlook in the visual clutter: gray would have been a better color choice for this category of vehicle. At least the explanatory legend is conveniently positioned inside the graph and sticks to simple solid colors instead of using dots and diagonal lines.

Source: US Department of Transportation, National Highway Traffic Safety Administration. *Traffic Safety Facts 2014*. DOT HS 812 261. Washington, DC: GPO, 2016. p. 49.

Circle graphs.

- Note that the purpose of a circle graph (also referred to as a "pie chart") is to display the number and relative size of the divisions of a subject.
- Restrict the number of segments in a circle graph to seven or eight. More will create confusion about their relative sizes. If necessary, combine several smaller segments and create a second circle graph to display the composition of that combined segment.
- Watch for possible distortion when you use three-dimensional circle graphs.
- Clearly label all segments. Whether they are placed inside or outside the circle, labels should be horizontal for easier reading. Avoid using explanatory legends (or keys).
- As you segment the graph, begin with the largest section in the upper right-hand quadrant. The remaining segments should be arranged clockwise, in descending order. Color the sections clockwise from darker to lighter so that color reinforces size.

See Figure 6–8 for an example of a circle graph.

Line graphs.

- Note that the purpose of a line graph is to show the degree and direction of change relative to two variables.
- Label each axis clearly. Like bar and column graphs, line graphs must have clearly labeled scales to show the variables you are measuring. Ordinarily, the independent variable is placed on the horizontal, or x, axis and the dependent variables are placed on the vertical, or y, axis (and, in a three-dimensional graph, on the diagonal, or z, axis).
- Choose the scale of each axis to show the appropriate steepness of the slope of the line. Typically, the scales start at zero, with the intervals equally spaced on each axis.
- The major difficulty in designing line graphs lies is choosing the spacing for each axis so that the steepness (slope) of the line accurately measures the actual trend suggested by the data. Typical software applications will allow you to adjust the intervals on the x axis and the y axis, but your job is to decide whether the slope of the graph accurately depicts your data or gives a distorted impression.
- Avoid using more than three data lines in one graph unless they are separated by visible space. Graphs with lots of overlapping and intersecting data lines are usually difficult to interpret.
- Keep the data lines in your graph distinctive by using either different colors or different styles for each line.
- If possible, label each data line. Avoid explanatory legends (or keys) because these will slow reader comprehension.

See Figures 6–9 and 6–10 for examples of line graphs.

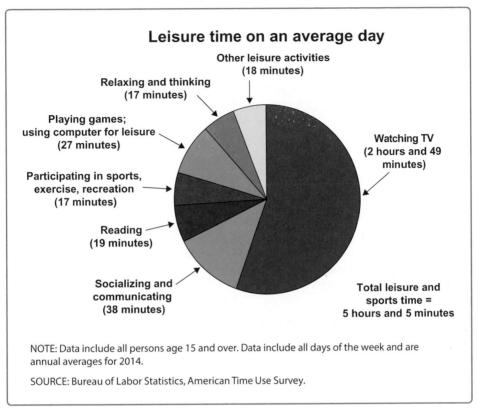

Leisure time on an average day

Other leisure activities
(18 minutes)

Relaxing and thinking
(17 minutes)

Playing games;
using computer for leisure
(27 minutes)

Participating in sports,
exercise, recreation
(17 minutes)

Reading
(19 minutes)

Socializing and
communicating
(38 minutes)

Watching TV
(2 hours and 49
minutes)

Total leisure and
sports time =
5 hours and 5 minutes

NOTE: Data include all persons age 15 and over. Data include all days of the week and are annual averages for 2014.

SOURCE: Bureau of Labor Statistics, American Time Use Survey.

• **FIGURE 6–8** Circle Graph This circle graph uses clear labels for each of its seven sections and consistently positions the labels outside the circle with arrows pointing to the pertinent sections. Circle graphs are easiest to decipher if their sections progress clockwise from largest to smallest, but this graph starts with the two largest sections and inexplicably shifts to arbitrary ordering. The coloring of the sections is also entirely decorative: instead of using increasingly lighter versions of a single color to reinforce the progression in size from largest to smallest, or using seven different colors to emphasize the distinctiveness of the seven sections, here the colors are assigned arbitrarily, including three variations of blue instead of blue, yellow, and brown.
Source: US Department of Labor, Bureau of Labor Statistics. *American Time Use Survey*. 2015. http://www.bls. gov/tus/charts/chart9.pdf

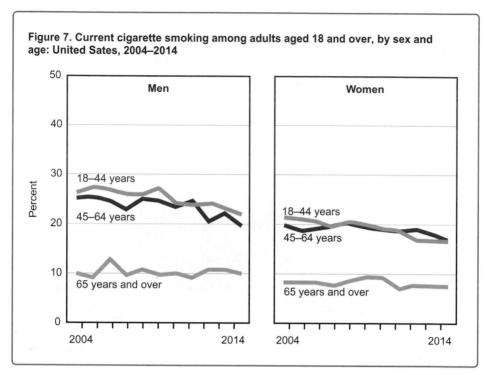

Figure 7. Current cigarette smoking among adults aged 18 and over, by sex and age: United Sates, 2004–2014

● **FIGURE 6–9** Line Graph This pair of line graphs uses clear labeling to identify and differentiate three data lines. It also uses color to allow easy comparison and contrast of the data lines across the two graphs, making evident that men have a higher rate of cigarette smoking at all ages and that the rate is dropping for men and women aged 18–44 and 45–64. Horizontal grid lines assist viewers in deciphering the relative positions of the paired data lines, while vertical grid lines prove unnecessary for the 10-year period.

Source: US Department of Health and Human Services, National Center for Health Statistics. *Health, United States, 2015: With Special Features on Racial and Ethnic Health Disparities*. DHHS Publication No. 2016-1232. Washington, DC: GPO, 2016. p. 13.

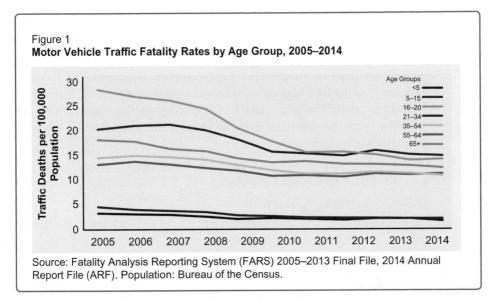

Figure 1
Motor Vehicle Traffic Fatality Rates by Age Group, 2005–2014

Source: Fatality Analysis Reporting System (FARS) 2005–2013 Final File, 2014 Annual Report File (ARF). Population: Bureau of the Census.

● **FIGURE 6–10 Ineffective Line Graph** This line graph offers several challenges, starting with the use of a color-coded legend for its seven data lines. If the graph were designed with more horizontal space, direct labeling of each data line would be possible and the legend could be eliminated. This change would also minimize the crowding of the data lines, especially in the later years. The graph does uses distinctive colors for the data lines, but the assignment of color is arbitrary, neither lighter to darker nor darker to lighter by age group. The non-white background for the graph only diminishes the contrast of the colors. In addition, a key point of the graph is that driver fatalities in all age groups dropped from 2005 to 2014 but most precipitously in the 16–20 age group. Designating this group with a black data line would direct attention to this important fact.

Source: US Department of Transportation, National Highway *Traffic Safety Administration. Traffic Safety Facts 2014: Older Population.* DOT HS 812 273. Washington, DC: GPO, 2016. p. 1.

Organization charts.

- Note that the purpose of an organization chart is to map the various divisions and levels of responsibility within an organization.

- Make the chart as simple as the organization itself, with the levels of hierarchy organized highest to lowest and positioned on the chart from top to bottom. The more levels of hierarchy in the organization, the more vertical the chart; the more divisions in the organization, the more horizontal the chart.

- Use the same shape in the same size for all divisions of the organization that are at the same level in the hierarchy.

- Label each division of the organization.

- If space allows, put the labels directly on the division; if not, attach the label to the subject with thin rules (never arrows).

- Position all the labels on the horizontal so that the viewer doesn't have to rotate the page or screen to read the labels.

- Connect each level of the organization to the higher and lower levels with a clear line (never arrows).

- Connect optional, informal, or temporary relationships with a dotted line.

See Figure 6–11 for an example of an organization chart.

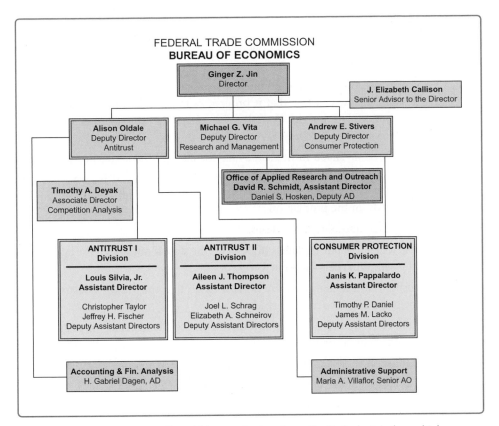

FEDERAL TRADE COMMISSION
BUREAU OF ECONOMICS

Ginger Z. Jin
Director

J. Elizabeth Callison
Senior Advisor to the Director

Alison Oldale
Deputy Director
Antitrust

Michael G. Vita
Deputy Director
Research and Management

Andrew E. Stivers
Deputy Director
Consumer Protection

Timothy A. Deyak
Associate Director
Competition Analysis

Office of Applied Research and Outreach
David R. Schmidt, Assistant Director
Daniel S. Hosken, Deputy AD

ANTITRUST I
Division

Louis Silvia, Jr.
Assistant Director

Christopher Taylor
Jeffrey H. Fischer
Deputy Assistant Directors

ANTITRUST II
Division

Aileen J. Thompson
Assistant Director

Joel L. Schrag
Elizabeth A. Schneirov
Deputy Assistant Directors

CONSUMER PROTECTION
Division

Janis K. Pappalardo
Assistant Director

Timothy P. Daniel
James M. Lacko
Deputy Assistant Directors

Accounting & Fin. Analysis
H. Gabriel Dagen, AD

Administrative Support
Maria A. Villaflor, Senior AO

• **FIGURE 6–11** Organization Chart This organization chart effectively depicts the multiple levels of authority within this agency: for example, assistant directors report to a deputy director who reports to the director versus the advisor who has direct access to the director. And while it is typical to display only position titles, here the names of the individuals in each position bring a sense of humanity to the organization. The use of color, however, could cause confusion: yellow is consistent for all the divisions, but orange is assigned to the agency's director and to a unique office that reports to a deputy director, and grey is assigned to deputy directors as well as to supporting administrators (the variation of double-ruled versus single-ruled boxes might easily go unnoticed). *Source*: US Federal Trade Commission. Bureau of Economics. https://www.ftc.gov/system/files/attachments/ bureau-economics-organization-chart/be_org_chart-jul16.pdf

Flow charts.

- Note that the purpose of a flow chart is to show the sequence of steps in a process or procedure.
- Make the flow chart as simple as the process itself. If a process is simple, design the flow chart so that it progresses in a single direction, usually top to bottom or left to right. Complicated designs that spiral and zigzag imply a more complicated process.
- Use the same shape in the same size for all equivalent steps or phases but different shapes for steps or phases of a different kind (e.g., circles for the stages in researching a document, squares for the stages of writing, and diamonds for the production stages of printing and binding).
- Label each of the steps or phases.
- If space allows, put the labels directly on the step; if not, attach the label to the step with thin rules (never arrows).
- Position all the labels on the horizontal so that the viewer doesn't have to rotate the page or screen to read the labels.
- Connect each step or phase in the sequence to the next step or phase with a clear directional arrow.
- Connect reversible or interactive steps or phases with double-headed arrows.
- Connect recursive or cyclical steps or phases with circular arrows.
- Connect optional steps or phases with dotted-line arrows.

See Figure 6–12 for an example of a flow chart.

● **FIGURE 6–12** Flow Chart This flow chart uses several levels of redundancy to simplify a process that might otherwise mystify or intimidate a prospective filer of a patent application. The process is organized in three columns: the number of each step, the agent of action, and the corresponding action itself. Actions are displayed in blue rectangles, while decision points are displayed in white diamonds. Arrows direct the viewer to the next step in the process and reinforce the numbering. Links to more information are included in almost every step but always opening as new pages in the viewer's browser to make it easy to preserve this overview of the entire process.

Source: US Department of Commerce, Patent and Trademark Office. *Process for Obtaining a Utility Patent.* 2013. http://www.uspto.gov/patents/process/index.jsp

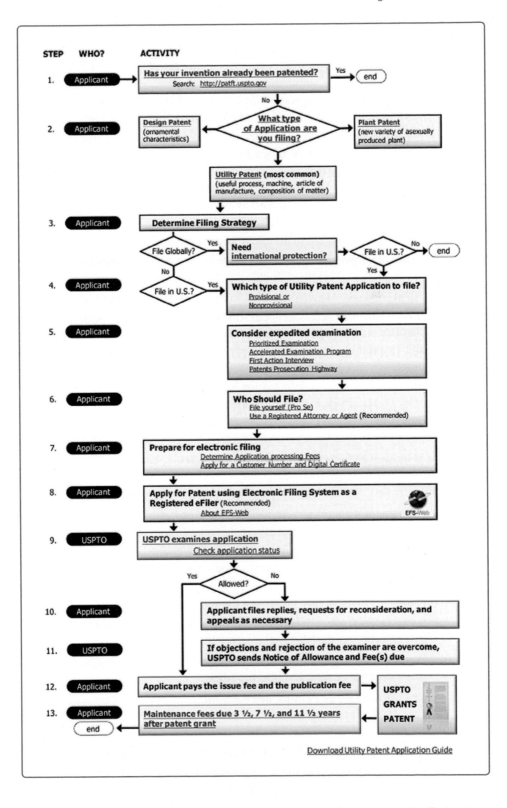

STEP   WHO?   ACTIVITY

1. Applicant   **Has your invention already been patented?** Yes → (end)
    Search: http://patft.uspto.gov
    No ↓

2. Applicant   **Design Patent** (ornamental characteristics) ← **What type of Application are you filing?** → **Plant Patent** (new variety of asexually produced plant)
    ↓
    Utility Patent (most common) (useful process, machine, article of manufacture, composition of matter)

3. Applicant   **Determine Filing Strategy**
    File Globally? Yes → **Need international protection?** → File in U.S.? No → (end)
    No ↓   Yes ↓

4. Applicant   File in U.S.? Yes → **Which type of Utility Patent Application to file?**
    Provisional or Nonprovisional

5. Applicant   **Consider expedited examination**
    Prioritized Examination
    Accelerated Examination Program
    First Action Interview
    Patents Prosecution Highway

6. Applicant   **Who Should File?**
    File yourself (Pro Se)
    Use a Registered Attorney or Agent (Recommended)

7. Applicant   **Prepare for electronic filing**
    Determine Application processing Fees
    Apply for a Customer Number and Digital Certificate

8. Applicant   **Apply for Patent using Electronic Filing System as a Registered eFiler** (Recommended)
    About EFS-Web   EFS-Web

9. USPTO   **USPTO examines application**
    Check application status
    Yes ← Allowed? → No

10. Applicant   **Applicant files replies, requests for reconsideration, and appeals as necessary**

11. USPTO   **If objections and rejection of the examiner are overcome, USPTO sends Notice of Allowance and Fee(s) due**

12. Applicant   **Applicant pays the issue fee and the publication fee** → **USPTO GRANTS PATENT**

13. Applicant   **Maintenance fees due 3 ½, 7 ½, and 11 ½ years after patent grant** ←
    (end)

Download Utility Patent Application Guide

Project schedule charts.

- Note that the purpose of a project schedule chart (also referred to as a "Gantt chart," after its creator, Henry Gantt) is to display the start and stop dates for all consecutive and simultaneous tasks related to a project. It offers a quick picture of tasks in progress at any given point in time, as well as how much work has been completed on each task and how much work remains.
- Make the project schedule chart as simple as the project itself, with one horizontal bar for every major task or activity and vertical grid lines for each reporting period (e.g., daily, weekly, monthly). Excessive division of the project into a multitude of minor tasks or milestones will create a chart that is difficult to decipher. Also avoid abbreviations and symbols and their explanatory legends (or keys).
- List the tasks from top to bottom in order of their start date.
- Extend a horizontal bar for each task from its start date to its anticipated date of completion or specified deadline.
- Fill in (or color) each bar from left to right as work on the task proceeds.
- If you don't have access to applications that will generate a project schedule chart, create a table in your word processing program with a row for each task and a column for each reporting period and fill in cells from left to right as progress is made.
- Incorporate additional information for each task as necessary and appropriate (e.g., names of task managers, facilities and resources assigned for each task, funds budgeted and expended in each reporting period), but realize that every extra level of detail complicates the illustration and slows interpretation.

See Figure 6–13 for an example of a project schedule chart.

Diagrams.

- Note that a diagram should identify parts of a subject and their spatial or functional relationship.
- Keep the diagram as simple as possible, avoiding unnecessary details or distracting decorations and focusing the viewer's attention on the key features or parts of the subject.
- Label each of the pertinent parts of the subject.
- If space allows, put each label directly on its corresponding part or adjacent to it; if not, connect the label to the part with a thin rule or arrow.
- Position all the labels on the horizontal so that the viewer doesn't have to rotate the page or screen to read the labels.

See Figure 6–14 for an example of a diagram.

Photographs.

- Note that photographs show what a subject looks like in realistic detail.
- Keep the photograph as simple as possible, focusing the viewer's attention on the key features or parts of the subject.

Task	Sep 2015	Oct 2015	Nov 2015	Dec 2015	Jan 2016	Feb 2016	Mar 2016	Apr 2016	May 2016	Jun 2016	Jul 2016	Aug 2016	Sep 2016
Task 1 – Program Mgt.													
Task 2 – Concept of Operations													
Task 3 –Security Concept													
Task 4 – Safety Plan													
Task 5 – Performance Measurement													
Task 6 – System Requirements													
Task 7 – App Planning													
Task 8 – Human Use Approval													
Task 9 – Training Plan													
Task 10 – Partnership													
Task 11 – Outreach Plan													
Task 12 – Deployment Plan													
Task 13 – Readiness Summary													

● Public Webinar

• **FIGURE 6–13** Project Schedule Chart This project schedule covers 13 tasks across 13 months, using bright orange to make readily apparent the months of activity for each task. The viewer will notice immediately that the project starts slowly in September 2015, picks up in February, and shifts to high activity in January, staying that way through May, easing a bit through August, and coming to a stop in a still hectic September 2016. A blue circle indicates the approximate dates of three public webinars with the explanatory legend for the blue circle displayed in unusually large type at the bottom of the chart, inviting the viewer's unrivaled attention. The implication of size here is that the webinars are the three principal activities of the project (i.e., all effort leading to or from the creation of the webinars) instead of noteworthy points or milestones in a process. Two possibilities for a quick fix: 1) using a smaller circle in the legend and the chart and a smaller type size in the legend or 2) substituting the word *webinar* in the smallest readable type size for the blue circle in the chart and eliminating the distracting legend altogether.

Source: US Department of Transportation. *Connected Vehicle Pilot Deployment Program: ICF/Wyoming.* http://www.its.dot.gov/pilots/pdf/CVP-ICF-yomingBriefing_v6.pdf

- Exercise caution in editing or polishing the photograph. Never insert or delete images. Viewers typically expect a photograph to be a representation of reality, and ethical communicators strive to meet that expectation. Cropping a photograph in order to close in on a subject and eliminate distractions in the background is a standard practice, but never insert or delete objects or change the size or color of objects in a photograph.

- If appropriate or necessary for the viewer's understanding, apply labels for each of the pertinent parts of the subject to direct the viewer's attention.

- If space allows, put each label directly on its corresponding part or adjacent to it; if not, connect the label to the part with a thin rule or arrow.

- Position all the labels on the horizontal so that the viewer doesn't have to rotate the page or screen to read the labels.

See Figure 6–15 for an example of a photographic display.

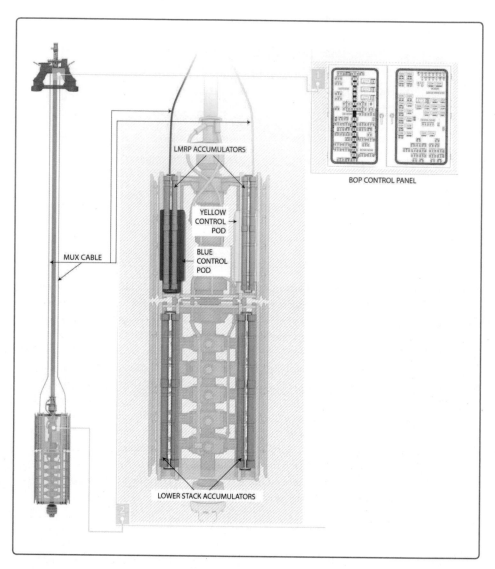

• **FIGURE 6–14** Diagram This diagram focuses on the details pertinent to the investigation of the explosion and fire that killed 11 and injured 17 at the Deepwater Horizon oil-drilling facility in the Gulf of Mexico. The image on the left depicts the spatial relationship of the rig at the surface to the blowout preventer (BOP) stack below it and the multiplexed (MUX) cable connecting the two, while the magnified images to the right focus on key components related to the investigation. Each of these components is labeled, and the figure caption explains their operation. While the report itself includes a glossary of 45 abbreviations, the caption here could offer convenient parenthetical definitions of the three abbreviations used in this diagram (e.g., LMRP: Lower Marine Riser Package).

Source: US Chemical Safety and Hazard Investigation Board. *Explosion and Fire at the Macondo Well, Deepwater Horizon Rig, Mississippi Canyon Block #252, Gulf of Mexico.* Investigation Report, Volume 2. No. 2010-10-I-OS. Washington, DC: GPO, 2014. p. 24.

• **FIGURE 6–15** Photographic Display This display of four photographs offers the viewer a realistic perspective from inside a research vehicle while the green circles and lines direct attention to the adjacent magnified images of warning signals that would otherwise be unreadable. The effectiveness of this photographic display is jeopardized, however, by the extraneous materials in the research environment that are visible through the windshield (e.g., the shelving unit on the left side) and in the rear-view mirror. Positioning the vehicle in front of a blank wall and tilting the mirror would eliminate these potential distractions.

Source: US Department of Transportation, National Highway Traffic Safety Administration. *Enhanced Seat Belt Reminder System Features for Teenagers*. DOT HS 812 173 Washington, DC: GPO, 2015. p. 10.

Infographics.

- Note that infographics make information easier to understand and easier to remember.

- Keep the pictorial images of the infographic as simple and straightforward as possible, avoiding unnecessary details or distracting decorations and focusing the viewer's attention on key traits.

- Choose pictorial images appropriate to your likely viewers (especially their age, level of education, and culture). Make sure the image clarifies without insulting or irritating your viewers.

- Choose pictorial images appropriate to the gravity of your subject (especially for issues related to safety). Make sure the image neither trivializes nor distorts the subject.

- Reinforce the pictorial image with a minimum of explanatory text, labels, or captions.

See Figure 6–16 for an example of an infographic.

Video Clips.

- Note that video clips illustrate a process, operation, or incident. Film clips depict activity in realistic detail. Animation clips simulate activity that would be impossible, imprecise, or impractical to film.

- Keep the video clip as simple and straightforward as possible, avoiding unnecessary details or distracting decorations and focusing the viewer's attention on the key actors, equipment, and steps or stages in the process, operation, or incident.

- If appropriate or necessary for the viewer's understanding, insert audio narrative or titles and captions that direct the viewer's attention to pertinent information.

- Choose animation images appropriate to your likely viewers (especially their age, level of education, and culture) as well as the gravity of your subject (especially for issues related to safety). Make sure the animation clarifies without insulting or irritating your viewers and neither trivializes nor distorts your subject.

- Include closed captioning for the audio portion of your video clip (i.e., for narration, dialogue, sound effects, and music).

- Exercise caution in editing or polishing a film clip. Viewers typically expect a film clip to be a representation of reality, and ethical communicators strive to meet that expectation. Editing a film for extraneous material is a standard practice, but never change the speed or sequence of the process, operation,

• **FIGURE 6–16** Infographic This array of infographics replicates a smartphone display, despite the error of the extraneous cursor. The design choice is clever and engaging as this device is a familiar source of information for the target audience for this message—adults in their child-bearing years. The illustration is easy to read and easy to navigate with intuitive images and clear verbal and numerical labels.

Source: US Department of Agriculture, Center for Nutrition Policy and Promotion. *The Cost of Raising a Child*. 2014. http://www.cnpp.usda.gov/sites/default/files/expenditures_on_children_by_families/CRC2013InfoGraphic.pdf

It's all done before the dirt is loaded into the hopper.

• **FIGURE 6–17** Film Clip This 6-minute video explains the process that geophysicists use to study debris flows from landslides and volcanic eruptions. The use of film offers a realistic impression of the substantial and detailed efforts involved in setting up the specialized equipment and running the experiments. Titles and diagrams are inserted periodically during the film to identify the researchers involved and to reinforce points made in the audio narration about the activities depicted. Closed captions make the audio narration accessible to viewers with limited hearing. *Source*: US Department of the Interior, US Geological Survey. *Catching the Quakes.* Video. 2016. https://www. usgs.gov/media/videos/catching-quakes

or incident (unless the change is altogether obvious), and always make sure that your editing creates no distortions or misrepresentations. In addition, treat each frame of the film as you would a photograph: never insert or delete objects or actors and never alter the size or color of the items in a frame.

See Figures 6–17 and 6–18 for screen shots of sample video clips.

Designing Illustrations Ethically

Displaying information ethically requires that you make careful choices about the design of your illustrations. For example, the scale of the *x* and *y* axes on a line graph has a significant impact on the data display. In designing a graph, you ordinarily start the *x* and *y* axes at zero. Exceptions are possible if beginning at some other point ("suppressing the zero") will not distort information. If several graphs are positioned side by side on a page, thus inviting comparison and contrast of the data, it would be unethical to use differing vertical scales because readers would likely come to incorrect conclusions.

Because the worker was using a highly conductive metal ladder

• **FIGURE 6–18** Animation Clip This brief video is about the safe use of ladders in the vicinity of overhead power lines. The use of animation allows the simulation of a cautionary case—a fatal accident in which a worker's aluminum ladder made accidental contact with a power line and conducted electricity (here depicted in purple) from the line to the worker, resulting in a tragic electrocution. The animation proceeds to demonstrate how the use of fiberglass ladders and safe practices in the same work environment would avoid similar accidents. Closed captions make the audio narration accessible to viewers with limited hearing.
Source: US Department of Labor, Occupational Safety and Health Administration. *Electrocution/Work Safely with Ladders Near Power Lines*. Video. 2014. https://www.osha.gov/video/index.html

Using distorted graphs, however, isn't the only error that will result in the creation of unethical illustrations. It is unethical to create a drawing that puts features on a product it doesn't really have. It is unethical to design a flow chart that disguises the complexity of a process by making it look relatively simple. It is unethical to stage or doctor a photograph to create a positive or negative impression of your subject that isn't fully justified. It is unethical to edit a film so that a 5-minute operation seems to take only 4 minutes. It is unethical to incorporate illustrations that might be easily misinterpreted.

In addition, if you illustrate information regarding people, you must strive to be sensitive to their humanity. A thoughtless or insensitive illustration could be widely and quickly circulated on the Internet, jeopardizing your organization's reputation for caring about people, especially the people in the communities in which it operates. For example, if you choose a conventional bar graph to depict the human beings killed in mining explosions, you make simple objects of real people who have surviving families and friends. In mapping the damage from a tornado, if you ignore the locations of the people who were killed and focus only on demolished buildings, you both diminish the loss of life and give a distorted picture of the disaster. Always consider ways that you might bring a sense of humanity (and thus greater accuracy) to your illustrations. Notice, for example, how the human dimension of safety on construction jobs is made evident in Figure 6–19 by the inclusion of icons of the workers.

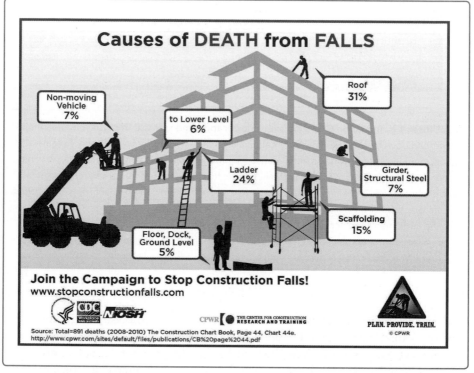

• **FIGURE 6–19** Infographic Sensitive to Potential Victims This infographic identifies the relative frequency of seven different kinds of fatal falls at construction sites. It does so by positioning human icons at the job site engaged in typical but potentially dangerous activities and attaching a clear numerical and textual label to each icon that identifies the activity and its frequency as the cause of a fatal fall. It is impossible to view the data here without thinking about actual construction workers and recognizing the dangers they expose themselves to every day in their occupation.
Source: US Department of Health and Human Services, Centers for Disease Control and Prevention, National Institute for Occupational Safety and Health. 2016. http://www.cdc.gov/niosh/construction/pdfs/cpwr-infographic-2-508c.pdf

Testing Your Illustrations

You might think your tables and figures are lucid and engaging, but some in your audience could be unfamiliar with the kinds of illustrations that you are using and unsure of how to decipher a graph or interpret a diagram. Others might be confused by the headings in a table or puzzled by the caption for a photograph. And still more could find the video you included to be entertaining but uninformative. Testing your illustrations on representative readers will help you to make sure that your tables and figures are as effective as possible.

To conduct your usability test, give each of your representative readers a copy of your illustrations and ask him or her to find the answers to several questions about the subject and about the illustration itself. For example, in a circle graph about childhood diseases, you might ask readers to find the answers to the following questions:

- Which childhood disease occurs with the third greatest frequency?
- What is the relative frequency of measles?
- What is the key message of this circle graph?

You would observe each representative reader as he or she examines your illustration and assess the time it takes to answer each question as well as the accuracy of the answers.

You could thereafter also interview the representative readers, either individually or jointly, about their impressions of your illustrations: Which did they like? Which did they dislike? What elements in each did they find helpful? What elements did they find unclear, distracting, or unnecessary? What do they think would make your illustrations more effective?

The results of this usability test will guide you in revising your illustrations according to the needs, interests, and abilities of your audience.

✐ Illustration Checklist

Planning

☐ What kinds of illustrations is your audience familiar with?

☐ Do you have information that could be more easily or quickly communicated to your audience visually or in a combination of words and images?

☐ Do you have definitions that could be displayed visually in whole or in part?

☐ Do you have any processes or procedures that could be depicted in a flow chart?

☐ Do you have information on trends or relationships that could be displayed in tables and graphs?

☐ Do you have masses of statistics that could be summarized in tables?

☐ Do you need to depict objects? If so, what do you need to display about the objects? Do you need to focus attention on specific aspects of the objects with diagrams? Do you require the realistic detail of photographs?

☐ Do you need to depict incidents or operations? If so, what do you need to display? Do you need to focus attention on specific aspects through animation? Do you require the realistic detail of film?

☐ What are the design conventions of your illustrations?

☐ Are there suitable illustrations you could borrow or adapt? Or will you need to create them yourself?

☐ How will you make your illustrations accessible to individuals with limited vision or hearing?

Revising

☐ Are your illustrations suited to your purpose and audience? Do your illustrations communicate information ethically? Are your illustrations effectively located and easy to find?

☐ Are your illustrations accessible?

☐ Are your illustrations numbered and labeled?

☐ Do your verbal and visual elements reinforce each other?

☐ Are your illustrations genuinely informative instead of simply decorative? Have you helped your audience to interpret your illustrations with commentary or annotations?

☐ Have you acknowledged the sources for borrowed or adapted tables and figures?

☐ Have you tested your illustrations with representative users?

 EXERCISES

1. Examine the illustrations used in three journals or magazines in your major field of study. Which types of illustration do you ordinarily find? Which types don't you find? Which occur most often? Which occur least often? What are the conventions for tables and figures in your field? Summarize your findings in a brief oral presentation with slides that display representative illustrations from your field.

2. Zeitgeist Corporation designs games for mobile devices and is looking for individuals with the experience and training to conceive and design state-of-the-art products. It has been a small private corporation for 5 years and hopes to achieve a major expansion. The president of Zeitgeist, Miyuki Sasaki, has asked

you to develop recruiting materials. You've compiled the following information that you believe will be pertinent:

History	Founded in 2012, in Long Beach, California, by Miyuki Sasaki, BS in Computer Science, 2010, University of Ohio; MBA, 2011, University of Los Angeles
Sales	2013: $250,000; 2014: $750,000; 2015: $1 million; 2016: $4 million; 2017: $7 million
Employees	With specialists in design (3), graphics and animation (4), programming (5), video editing (3), production (2), quality assurance (1), documentation (1), sound (1), music (1), closed captioning and subtitles (1)
Products	5 current products, looking to diversify, especially in action–adventure, newest product: Street Maniac (driving game); biggest seller: Ninja Zombie Diaries (fantasy role-playing)
Facilities	State-of-the-art equipment, serves both Android and iOS systems
Location	1221 Alamitos Avenue, Long Beach, California: new environment-friendly building, spacious offices
Salary/benefits	Competitive with industry; exceptional medical coverage, 4 weeks of vacation each year, 2 weeks of paid community service time each year

Design recruiting materials for Zeitgeist that incorporate as wide a variety of illustrations as possible.

3. In designing your recruiting materials for Zeitgeist, you are surprised by a dilemma. Miyuki Sasaki is willing to hire people with disabilities and would like you to include that information. At this time, Zeitgeist has no employees with disabilities, but Jacqueline Bailey, a programmer, injured her back in a skiing accident and is temporarily restricted to a wheelchair. Sasaki proposes including a photograph of various Zeitgeist employees (including Bailey in the wheelchair) to indicate to prospective employees that people with disabilities would find a supportive environment at Zeitgeist. You mention the idea to staff photographer Kishor Mitra, but he objects, claiming such a picture would be deceptive. As project leader, you must decide who is right and compose one of two e-mail messages: either to convince Sasaki that such a photograph would be unethical or to direct Mitra to take the photograph in spite of his objections. How do you proceed?

4. Examine Figure 6–20. If you were given 15 minutes on the job to revise this one-page document with more effective illustrations, what changes would you make? Because you won't have time to do everything, focus your efforts on the additions, deletions, and substitutions that you think will have the greatest impact on the document's usability.

INTERNET OF THINGS
TIP CARD

The Internet of Things refers to any object or device that sends and/or receives data automatically via the Internet. This rapidly-expanding set of "things" includes tags (also known as labels or chips that automatically track objects), sensors, and devices that interact with people and share information machine to machine.

WHY SHOULD WE CARE?

- Cars, appliances, wearables, lighting, healthcare, and home security all contain sensing devices that can talk to another machine and trigger other actions. Examples include: devices that direct your car to an open spot in a parking lot; mechanisms that control energy use in your home; and other tools that track your eating, sleeping, and exercise habits.

- This technology provides a level of convenience to our lives, but it requires that we share more information than ever. The security of this information, and the security of these devices, is not always guaranteed.

- Though many security and resilience risks are not new, the scale of interconnectedness created by the Internet of Things increases the consequences of known risks and creates new ones.

SIMPLE TIPS

Without a doubt, the Internet of Things makes our lives easier and has many benefits; but we can only reap these benefits if our Internet-enabled devices are secure and trusted. Here are some tips to increase the security of your Internet-enabled devices:

1. **Keep a clean machine.** Like your smartphone or PC, keep any device that connects to the Internet free from viruses and malware. Update the software regularly on the device itself as well as the apps you use to control the device.

2. **Think twice about your device.** Have a solid understanding of how a device works, the nature of its connection to the Internet, and the type of information it stores and transmits.

3. **Secure your network.** Properly secure the wireless network you use to connect Internet-enabled devices.

Stop.Think.Connect.™ is a national public awareness campaign aimed at increasing the understanding of cyber threats and empowering the American public to be safer and more secure online. The Campaign's main objective is to help you become more aware of growing cyber threats and arm you with the tools to protect yourself, your family, and your community. For more information visit www.dhs.gov/stopthinkconnect.

 Homeland Security   www.dhs.gov/stopthinkconnect   STOP | THINK | CONNECT

• **FIGURE 6–20** Document for Exercise

PART **TWO**

Applications

E-mails, Texts, Memos, and Letters

On the job, you will write and receive more e-mail messages, text messages, memos, and letters than probably any other kind of document. And you will likely write and receive them more often than any other kind of document. Because of the sheer quantity and frequency of correspondence that people must manage, they can often overlook, forget, or misread important items of information.

 Quick Tips

- Prepare readers for the key information in your message by using a clearly worded subject line or opening sentence.
- Put the most important information in the opening paragraph.
- Format the message to help readers recognize its organization and major points.
- Keep sentences and paragraphs short. Readers should not have to read your message more than once to grasp your meaning.
- Try to keep your message to a single page or screen. If your message exceeds a single page or screen, use headings to guide your reader through it.
- Keep a copy of every message you send and every message you receive for at least 3 years. When others forget what they wrote to you or what you wrote to them (or the fact that you wrote at all), your ability to retrieve the message in question will save time, avoid arguments, correct mistakes, and reinforce your reputation for efficiency.

E-mail and Text Messages

E-mail and text messages constitute the majority of job-related communication. You will often find yourself reading or writing at least 100 e-mail and text messages every day. Text messages are by definition brief and, thus, ideal for urgent notifications and quick acknowledgments—information that can't wait for e-mail. Appropriate e-mail and texting etiquette continuously evolve, but both kinds of messages require conciseness and readability. Keep in mind that typical mobile devices display the list of incoming e-mail messages with the opening five lines of each message and the list of incoming text messages with the opening two lines. You will assist your recipients in navigating their incoming messages if you make sure to put your key point in the opening lines.

Note also the continuing error of people e-mailing or texting thoughtlessly, without considering the issues that could arise if their messages were later forwarded to unintended recipients or made public. Exercise discretion in all cases and with all audiences. Keep in mind that everything digital is universally and infinitely available. As thousands of government and corporate officials across the world discover again and again to their great dismay, if you or anyone you know is ever investigated for any kind of impropriety, all your e-mail and text messages could be exposed for humiliating scrutiny.

Memos and Letters

Memos are the relatively short and informal documents that circulate within organizations and address internal decisions and operations. Memos can be posted to the organization's website (including bulletin boards, blogs, and wikis), sent as attachments to e-mail messages, or printed and distributed.

Letters are typically intended for individuals outside the writer's organization, but can be used as official documents within the organization. Like memos, letters can be sent as e-mail attachments.

Letters and memos almost always benefit from the use of headings. Even a one-page document will be easier for readers to skim and navigate if it includes headings. And any memo or letter that exceeds one page must include headings to assist readers in identifying the topics covered and in later retrieving key items of information. Notice how the headings and subheadings in Figure 7–1 help the reader to assess the situation and take appropriate action in the limited time available.

Guidelines for Ensuring Quality

The following seven guidelines will help you to communicate effectively by e-mail, text, memo, and letter:

1. Make sure you look and sound professional as you choose words and put together sentences. Review and proofread messages carefully before you send them.

USDA

United States Department of Agriculture

Farm and Foreign Agricultural Services

Risk Management Agency

1400 Independence Avenue, SW
Stop 0801
Washington, DC
20250-0801

July 29,2016

INFORMATIONAL MEMORANDUM: IS -16-002

TO: All Approved Insurance Providers
All Risk Management Agency Field Offices
All Other Interested Parties

FROM: Robert Ibarra   */s/ Robert Ibarra*   *7/29/2016*
Deputy Administrator for Insurance Services

SUBJECT: Valuing Tart Cherry Crop Production on Diverted Acreage

BACKGROUND:

In June, the Cherry Industry Administrative Board (CIAB) that administers the Federal Tart Cherry Marketing Order estimated a large 2016 tart cherry crop and announced diversion provisions on the crop for supply control.

On July 8, 2016 fruit crops in northern Michigan were impacted by a significant hail storm. Reports indicate varying degrees of damage has occurred to the tart cherry crop and has resulted in questions regarding how production on diverted acreage of tart cherries is valued.

ACTION:

Insureds who suffer a crop loss should notify their crop insurance agent immediately. The Approved Insurance Provider (AIP) can send a loss adjuster to discuss the insured's options and work the claim.

For tart cherry acreage that may be diverted:
The Actual Revenue History (ARH) Tart Cherry Crop Pilot Provisions require the insured to notify the AIP within 2 days of any decision to divert any part of your crop under the in-orchard diversion provisions of the Federal Tart Cherry Marketing Order for supply control. The insured's AIP has the option to accept the CIAB assessment of marketable tart cherries or complete their own appraisal for tart cherries with diverted acreage.

Annual price for diverted production:
The marketable in-orchard diverted production will be valued at 80 percent of the annual price. The annual price is defined in the Actual Revenue History (ARH) Tart Cherry Pilot Crop Provisions.

DISPOSAL DATE:

November 1, 2016

USDA is an Equal Opportunity Employer and Provider

• **FIGURE 7–1** One-page Memo Using Headings and Subheadings

Every digital message you send will exist indefinitely and could easily be forwarded to additional readers across the world or obtained by unexpected audiences for years to come. In addition, every message you send or receive through your company e-mail account (as well as all attached documents) and every text message you send or receive through your company's message service are subject to review at any time by officials of your company and their legal representatives. Your privacy is limited to messages sent or received through your private accounts on your personal equipment, but even these messages could be subject to investigation if you are discussing company or government business.

2. Keep the six *C*'s in mind as you develop e-mails, texts, memos, and letters: be *concise, concrete, complete, correct, courteous,* and *clear.* Be attentive to how your words can be perceived by your reader. Rephrase anything that you think might be construed in ways you do not intend.

3. Be aware of how your message looks. A letter or memo with typographical or formatting errors makes an unfavorable statement about you and your organization. Design e-mail messages with equal care because these may be printed and copies made and distributed.

4. Watch for spelling errors as well as errors in sentence structure and word usage. For especially important e-mail messages, consider printing copies before you send them. Editing printed material often reveals errors you may miss as you compose or proofread on a screen. Also use the spellcheck function available on your e-mail application to help you find errors, but keep in mind that spellcheck is no substitute for careful reading.

5. For clarity in international communication, spell months of the year instead of using numbers: that is, 3/11/17 would likely be read as March 11 in the United States but as November 3 in Europe and Asia, and readers won't necessarily know whether you are using your practice or trying to be sensitive to cultural differences and adopting theirs. Consider also the pertinent cultural practices related to weights and measures, titles of address, spellings of names, monetary currencies, etc. If identifying a price, indicate the currency of the denomination (e.g., $15 USD).

6. For text messages on the job, avoid slang and abbreviations unless you know your recipient personally. Always keep your messages on point and professional. Always assume that your message could be forwarded to your boss, published in the newspaper, or distributed on Twitter.

7. For memos, letters, and e-mail messages, use the following structure:

First Paragraph	State the purpose of the message or the main information the reader needs. If your message delivers disappointing news, you may want to cushion the reader's disappointment with a supportive opening and reserve the news for the following paragraph. If you are communicating

	across international cultures you might choose to acknowledge this situation with a special greeting in your reader's language.
Middle Paragraphs	Support or develop the main topic stated in the first paragraph. Limit each paragraph to one idea
Final Paragraph	Tell the reader what to do or what position to take. If appropriate, include closing comments (e.g., offer to explain in greater detail or answer questions).

Appropriate Tone in E-mails, Texts, Memos, and Letters

Be careful of how you come across to your readers. What you write always carries with it a perceptible attitude toward the audience and the subject. As they digest your message, readers often vocalize your sentences and phrases. Anticipate the intonation that your readers will attach to what you write. Adopt a respectful, positive perspective toward your reader—a "you" attitude that addresses the subject from your reader's point of view—and avoid wording that might sound rude, sarcastic, or irritable. Remember: You want to be clear, but you also want to maintain a favorable relationship with your readers.

Avoid phrases that suggest that the reader is careless or unintelligent:

You neglected to . . . You failed to . . .
I simply cannot understand how you could . . .

Also avoid phrases that imply that the reader might be lying or exaggerating:

You claim that . . .

Omit excessive and insincere flattery:

It is indeed a profound privilege for us to work with you on your immensely important project.

I look forward to the extraordinary opportunity to submit to you my full proposal. Your firm is known worldwide for its incomparable commitment to excellence, which in all humility I aspire to emulate.

Keep in mind that your effort to be formal and objective might come across as impersonal:

The changes agreed upon per our conversation are herein included. Your immediate reply will be greatly appreciated.

Instead, write with a conversational tone that lets your reader know that you are grateful for his or her attention:

The changes included here are the ones we discussed during yesterday's telephone conversation. Please let me know your decision as soon as you can.

Often, in an effort to be clear, writers may sound tactless:

> If your employees had actually read the procedures before trying to install the pump, they would have seen that the installation requires that the sealant be allowed to set for 1 hour before beginning the second stage of the process. Not following this procedure causes the diagnostic to shut down because the sealant has not dried enough to absorb the test stress. To avoid future installation problems, please see to it that your employees read the procedures and read them thoroughly.

Thinking about the emotion that your writing may elicit could help you to phrase the same idea in a less caustic way:

> Please note in the procedure manual that the sealant must set for 1 hour before the second stage of the installation process can begin. This amount of time is necessary for the sealant to dry enough to absorb the test stress. If the second stage of the installation process is attempted before the sealant has dried, the diagnostic will shut down.

Following are two additional examples of tactless phrasing and their revisions:

> Regrettably, equipment expense reimbursements are not allowed.

versus

> My office can reimburse you for travel-related expenses only. Your invoice for equipment may be submitted to the VP of Operations.

> We cannot fulfill your request for a list of company positions and annual salaries as this is highly confidential information and only company executives are allowed to have access to it.

versus

> The list of positions and corresponding salaries that you requested is confidential information for use by our company personnel only. For your research, however, we can send you related information on industry-wide trends in similar positions.

Guidelines for Dealing with Tone

You cannot anticipate every possible meaning that your message will have for your reader, but you can achieve clarity and maintain goodwill if you keep the following guidelines in mind as you compose your letter, memo, text, or e-mail:

- Allow more time for designing messages about sensitive issues or that might be subject to misreading. Try to avoid sending bad news by text or e-mail.
- Anticipate the emotional impact of messages that will convey negative or disappointing news. People often take bad news personally and may be more inclined to take offense or respond with anger, confusion, or suspicion.

Dissatisfied readers could attach unintended meaning or motives to individual words and phrases.

- Always analyze your reader as carefully as possible, particularly the reader's frame of reference and attitude toward both you and the subject. Always be courteous.

- Read aloud what you have written. When you hear what you have written, you may often detect words and phrases that don't convey the attitude you intend.

- For any especially difficult message, compose the message, save the draft, and then let it sit for a while. Later, read the message aloud to check for clarity and courtesy.

Writing for Social Media

You could be assigned on the job to communicate with the public through social media about your organization's products, services, or projects. Keep in mind that all the principles of effective writing apply. You must have a clear sense of your audience and your purpose in writing, you must write in a way that is easy to read and easy to understand, and you must come across as a likable and credible representative of your organization. To achieve your objectives, consider the following:

1. **Write in a Distinctive and Consistent Voice:** Your style of writing (from formal to informal) and your attitude toward your subject and readers (e.g., serious and authoritative, caring and supportive, ironic and lighthearted) will characterize your organization in the minds of your readers. The more distinctive and consistent you are, the more memorable and persuasive will be your messages. Make sure your voice is appropriate for your organization and creates a positive impression of its products, services, and projects.

2. **Ask a Colleague to Review and Edit Your Messages:** Because your messages on social media could be distributed universally and infinitely, make sure that their meaning is clear and that misinterpretation is almost impossible. The easiest and surest way to assure the clarity of your intended meaning is have a judicious colleague review your messages prior to their distribution. Make sure your reviewer gives extra attention to every message that you think is hilarious: errors here could make your organization look insensitive, foolish, or sordid instead of perceptive, inventive, and skilled.

3. **Keep Your Messages Concise:** The principle of "brief and to the point" prevails across all social media: that is, readers expect maximum information in minimum words. Look for opportunities to substitute shorter and simpler words or to delete unnecessary words. Consider emojis and abbreviations to save words but only if this is appropriate to your voice.

4. **Give Your Reader Useful and Usable Information:** Readers will read your message and look for subsequent messages from you if you provide information they need in order to take actions and make decisions about their jobs, their professions, their industries, and their communities. This information must be accessible, typically on a mobile device, with operating links to more

information from credible online sources. Check the links in your message to verify their operation and make sure this information is genuinely pertinent to your reader's needs. Never include a link to a site that you have never visited.

5. **Encourage Sharing of Your Message:** A key objective of social media is the sharing or forwarding of messages. The more engaging your message, the more likely it is a reader will forward or share it. To make your messages more engaging, focus on actions, invite answers with questions, and include the keywords that might occur in a topic search of related messages.

For example, instead of

ABC has 3 new courses in Graphic Design for students to enroll in today at abc.edu/courses.

write

Are you a graphic artist looking for 3 new design courses? Enroll today at ABC.edu/courses.

Also keep in mind that your job is to activate readers with interesting messages but never with hostile or offensive messages: again, a conscientious colleague will alert you to potential disasters in the making.

Planning and Writing Correspondence

The following cases (7–1 through 7–6) introduce typical situations on the job in which writers must write a letter, memo, or e-mail message.

Review of Principles

As you plan your e-mails, texts, memos, and letters . . .

10 Questions to Ask Yourself

1. How well do I know the readers?
2. How much do they know about the topic discussed in the message?
3. How will they respond to what I will need to say? In what ways can I use this communication to build rapport for my organization?
4. What exactly am I trying to accomplish with this message?
5. What is their level of knowledge about the subjects discussed?
6. What is their attitude toward me and my organization?
7. What previous business dealings have I/we shared with them?
8. How much and what kind of information should I include, based on their profile?
9. How technical can I be in presenting my message?
10. What strategies can I use to make this message easy to read and understand?

CASE 7-1	**Informational E-Mail Message**

David Reilly, an administrative assistant to HCI's chief financial officer (CFO), must notify 14 division managers of a meeting to deal with looming budget cuts. David decides to attach three documents that those attending will need to have read by the meeting. The e-mail message explains the reason for the meeting, the documents attached, and the proposed agenda. As David explains, everyone should bring the attached documents to the meeting.

This e-mail message exemplifies effective design (Case Document 7–1). Note the clearly phrased subject line, the action-required statement in bold type, and the placement of the main information—the meeting and required attendance—in the opening sentence. The meeting agenda appears as a list, and the managers are told exactly what they must do with the attached materials. The message is concise and visually accessible.

CASE DOCUMENT 7–1

From: David Reilly [david.reilly@hci.com]
Date: February 23, 2017
To: Operations Support Staff
Subject: Agenda for March 13 Meeting on Proposed Budget Cuts

Attachments: CFO 1-31.pdf, NSP bid.pdf, OSHA oxygen proc rev.pdf, SHR mntnc cost.pdf

ACTION REQUIRED: Prepare New Cost Figures for Your Projects by March 12

All division managers are required to attend a March 13 meeting to decide how to cut the third and fourth quarter budgets. Please refer to the CFO's memo of January 31 (see attached) for guidelines on cuts.

You will receive a separate e-mail message later today asking you to put this mandatory meeting on your calendar. Please clear your calendar now of any conflicting appointments.

Time and Location: March 13, 8:00–11:00 a.m., Third Floor Conference Room

Please have four copies of your budget prepared with the 15% cut suggested by the CFO's 1/31 memo.

Meeting Agenda
1. Budget presentations and discussion
2. Proposed bid development on the North Shore Power project (see attached)
3. Revision of Procedures to Meet OSHA guidelines on oxygen tanks (see attached)
4. Cost overrun problem with maintenance contractors on the Signal Hill Reclamation project (see attached)

Please review all attached documents and be prepared to discuss viable solutions.

If you have problems or questions, please let me know.

DR

David Reilly
Administrative Assistant Office of the CFO
HCI, Inc.
889–742–2500

CASE 7-2 **Instructional Memo**

The following memo (Case Document 7–2) announces Quaker Pharmaceutical's new agreement with a local rental car dealership and provides instructions for renting cars for business use. The writer, Luning Jiang, suggests that readers may want to keep the memo for reference.

CASE DOCUMENT 7–2

MEMORANDUM

TO: Quaker Pharmaceutical sales representatives
FROM: Luning Jiang
SUBJECT: New Benefits from Horizons Rent-A-Car
DATE: 10/17/17

Quaker Pharmaceutical signed an agreement today with Horizons Rent-A- Car offering a 20% discount for all business-related travel. Please retain this memo for your files.

Business Vehicle Guidelines
- 1 or 2 employees traveling: use compact car; 3 or more employees traveling: use sedan or minivan.
- Luxurious vehicles for transporting clients or potential clients must be approved by your supervisor.
- Please lease the most economical vehicle available.

Regulations

- Employees are expected to follow all traffic laws.
- Employees under 25 cannot drive unless accompanied by a fellow employee over 25.
- Vehicle cannot be used to transport persons or property for hire.
- Vehicle cannot be used to tow anything.
- Horizons Rent-a-Car and **Quaker Pharmaceutical** are not responsible for loss or theft of personal belongings left unattended in the vehicle.

Reservations

To reserve your car for business travel, go online at www.rentcar.com or call 1–800–RENTCAR.

- Valid driver's license needed for reservation
- No deposit required

Online Instructions:

1. Go to http://www.rentcar.com
2. Locate login box for Gold Corporate Customers (GCC)
3. Enter username: QPrental (case-sensitive)
4. Enter password: your last name (all lowercase)
5. Follow remaining steps as prompted

For questions, contact Shirley Kimball, the Office Services Manager, at 898–757–3478 by call or text.

CASE 7-3   Letter Requesting Information

Juanita Márquez serves as local arrangements chairperson for the American Society of Engineering Education's (ASEE's) regional meeting. One of her first responsibilities is to locate a hotel in which the regional meeting can be held. She decides to send a letter to seven hotels near the sports arena so that conferees can enjoy sports events during the evenings of the 3-day conference. Juanita writes the same letter to all seven hotels requesting information about their facilities.

When Juanita writes to each of the seven hotels, she has several questions about availability, facilities, and equipment. Juanita needs each hotel to respond by a specific deadline, and she wants to be sure that every hotel supplies all the information she requests. With these answers, she can compare costs and services of hotels that are interested in hosting the convention. She uses a *block letter format* with all elements aligned on the left margin. She chooses to conduct this correspondence by formal business letter instead of e-mail, to make sure she obtains signed paper copies of all promises made by the hotel representatives.

CASE DOCUMENT 7–3

Pittsburg City College
200 Rosser Hall
Department of Civil Engineering
Pittsburg, KS 66760 (date)

(name of convention manager) (name of hotel)
(mailing address) (city, state, zip)

Dear (   ):

The South Central Chapter of the American Society of Engineering Education will have its annual meeting in Kansas City, April 23–26, 2018.

With the convention 15 months away, the local arrangements committee is seeking a hotel that will serve as convention headquarters. Our committee would like to consider your hotel as a possible site because of its location and its reputation in handling conventions.

By March 1, 2017, I will need answers to the following questions, if you are interested in hosting our convention. I will contact you by telephone no later than March 15, 2017, to arrange an appointment to discuss our needs and your facilities in further detail.

Number of Conferees Expected
Based on registration from past conventions, we expect approximately 500 people to attend. Of that number, approximately 450 will need rooms. Of that number, approximately 250 will require double occupancy rooms.

Conference Accommodations Needed
We need you to provide answers to the following questions:

1. Can your hotel accommodate 500–600 people, April 23–26, 2018?
2. Since the conference will feature 3 days of concurrent sessions, do you have available conference rooms in the following configurations?

   • three rooms located in the same general area that will hold 100 people
   • three additional rooms that will hold 50 participants, with six persons per round table

   We will have three concurrent sessions at each time slot—two in the morning and two in the afternoon.
3. Can your hotel provide the following multimedia equipment for each room?

   • screen
   • projector that can be attached to a variety of mobile devices

- microphones
- lectern
- high-speed Internet connection (please specify wired or wireless)

4. Can the seating arrangements be altered in each room between sessions?
5. Can the hotel provide refreshments during the morning break and after-noon break? Can these refreshments be made available at a location that is convenient to the meeting rooms?

Response Deadline
I will need a written response to each question by *March 1, 2017.* Also, please include in your reply a price list and menu for refreshments available for conferences.

If you have any questions, please contact me at 620-231–7000 Monday– Friday. If I am away from my desk, just leave a message on my voice mail and a time when I may return your call.

Sincerely,

Juanita Márquez

Juanita Márquez
ASEE Convention Arrangements Chair

CASE 7-4   **Unfavorable News Letter**

Nancy Krueger is a highly experienced safety engineer who was interviewed by Novel Engineering. Ms. Krueger was impressive during the visit to corporate headquarters, but the interview team recommended another finalist for the position. Despite Ms. Krueger's exceptional credentials, the second candidate had qualifications that more clearly matched the company's needs. Ritu Srivastava, director of personnel, must write to Ms. Krueger to explain that a job offer will not be forthcoming.

In this situation, Ritu may call Ms. Krueger to give her the disappointing news, but she will also follow the call with a formal letter. Avoid using e-mail or text messages to deliver bad news. Note that the letter gives reasons for Ms. Krueger not being selected, but the news is not explained in a harsh, critical way. It is intended to announce the bad news in as positive and supportive a way as possible. Notice also that Ritu cushions the bad news with a friendly opening and appropriate compliments. Because Ritu was heavily involved with Ms. Krueger during the application and interview process, she addresses her as "Nancy." Do not use first names unless you really know the person you are addressing. Ritu uses a *modified block format*, which uses paragraph indentions and shifts the date, closing, and signature line to the right side of the page (Case Document 7–4), because she thinks this format looks a little less stiff and formal.

CASE DOCUMENT 7–4

Novel Engineering, Inc.
4415 Orange Avenue
Syracuse, NY 13233

January 7, 2018

Ms. Nancy Krueger
1248 East 33rd Street
Cleveland, OH 44125

Dear Nancy:

All of us here enjoyed the two days you visited with us two weeks ago. Your perception of our clients' needs indicates that you have a firm understanding of the role safety engineering plays in the local contracts we win and manage.

While your background would be invaluable to us, we have only one position available, and we have selected an applicant with international project experience.

As we discussed during your visit, our range of clients has expanded. We now receive RFPs from countries in Europe, South America, and Asia. Developing responses to their needs requires us to expand our team, particularly when we can do so with professionals with international experience.

Thank you for considering us. We were genuinely impressed with your professionalism, your analytical skills, and your record as a team player. And I was immensely pleased to have the opportunity to meet you. I expect to hear great things of you in this profession.

Sincerely,

Ritu Srivastava

Ritu Srivastava
Manager, Personnel

CASE 7-5 **Claim Letter**

Catherine Leigh, a senior at Astoria College, discovers that she has been denied a tuition refund for a course she thought she dropped. As though the monetary penalty weren't enough, she also discovers that she has been given a grade of WF (Withdraw Failing) for the course. This withdrawal on her transcript will hurt the exceptional grade point average (GPA) she has been building for 3 years and makes admission to veterinary medicine almost impossible. She takes her complaint to the dean of Academic Affairs and follows up the face-to-face meeting with a formal letter that explains the problem in detail as well as the solution she is proposing.

Catherine knows that the dean of Academic Affairs has a lot to do and may give the letter to an assistant to investigate the complaint. She therefore doesn't assume that her reader remembers (or even knows) anything more about the case than what she provides in her letter. And she uses headings to help the reader find information quickly. She adopts a block format to make herself look judicious and methodical, and she proofreads carefully so that she doesn't come across as the kind of student who easily makes mistakes.

CASE DOCUMENT 7–5

Catherine Leigh
601 Ocean Avenue
Astoria, FL 34103

September 8, 2017

Dr. Neville Hodges
Dean of Academic Affairs
Astoria College
Astoria, FL 34103

Dear Dean Hodges:

I am writing to give you a written record of the situation I discussed in your office on Friday, September 5. I enrolled in the summer session of Math 3325, in section 001, under the instruction of Dr. Voskovic. I dropped the course before classes started but nevertheless was denied a tuition refund and given a grade of WF.

Description of Events
I registered for Math 3325 in February, but later my plans for the summer changed. I believed that all I had to do was go online and drop the course. I dropped the course online and did not attend any of the class sessions. During the first week of June, I called the registrar for information about my refund check. I asked the woman who answered my call if I had taken care of everything I needed to in order to get my refund check. She asked for my ID number, which I gave her. She then informed me that I should receive my check in the mail shortly. I waited for two months.

I returned to the registration office in person during the first week of August to inquire about the refund check. It was then that I was informed that I was dropped from the course due to excessive absences and I could not receive a refund. However, I was assured that I would get a W on my transcript. I was upset that my tuition for the course would not be refunded, but I was relieved that the incident would not hurt my GPA.

This relief, however, was short-lived. The following day, while I was working on my application to the Florida College of Veterinary Medicine, I discovered that my unofficial transcript online has a grade of WF for Math 3325. This grade will seriously hurt my chances of admission.

My Request

I request that the WF on my transcript be changed to a W and that my tuition for the course be refunded. I sincerely believed that I had dropped the course. My telephone conversation with the registration office only confirmed this belief.

I am a senior honors student with a GPA of 3.63. I fully understand that once a student enrolls in a course, he or she must attend class. I came to Astoria to prepare for a career in veterinary medicine: I would never enroll in a course and not attend it. I hope you will agree that I am conscientious about my studies. Please do not allow this one error to ruin my chances of acceptance to the Florida College of Veterinary Medicine.

I know that the registrar's office is hectic, and I regret that this error occurred. However, given the situation I have described, I do not know how I could have proceeded in a different way. I believe that I did everything I could to drop the course correctly.

Sincerely,

Catherine Leigh

Catherine Leigh
Astoria ID: 418556

CASE 7-6   **Letter of Reply**

Gabriel Dominguez owns an agricultural equipment business. He has just received an order and a check from Bruce Pickering, who wants to replace an irrigation pump.

Gabriel needs to notify Mr. Pickering that the pump he has ordered is being discontinued and replaced by two new pumps, both of which are more expensive but are designed to offer better performance. Gabriel wants to be sure that Mr. Pickering doesn't think that Gabriel is simply trying to sell him a more expensive pump when the current model will do the job.

In designing this letter, Gabriel wants to be sure that he does not suggest to Mr. Pickering that he is the object of a bait and switch. Thus, he opens the letter by telling Mr. Pickering that he has the pump and can ship it immediately (Case Document 7–5). However, he wants Mr. Pickering to know that several improved models are available and that he may wish to choose one of the new models. Gabriel then presents all the information that he believes Mr. Pickering will need to make his decision and invites Mr. Pickering to contact him.

Gabriel could call Mr. Pickering and talk to him, but in this case a business letter that expresses concern for Mr. Pickering's investment and presents exact information about the alternative pumps will make Mr. Pickering's decision easier. Note that Gabriel chooses the block letter format: he thinks its formal and straightforward appearance will reinforce his credibility and indicate respect for his customer and their business relationship.

To give Mr. Pickering the necessary information as soon as possible, Gabriel will prepare the letter with his electronic signature and e-mail it as an attachment; for this important sale, he will also put a paper copy of the letter with his ink signature in 2-day delivery so that Mr. Pickering is confident that everything is official and legal. (In the e-mail message, Gabriel advises Mr. Pickering of the attached electronic copy and the forthcoming paper copy of the letter.)

CASE DOCUMENT 7–6

Dominguez Irrigation Manufacturing & Supply
2101 Cerritos Drive
Agoura Hills, CA 91301
817-569-3766

February 23, 2017

Mr. Bruce Pickering
Route 1, Box 41
Silsbee, TX 77656

Dear Mr. Pickering:

We have received your order and check for $698 as payment for our 25-hp Model XM 213 auxiliary pump. Although we do have several in stock and we can ship one to you immediately, manufacture of this model has ended. We are holding your payment and order until we hear from you regarding the following alternatives.

Here is the situation. Purchasers of the 25-hp model have indicated that the low output of this pump is generally insufficient to run the new drip irrigation systems currently being manufactured. This particular pump is not designed to handle extensive operation of these larger systems. As a result, we have discontinued the model. We have reduced the price to $548 due to our inability to warranty the motor for more than 1 year or supply replacement parts for more than 2 years.

We now manufacture two larger pumps, a 50-hp model and a 75-hp model. The 50-hp model will power a #70 system. The 75-hp model will power a #90 system. Here is a comparison of these three models.

Hypo	Pumps			
Motor	Model	Rating	Warranty	Price
25 hp	XM2130	30	1 year	$ 548
50 hp	XM3170	70	5 years	$1,795*
75 hp	XM4190	90	5 years	$2,350*
*Additional warranty available for $125/yr.				

While we believe that you will be happier with the reliability and efficiency of either of the larger pumps, if you choose to purchase the 25-hp model, we will ship it immediately and refund the $150 difference in price. I am sending brochures on both new motors. Please look them over and call me at our 800 number, extension 145, if you have any questions. As soon as we hear from you, we will ship you the motor of your choice within 2 working days.

Sincerely,

Gabriel Dominguez

Gabriel Dominguez
Owner

Correspondence Checklist

Planning

☐ What is your subject, and what is your purpose?

☐ What do you want to have happen as a result of your correspondence? What will you do to achieve your objective?

☐ Who are your primary readers? Who are your secondary readers? Do your primary and secondary readers have different needs? How will you satisfy the different needs of all your readers?

☐ Why will your readers read your correspondence?

☐ What is the attitude of your readers toward you, toward your subject, toward your purpose?

☐ If you are addressing international readers, do you understand their cultural practices? What adjustments in your correspondence will their cultural practices require?

☐ Will you write an e-mail, text, memo, or letter?

☐ How will you make your message accessible for all of your likely readers?

EXERCISES

1. Civil Engineering Associates (CEA) of Lubbock, Texas, was established in 2010 by Robert Davidson and Walter F. Posey, both graduates of the University of Amarillo. Business for the company was good, and CEA took on six additional partners between 2013 and 2017: Alvin T. Bennett, Wayne Cook, Frank Reynolds, John W. Castrop, George Ramirez, and Richard M. Burke—all graduates of the University of Amarillo.

   For the last 5 years, the partners of CEA have met every Friday for a working lunch at Coasters, a local restaurant and bar that features attractive young waitresses wearing provocative swimsuits. The customers are almost exclusively men, and the interaction between customers and waitresses often appears flirtatious.

   This year CEA hired Elizabeth Grider, a Missouri State University graduate, as a new partner in the firm. She has attended two of the working lunches at Coasters and finds herself uncomfortable in this environment. She does not feel, however, that she can just skip the lunches because these are the only regular times at which all the partners gather. Projects are often discussed and work assignments decided at these Friday meetings. In addition, the lunches offer the opportunity to establish a cordial and friendly working relationship with the partners. She realizes that "Friday at Coasters" has become a long-standing tradition at the firm, and she hesitates to upset the status quo. However, she really wishes they could find a better place for their meetings.

   As Elizabeth Grider, write a memo to the senior partner, Robert Davidson, explaining the issue with Coasters and recommending one or more solutions. Of course, Grider will speak to Davidson directly but thinks that writing a memo will help her organize her thoughts, anticipate objections, and prepare for a likely difficult conversation. And after their meeting, she will leave the memo with Davidson as a written record of her recommendations.

2. You are the manager of Activ8 Sports in Cleveland, Ohio. Activ8 Sports, a regional chain of nineteen athletic equipment stores, has headquarters in Detroit. This morning you received a call from a vice president of the company. The

corporate office's social media specialist was doing the usual daily search on Twitter, Yelp, Facebook, etc., for customer comments about Activ8 Sports. He discovered that several of the salespeople at the Cleveland location were using Twitter to make derogatory comments about a coach at a local high school. The coach is a regular and frequent customer at the store and is sometimes abrupt and argumentative with the salespeople. The vice president directed you to "fix this Twitter thing immediately" and implied your job was at risk. You decide to reprimand each of the salespeople in question. You will also issue a memo for their files as an official warning with a reminder of the corporate policy that proscribes discussing company business in public, including social media.

As the manager of Activ8 Sports, write this memo: try to keep your salespeople cooperative and productive but also make clear that this is their only warning.

You also know that you must compose a letter of apology to the high school coach before she hears about the indiscretion of your salespeople. You know that sooner or later she will find out as the comments could easily go viral at the high school. The coach has been a loyal customer, purchasing all uniforms and equipment thorough your store year after year. You certainly can't afford to lose her business.

As the manager of Activ8 Sports, write to the coach to explain the situation and apologize for the comments of your salespeople on Twitter.

3. Examine Figure 7–2. How could you make this memo easier to read and more likely to achieve its objective of generating nominations. Consider changes to the wording and the design of the memo. Compose the revised memo as well as a letter to the National Labor Relations Board to encourage it to adopt more effective communication practices. Include a copy of your revised memo with your letter.

4. Using the information in Figure 7–2, develop a series of five engaging text messages that could be distributed over a 30-day period as reminders to recipients to nominate their colleagues for the NLRB's bilingual awards. You will have to decide which information in the memo is vital for a nominator to know as well as how to explain this information briefly and persuasively. You will also have to determine a logical sequence for the five messages.

5. The following messages were composed by the U.S. Fire Administration for the public to share through social media (www.usfa.fema.gov/prevention/outreach). How effective is each message? Which of the five would you be likely to share?

- College students: take a moment to watch this campus #firesafety video http://www.youtube.com/watch?v=yMQQhMztyuw-these tips can save your life!

- Adults 65+ are 2.6x more likely to die in fires. Watch this short video for older adult #firesafety tips https://www.youtube.com/watch?v=3xv8bdKhBno

- When fire strikes, deadly smoke fills your home in minutes. Watch this for fire escape planning tips https://www.youtube.com/watch?v=C9KSFRq4rXA
- Electrical failure/malfunction is a leading cause of home fires. Learn abt electrical fire safety at https://www.usfa.fema.gov/prevention/outreach/electrical.html
- Upcoming travel or vacation plans? For a list of hotels that meet fire safety standards visit https://apps.usfa.fema.gov/hotel/

OFFICE OF THE GENERAL COUNSEL
Division of Operations-Management

MEMORANDUM OM 16-18   June 13, 2016

TO: All Regional Directors, Officers-in-Charge,
                and Resident Officers

FROM: Anne Purcell, Associate General Counsel

SUBJECT: Nominations for Bilingual Awards

      Pursuant to Article 12, Section 2(b) of the Collective-Bargaining Agreement (Agreement) between the General Counsel and the National Labor Relations Board Union (NLRBU) covering field office professional and support staff employees, this memorandum announces the nomination period for FY 2016 bilingual awards. Nominations for bilingual awards shall be submitted to the Regional Director, who shall forward approved nominations to his or her Assistant General Counsel in the Division of Operations-Management no later than July 15, 2016. Nominations may be submitted to the Regional Director (through appropriate supervisory channels) by any employee, the NLRBU, or any NLRBU local. All nominations must be in writing.

      To be eligible for a bilingual award, an employee must meet the following criteria: (1) the employee must, on a regular basis rather than occasionally, utilize his or her bilingual skills in the performance of Agency work; (2) the employee's current summary rating must be at the "fully successful" level, or higher; and (3) the employee cannot be otherwise compensated for the use of his or her skills, either through salary or some other type of incentive award.

      The awards approved by the Regional Director shall be designated as Level 1, Level 2, or Level 3, depending on the employee's contribution to Agency work, with Level 3 being the greatest contribution. Employees who are approved for an award will be granted a monetary award as set forth in the Agreement. In addition, these employees will receive a certificate of commendation from the General Counsel.

      Questions about this program should be directed to your AGC or Deputy.

                                   /s/
                                   A. P.

cc: NLRBU

• **FIGURE 7–2** Document for Exercises 3 and 4

8

Technical Reports

Reports, like correspondence—e-mails, texts, memos, and letters, discussed in the previous chapter—remain the most commonly written workplace documents. Reports serve a variety of functions: they provide information, instructions, analysis, conclusions, and recommendations based on analysis.

Given the quantity of information generated and shared in most work environments, getting your report read has become an increasingly critical consideration. In this chapter we will explain the basics of designing clear and accessible reports that readers are more willing to take the time to read.

 Quick Tips

None of your readers will read your entire report unless it's a single page. Different readers will read different sections of your report according to their needs. Almost all your readers will read the summary—make sure this section exemplifies clarity and conciseness. The busier your reader, particularly those who receive dozens and hundreds of documents daily, the more important are directness, clarity, and conciseness if you want this reader to read your document.

If a report doesn't have a concise, readable summary, readers will flip to the end to find the conclusions. If they can't find those easily, they likely will not read any portion of the report.

Kinds of Reports

Many routine reports provide information, but others go beyond simple reporting. They categorize and then analyze information. From the analysis, the writer may evaluate evidence, draw conclusions, and perhaps recommend action based on those conclusions. Analytical reports often defy rigid classification, but for the purpose of learning to write analytical reports, we can generally identify the following types:

- If the analysis focuses on a recommendation, the report may be called a *recommendation report*.
- If the analysis emphasizes evaluation of personnel, data, financial options, or possible solutions to problems or avenues for exploration, the report may be called an *evaluation report*.
- A *feasibility report* analyzes a problem, presents possible solutions to the problem, determines criteria for assessing the solutions, assesses the solutions against the criteria, and then shows the best solution(s) based on the reported analysis of the solutions. The analysis underpins the conclusions and any recommendations. Based on available resources, the writer determines criteria against which possible solutions may be measured and methods of applying criteria and analyzing solutions against criteria selected. Many feasibility reports are long because they represent the way a problem has been defined, studied, and resolved. Results of feasibility studies can lead to major financial expenditures.
- A *progress report* or *status report* both informs and analyzes (see Chapter 9). Status reports describe and evaluate the work done on a project, the money expended, and problems encountered.
- A *trip report* documents information gathered on a trip, evaluates this information, and may suggest actions based on findings.
- A *personnel report* describes an employee's performance, analyzes effectiveness, suggests methods of improvement, and estimates the employee's potential for promotion.
- An *economic justification report* explains the cost of a project or action and then argues for the cost-effectiveness of the project.

Report Categories—Informal and Formal

Reports can generally be divided into two categories: informal reports, used for daily internal communications in business organizations, and formal reports, generated for readers outside the organization or for internal use on major projects. The two kinds of reports share similarities in structure and major components. They differ principally in length and in format.

Informal Report Headings

Internal reports use a memorandum heading that may include many of the following items. Organizations usually have their own specific requirements for internal reports and use a distinctive memo template. Common elements of memo headings usually include the following:

Date:
To:
From:
Subject:
Reference:
Action Required:
Distribution List:

Subject line. The subject line should state clearly and concisely the content of the report. The clearer the subject line, the better the odds that your readers will read your report. Your subject line should make your readers want to read your report.

Reference. Many reports respond to previously written reports, to company policies, or to items such as engineering specifications and legal issues. Be sure your readers have what they need and understand how your report links to previous reports and documents.

Action required. If you need your reader(s) to respond to your report by a certain date, then say so here. For example, "Your approval needed by 5/10/2018." (See Case 7–1, which shows a memo with an action-required statement.)

Distribution list. This report segment, which can occur at either the beginning or the end of the report, indicates who will receive copies as well as the report's file name for later reference. Be sure to include on this list everyone who should receive a copy. Failing to include all appropriate recipients can create problems for you as those omitted may resent their being eliminated from the communication chain.

Parts of an Informal Technical Report

Introduction. Reports should begin with an introduction to the content covered. For routine memo reports, keep your introduction concise but long enough to ensure that readers understand what they are reading and why they are included on the distribution list. Also remember that a report will be kept in a file (virtual or paper) and will be accessible to readers many months and years after you have written it. The introduction should provide enough background information to ensure that all readers understand what follows.

Thus, a memo report introduction should contain the following elements. How much you develop each item will depend on the particular report situation—what you think readers will know.

Subject or report topic. What is your report about?

Report purpose. What do you want your report to accomplish? Often, you can combine the report topic and purpose into one sentence. If possible, begin the report with a sentence or two that states the subject and purpose. For many routine reports, a subject/purpose statement will give readers what they need.

> As you requested, I have prepared a brief report on the best options in alternative fuels. Current research over the past 5 years suggests that biofuels, compressed natural gas, and alcohol fuels may present the best alternatives to conventional fossil fuels. This report presents the advantages and disadvantages of each and should offer interesting discussion points for the bimonthly brown-bag lunch meetings.

Background or rationale. What issues led to the need for this report? Be concise. If you must have more than a full paragraph, place the background or rationale in a separate paragraph. If your main readers don't need this information, then they can skip this paragraph. Readers on the distribution list or future readers will likely appreciate the information.

> This report recommends that the university develop online versions of our high-demand courses, particularly those most requested by students who leave campus for the summer.
>
> The university loses substantial funding when students do not enroll for our summer classes. Many would enroll for online versions of these courses if we offered them. The loss in revenue has increased over the past 3 years. To make a well-informed decision, we must determine (1) the courses we could develop for online/hybrid delivery, (2) the cost required for redesign, and (3) the time required for effective conversion. Discussion of each point follows.

Report development. What topics will you cover? Simply state how you will develop the report, mentioning each topic you will cover in the order you will cover it. This quick overview lets readers know what to expect and helps prevent surprise or disappointment.

Summary. The summary determines the success of many reports. Many readers who receive copies simply as information will read only the introduction and the summary. The summary should tell your readers the essential points and findings in the report. Informal reports may combine the introduction and summary into one paragraph, consisting of the key points and the purpose.

Discussion. This section develops each point that you state you will discuss in the introduction.

Conclusion. The conclusion ties together what you have presented and states the main ideas you want readers to remember.

Recommendations. If you need to recommend action, then include that here. To improve readability, list and number your recommendations. If you need to discuss them, you can do so following your list.

Attachments. Attachments include any material needed to support your discussion, conclusions, or recommendations. Attachments may include additional data, calculations, or previously written documents on the same subject.

Figure 8–1 exemplifies an effective informal report prepared by the chair of a university's marketing committee. Addressed to the university president, the report covers the committee's initial findings about promoting the institution to the parents of prospective students.

Developing Reports

How reports develop varies, and here we can provide only guidelines and a few examples. In the introduction, you state how you will present your report and the topics that your reader will find. Thus, you can have a main heading for each topic, and readers who have interest in one or two particular issues can go immediately to those issues.

 Note: You want to develop a reputation for writing readable, informative, and accurate reports. Don't make readers hunt for information—they won't. If your reports don't meet these expectations, readers will lose trust in you as a writer and perhaps as a colleague.

How you arrange the main components should depend on what you believe your readers will want. Note that in Figure 8–1 the writer begins with the main marketing recommendations. The introduction is brief, as the president of CSU expects it to be.

Richard Harrington

TO: Justin Perry, President   Date: June 4, 2017
        Cambridge State University

FROM: Richard Harrington, Professor of Finance
        Chair, CSU Marketing Committee

SUBJECT: **Marketing CSU—Status Report**

Our Backgrounds

The CSU Marketing Committee, formed by Provost Davis in June of 2016, emerged from concerns that many students and their parents who visited our campus knew little about CSU. The committee, composed of parents of CSU graduates and faculty are grateful for the opportunity to share our views of the university. We solicited input from nearly 300 parents. Based on our combined experiences with CSU, we offer our recommendations, compiled as of now, based on focus groups, surveys, and our own experiences as former students from other universities. Much work remains, and we solicit your suggestions.

Marketing Recommendations

1. Improve CSU's visibility. CSU is the best-kept secret in quality small colleges. In "getting out the word" about CSU, the university needs to know the audiences it wants to reach and their concerns in responding to college advertising.

2. Leverage on the recent *Princeton Review* recognition of CSU. Highlight the features that the *PR* noted.

3. Don't allow what a marketing firm thinks to control marketing decisions. Stakeholders often have valid views based on their experiences, backgrounds, and education.

4. In attempting to build name-recognition and prestige, do so on the basis of qualities that CSU offers that other state schools do not offer.

Marketing CSU   Page 1

• **FIGURE 8–1** Effective Informal Report

5. Include, in developing a marketing plan, issues that concern parents, grand-parents (who often help with college costs), as well as high school counsel-ors. Business should also be a consideration, as they will hire CSU graduates and be approached for financial support:

Cost

1. Show how CSU's cost makes it attractive. Discuss the Wallace scholarships and the amount of money available. Provide comparisons between CSU and your main competitors.

2. Point out that car insurance and the cost-of-living index in our town and county are much lower than those costs urban areas.

3. Emphasize how teaching, rather than research, provides the foundation of a CSU degree.

4. Discuss how CSU has improved the quality of life for students with the new facilities available to all students: the new student center, the new residence halls, the refurbished dorms and on-campus apartments, and the recreation/exercise facilities. We plan a survey of graduating senior at the end of every term.

Safety
Note that parents expressed extreme concerns about on-campus safety. They read/listen to news and know that universities, particularly large ones or universities in urban areas, do not offer safe havens for students. Point out the low on-campus crime rate at CSU and the low crime rate of Carrington in general. Compare this rate with those in nearby urban areas. These statistics are easy to find and provide a powerful argument for the benefits of a small college community.

Values
We list the following recommendations in no particular order. Different target audiences value each differently, but all seem to us important based our interviews with parents:

1. Parents also want a university that has a culture that supports civic values and has rules for conduct in its housing and on campus..

Marketing CSU   Page 2

• **FIGURE 8–1** *Continued*

2. They also want an environment of zero-drug tolerance, where the athletic program has a zero tolerance policy on drugs and tests athletes regularly.

3. Parents also like a campus that is clean and inviting. On this point, CSU excels. Marketing material for CSU should have effective photos of the campus, particularly areas on either side of the mall. Show trees, flowers, some of the great classrooms, and the integrated campus concept.

4. Parents want a university focused on education as well as career preparation. Discuss student service projects that have required critical thinking.

5. Parents want to know that a degree from CSU will "mean something." Having testimonies from former students who have succeeded will be extremely helpful.

6. Parents want faculty who care about students and demonstrate that concern. Parents also value research that "makes a difference," but they want students placed first. Parents want a college that offers personal development opportunities, such as on-campus leadership development programs and internships. They also want undergraduate research programs for students who seek those opportunities. Discuss faculty whose creative work and research has been recognized. Teachers whose research supports their teaching is also a useful point to highlight.

We look forward to meeting with you and your team during Parents Week in September.

• **FIGURE 8–1** *Continued*

CASE 8–1

Matthew Lunsford has worked in Bradshaw Engineering's fuel analysis and development division for 8 years. He recently received an assignment from Jerry Bradshaw, one of the three principals of Bradshaw Engineering, during an office discussion: "Matt, the sustainability issue intrigues me. Are we moving in the right direction? What issues do you see on the horizon? What questions do we need to be asking? Write a concise report for me and possibly the rest of our engineering team. Make it readable. Where we go from there I don't know."

Based on the conversation, Matt writes a report that informs, analyzes, and recommends. He knows the opening summary is critical: although Bradshaw is interested in this subject, the summary could be the only section that the engineering team will read. Matt inserts headings in the summary and throughout the report: he knows these will help readers to identify and retrieve key ideas for later discussion.

CASE DOCUMENT 8-1

Transmittal Memo and Summary

TO: Jerry Bradshaw   DATE: November 12, 2017

FROM: Matthew Lunsford

SUBJECT: Report on Current Issues in Chemical Engineering - Sustainability

As we discussed in our last staff meeting, employees at BE need to understand the current issues facing chemical engineering, particularly those that affect the definition of our field. In response to your request, I have devoted the past two months to research on the direction that chemical engineering seems to be moving. My goal: to help BE in our decisions about hiring, continuing education, and company development.

In light of the changes in the field of chemical engineering, I recommend (1) that engineers at BE participate in continuing education in the areas of alternative fuels and biotechnology and (2) that BE look to recruit college graduates and experience engineers who have had education or experience in these areas. Following these recommendations would help BE adapt to changes in the field of chemical engineering and continue offering excellent engineering services.

As you suggested, I have prepared a summary of my report for our engineering team and a report for you and the BE management. I look forward to your comments and recommendations.

Current Issues in Chemical Engineering: Sustainability

by
Matthew Lunsford

Summary

Report Purpose

The purpose of this report is to inform BE of changes occurring in the field of chemical engineering. This report focuses on sustainability and the technologies that will help move us toward a sustainable future as well as the challenges that chemical engineers still face. Based on the report findings, I close with several recommendations for BE that will help us stay up to date with these emerging technologies.

Research Methods

To gather information and to obtain direction for this report, I interviewed Dr. Chen, a professor of chemical engineering at Texas A&M University, and researched the topic of sustainability using databases such as ScienceDirect. These databases were available to me through the Texas A&M library.

The Importance of Sustainability

Sustainability, meeting our needs presently without compromising the ability of future generations to meet their own needs, has become a major issue in chemical engineering. Chemical engineers have shifted from focusing solely on profits to focusing on being profitable within the context of sustainability. Three main realizations have led to this change.

1. The fossil fuels that we currently rely upon for energy and chemical production will eventually run out, leaving future generations without oil and gas.
2. Burning fossil fuels destroys the environment by causing acid rain and possibly contributing to global warming.
3. As the world's population increases and becomes more affluent, the demand for energy and chemicals will increase dramatically.

Possible Solutions and the Chemical Engineer's Role

Renewable energy is a growing source of electricity that will help us move toward a sustainable energy future. Wind, geothermal, hydro, and solar power are all examples of renewable energy. Renewable energy is attractive because it uses limitless natural resources such as winder, water, sunlight, and heat from the earth to produce energy. The high price of electricity from renewable energy sources is the main obstacle keeping renewable energy from being more widely produced. Engineers can overcome this challenge by developing more efficient equipment for renewable energy production.

Alternative fuels, especially bioethanol, also present a solution to the problem of meeting our energy needs sustainably. Bioethanol producers use a fermentation process to make ethanol from renewable resources such as crops and other plants. Finding a sustainable feedstock for making bioethanol is one challenge involved in ethanol production. Cellulosic biomass, which includes prairie grasses and some trees as well as sewage sludge and municipal waste, appears to be the most sustainable feedstock available. However, the process to produce ethanol from cellulosic biomass is too expensive. Engineers must decrease the cost of production by finding low-cost pretreatments and by optimizing the process.

Another valuable tool that we can use to reach a sustainable future, especially in the chemical industry, is biotechnology. Biotechnology refers to the use of living organisms to accomplish tasks. Some companies are already using microorganisms to produce products sustainably and efficiently. Understanding and manipulating complex metabolic pathways is a major challenge that researchers must overcome in this area. Engineers must also receive the education necessary to create, run, and optimize processes involving biotechnology.

Conclusion

Though energy and chemical producers must utilize each of the technologies discussed in order to achieve sustainability, cellulosic ethanol and biotechnology are particularly important and should be the focus of chemical engineering research.

Recommendations

To keep up with the changes in the field of chemical engineering, I recommend that:

1. Chemical engineers currently employed at BE participate in continuing education in the areas of alternative fuels and biotechnology
2. BE recruit college graduates and experienced professionals who have education or experience with alternative fuels and/or biotechnology

Introduction

Over the past 20 years, a growing awareness of our world's limited resources, our environmental impact, and our increasing energy needs has sparked a change in thinking among chemical engineers. Formerly, they focused solely on profits. Now, however, they are placing greater emphasis on preserving our world's resources for future generations. Researchers have coined the term "sustainability" to describe this idea. Sustainability is one of the largest challenges currently facing chemical engineers, and it will continue to be just that for years to come.

This report covers sustainability in the chemical and energy industries and how chemical engineers play a vital role in achieving a sustainable future. I have written this report to keep BE up to date on important developments in the field of chemical engineering. The remainder of this report will (1) discuss the importance of sustainability, (2) highlight solutions that could help move us toward a sustainable future, and (3) close with several conclusions and recommendations that apply to BE.

The Importance of Sustainability

Companies and media often toss the word "sustainability" around, and several valid definitions exist. The most common definition that I encountered during my research is that sustainability is "development that meets the needs of the present without compromising the ability of future generations to meet their own needs" [1]. This statement carries the idea that we must be looking ahead to the good of those coming after us as we currently seek to meet our needs, whether they are financial or energy related. Otherwise, our future children and grandchildren may be left without a way to meet their own needs.

A major issue that we must overcome if we want to reach sustainability is meeting our chemical and energy needs with resources that are fully sustainable, ones that will not run out or destroy the earth that we live on [1]. Even major oil companies such as Exxon, which has invested $600 million in alternative fuels research, are taking notice of this issue [12]. Universities have also changed their chemical engineering curriculum to include courses in biotechnology, which can be used in the production of sustainable fuels. For the following reasons, the focus of the chemical engineering field has shifted away from focusing solely on profits and towards being profitable within the context of sustainability.

1. One reason is that the resources that we rely heavily upon will run out. Currently, the world meets 80% of its energy needs with fossil fuels such as oil, natural gas, and coal. The chemical industry also relies on these same resources as raw materials. Analysis shows that remaining fuel reserves will last about 40, 60, and 150 years, respectively [2]. So, if current consumption continues without the development of sustainable energy resources, 80% of societies'

1|

energy needs will be left unmet 150 years down the road. Researchers and engineers must focus their knowledge and skills toward developing energy resources and raw materials that will last indefinitely, or future generations will suffer the consequences of energy shortages and a lack of natural resources.

2. Destruction of the environment is a second reason why striving for sustainable energy demands the attention of the chemical engineering field. The fossil fuels that we currently burn for energy release chemicals that harm the environment. Chemicals such as SOx and NOx are particularly harmful because they react with water vapor in the atmosphere to form acids. These acids then fall to the earth as acid rain and can harm ecosystems that are sensitive to changes in acidity [1]. These ecosystems include oceans, lakes, rivers, and forests [1]. Electricity generation, residential heating, and industrial energy use contribute about 80% of SOx emissions, and motor vehicle exhausts contribute about 48% of NOx emissions [1].

   The burning of gasoline and other fossil fuels also releases CO_2 into the atmosphere, which, many scientists suspect, contributes to global warming through a greenhouse gas effect. CO_2 helps to trap heat within the earth's atmosphere, which may be causing temperatures on the earth to rise. If this trend continues, temperatures on the earth could rise between 2 and 4 degrees Celsius in the next century, which would cause ocean levels to rise [1]. Rises in ocean levels could flood populated coastal areas and displace agriculture, threatening whole populations [1]. Engineers must find energy resources that do not continue to degrade the environment so that society in the future still has an earth to live on.

3. Finally, engineers must develop sustainable resources that can keep up with increasing demand. By the year 2050, researchers expect the population of the world to double [1]. Along with that, they expect energy demands to grow by 1.5–3 times [1]. Large countries such as China and India continue to develop and require more energy for industry and motor vehicles as their populations become wealthier. This ever-increasing demand for energy will put our limited fossil fuel resources under more stress as time goes on. Therefore, the need exists for sustainable energy resources and raw materials for chemicals that will be able to meet this large demand indefinitely.

Therefore, the challenge for chemical engineers is this: develop energy resources and raw materials for chemicals that (1) will not run out, (2) will not destroy the environment, and (3) will meet the demand of an increasingly large and affluent population.

Possible Solutions and the Chemical Engineer's Role

Chemical engineers lead the effort to develop sustainable energy resources and raw materials. One reason for this is that researchers have already amassed a large amount of base knowledge on the subject of alternative fuels and renewable resources. Scientists have already discovered the basic principles behind each of the possible solutions that I will discuss [1]. The challenge now is overcoming the economic barriers that keep these solutions from being competitive with the nonrenewable fuels that we currently use [1, 3]. Because researchers cannot easily quantify all of the benefits of using sustainable resources, high costs still keep sustainable resources from being used more widely [2]. Further research is also necessary in some areas, such as in biotechnology. The following paragraphs discuss a few possible solutions to the problem of developing sustainable energy and chemical resources and mention some challenges that engineers must still overcome.

Renewable Energy

Renewable energy is one solution that has and will continue to help move us toward a sustainable energy future. Wind, geothermal, hydro, and solar power all fall into this category of energy production. Renewable energy attracts attention as a solution for a sustainable future because it uses limitless natural resources such as wind, water, sunlight, and heat from the earth to produce energy. Renewable energy is already a growing source of electricity with growth of 55% for solar power and 21% for wind power between 2008 and 2013 [5]. However, renewable energy still composed a small portion of the world's total energy production in 2013 as shown in Figure 1.

Some sources of renewable energy are more sustainable than others. Wind energy, especially, shows great promise as a major energy contributor for the near future [4]. Though they can be loud and disturb some wildlife, wind turbines have a great upside. They do not pollute the environment, they are relatively efficient compared to some other renewable energy sources, and they produce electricity at a reasonable price [4]. Research shows that wind power is currently the most sustainable form of renewable energy [4].

Some other sources of renewable energy, such as hydropower and geothermal power, come at a higher price to people and to the environment. Damming rivers to produce hydropower often displaces people, animals, and agriculture, and geothermal power plants must carefully process wastewater in order to avoid releasing toxic chemicals such as hydrogen sulfide, ammonia, mercury, and arsenic into the environment [4]. Solar power is the least efficient and most costly of the renewable energy sources listed [4].

Challenges

The main challenge associated with renewable energy is the cost compared to fossil fuels. Currently, electricity from burning coal and gas still costs less than any of the sources of renewable energy, which prevents companies and countries from using renewable energy more widely [4].

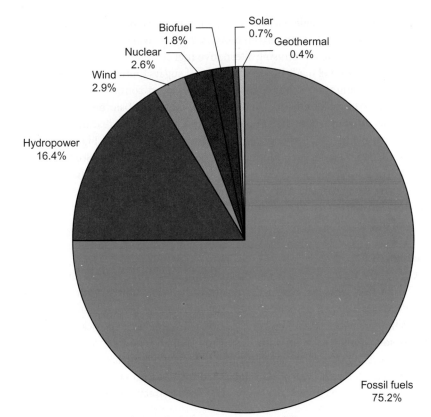

Figure 1 World Energy Supply 2013 (Information from source [5])

Possible Solutions

As engineers make more efficient designs for solar cells and wind turbines and as the renewable energy industry matures and grows, costs of the electricity produced will drop and renewable energy will become more competitive with conventional energy sources. Then, renewable energy will be a more significant contributor to a sustainable energy future.

Alternative Fuels

Though renewable energy will play a part in working toward a sustainable energy future, we cannot rely on it completely. Renewable energy resources such as rivers are not evenly distributed evenly throughout the earth, and places that do not have large amounts of renewable energy resources need other forms of sustainable energy that can be transported to them. Also, the majority of motor vehicles currently use combustible fuels, not electricity, for their energy supply. Therefore, if we want to move toward a sustainable energy future, we must develop sustainable alternative fuels that we can transport and use in motor vehicles. I will focus here on bioethanol because this is the most promising alternative fuel currently being produced.

Bioethanol is a fuel made by fermentation of sugars found in renewable sources such as plants and wastes. Bioethanol producers most commonly use either corn or sugarcane to make ethanol, but sorghum and cellulosic biomass also show promise as ethanol feedstocks. For environmental reasons, researchers see fuels made from plants as a very sustainable option. As shown in Figure 2, biofuels release CO_2 that plants such as corn take in during growth [3]. So, besides fossil fuels burned for feedstock growth and ethanol production, bioethanol does not pollute the environment. Also, because farmers can plant crops year after year, ethanol feedstocks are available indefinitely as opposed to oil, which will eventually run out.

Challenges

One of the major challenges associated with alternative fuels is finding a feedstock that is sustainable, one that will not destroy the agricultural environment and can keep up with demand. Countries such as the United States use corn as an ethanol feedstock. Corn, however, is not a sustainable feedstock because it degrades soil, cannot keep up with increasing demand, and causes food prices to increase [6, 7].On the other hand, sugarcane seems to be an attractive and sustainable feedstock. Brazil, which has established a successful ethanol industry that is now independent of government support, uses sugarcane to produce ethanol [2].

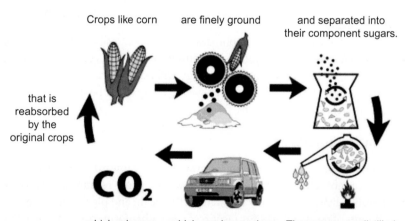

THE CARBON CYCLE

Crops like corn   are finely ground   and separated into their component sugars.

that is reabsorbed by the original crops

CO_2

which releases carbon dioxide   which can be used as an alternative fuel   The sugars are distilled to make ethanol

Figure 2 The Carbon Cycle for Bioethanol (Information from http://www.globalization101.org/biofuels-and-ethanol/)

The use of sugarcane avoids negative consequences, such as rising food prices and soil degradation, that are inherent in other feedstocks such as corn [2, 6]. However, not all countries have the climate necessary to grow sugarcane.

Another feedstock that will not compete with food or damage soil is cellulosic biomass. Cellulosic biomass includes a wide variety of plants from tree species such as willow, maple, and black locust to grasses like switchgrass and reed canary grass [8]. Wastes such as woodchips, sawdust, municipal residues, and sewage sludge can also be used as cellulosic biomass and made into ethanol [8]. Many researchers view cellulosic biomass as an attractive ethanol feedstock because of its low price and sustainability. But the cost of producing ethanol from cellulosic biomass currently keeps this feedstock from being used on an industrial scale.

For each feedstock the main step in the ethanol production process is very similar. During this step yeast or bacteria take the sugars obtained from the feedstock and produce ethanol in a fermentation process [9]. The key difference between the different raw materials is what occurs before the fermentation step. For grain feedstocks such as corn, ethanol producers must use enzymes, which are biological catalysts, to break down the starches into simple sugars before fermenting [8, 9]. This extra step adds additional cost to the process. Cellulosic ethanol requires even one more step. Before breaking down the feedstock into simple sugars, ethanol producers must first expose the carbohydrates, which are enclosed by a rigid shell called lignin, in a step typically called pretreatment [9]. This additional step causes cellulosic ethanol to be more expensive to produce than ethanol from other feedstocks.

Possible Solutions

For cellulosic ethanol to be produced and used on an industrial scale, chemical engineering researchers must determine ways to make it competitive economically with petroleum fuels. Decreasing production costs is a way that they can overcome this obstacle [9]. Researchers are currently giving a lot of attention to the pretreatment step, as this step comprises almost 40% of the total cost of production [10, 11]. Current pretreatment methods involve chemicals such as sodium or potassium hydroxide, which are too expensive [11]. Other pretreatment methods such as physically grinding or using steam are either too expensive or produce yields of carbohydrates that are too low [11]. Possibly viable pretreatment methods include the use of the dilute acids such as sulfuric acid or fungi, which degrade the lingnin [9]. However, using fungi for pretreatment is typically slow compared to other methods [9].

Chemical engineers have several other options available for decreasing production costs of cellulosic biomass. These options mainly consist of taking techniques currently used in the production of petroleum fuels and applying them to cellulosic ethanol production. Using simulation software to model and optimize the ethanol production process would help ethanol producers make the most ethanol at the lowest cost [10]. This technique is used pervasively in refineries. Another technique that chemical engineers can use to optimize production is heat integration [10]. Engineers are able to conserve energy by taking waste heat from one part of the process and using it to heat up another part of the process through the use of heat exchangers.

Biotechnology

Biotechnology is another powerful tool that chemical engineers can use to work toward a sustainable future. Biotechnology refers to the use of living organisms, such as fungi and bacteria, to accomplish tasks. Researchers can make genetically modified microorganisms to produce enzymes that are much more effective than traditional catalysts [3]. Through altering the natural metabolic pathways, researchers can also engineer bacteria that will produce desired products efficiently and from renewable resources [3].

This technology is especially useful in the chemical industry. For example, DuPont uses a genetically altered strain of *E. coli* to produce polymers from corn [3]. Normally this same polymer is made from fossil fuels. The chemical company Hoffmann-La Roche also uses a genetically altered strain of bacteria to produce vitamin B_2 [3]. This one-step process replaces an older process that involved six steps, resulting in savings on energy and decreased burning of fossil fuels [3]. Biotechnology has the potential to allow chemical companies to make products from renewable resources and burn less fossil fuel in the process.

As previously stated, ethanol producers also use biotechnology in the form of enzymes and bacteria.

Challenges

The main challenge associated with the use of biotechnology is the complexity of the microorganisms and the need for more knowledge and research. To produce a desired product using a microorganism, researchers often have to modify the metabolic pathways of that organism [13]. This process is carried out by modifying the organism's DNA [13]. Therefore, researchers must have a in-depth knowledge of the extremely complex metabolic pathways as well as the DNA that governs that pathway. Then, the researcher must know exactly how to alter the DNA so that the desired change in the metabolic pathway takes place.

For the engineer, the main challenge is to gain the knowledge necessary to efficiently run processes that involve biotechnology. Though some of the basic principles are the same for a petrochemical reaction and reaction involving enzymes, there are also many key differences. For example, while using biotechnology, an engineer must be concerned with not only keeping the microorganisms alive but also what stage of growth they are in, as this can affect what product they produce

[13]. In a petrochemical plant, chemical engineers have to worry about much different things such as the high temperatures and pressures often used.

Possible Solutions

The need for more research seems to be the limiting factor in the use of biotechnology in the chemical industry. Genetically engineering microorganisms is a relatively new art. So, if chemical producers want to use biotechnology to make more products efficiently and sustainably, researchers must continue to study and gain more knowledge on the metabolic pathways of microorganisms. Then engineers and researchers must develop processes using these microorganisms that can compete with petrochemicals. As the knowledge base grows, researchers will continue to develop strains of bacteria and fungi that can produce products sustainably. This developing technology could be the future face of a sustainable chemical industry.

For engineers to create, run, and optimize processes involving biotechnology, universities must prepare them to face the challenges of this relatively new field. Some universities have already begun to change their degree programs to reflect the move toward biotechnology. For example, UC Berkeley has changed the name of their chemical engineering department to the Department of Chemical and Biomolecular Engineering [14]. Also, MIT now offers degrees in both chemical engineering and chemical-biological engineering [15]. However, some universities such as Texas A&M University require very little biology-type courses in their undergraduate chemical engineering programs. Universities must continue to expand undergraduate education in bioengineering if engineers are going to be prepared to work in the changing chemical industry.

Conclusion

Despite posing a large challenge, sustainability in the energy and chemical industries seems to be a realistic future. No one technology, however, will bring us there. Energy and chemical producers must make use of a wide variety of options including renewable energy, alternative fuels, and biotechnology. As these technologies mature and develop, industry will call on chemical engineers to develop more efficient designs, processes, and microorganisms that will make sustainable energy and chemicals more competitive with those produced from fossil fuels.

Cellulosic ethanol and biotechnology are especially promising. Ethanol producers could potentially supplant much of the fossil fuel–derived energy resources with sustainable fuels. And biotechnology gives the chemical industry the opportunity to produce chemicals that are also made from renewable resources. These technologies are very important and should be the focus as we move toward a sustainable future, one where the needs of future generations will not be threatened.

Recommendations

In light of the changes that are occurring in the field of chemical engineering, I recommend that (1) chemical engineers currently employed at BE participate in continuing education in the areas of alternative fuels and biotechnology and (2) BE recruit college graduates and experienced professionals who have education or experience with alternative fuels and/or biotechnology. These measures will ensure that BE can offer the highest quality services to our clients who have needs in these developing areas.

References

[1]   Dincer, Ibrahim. "Renewable energy and sustainable development: A crucial review." *Renewable and Sustainable Energy Reviews* 4, no. 2 (2000): 157–175.

[2]   Goldemberg, José. "Ethanol for a sustainable energy future." *Science* 315, no. 5813 (2007): 808–810.

[3]   Rogers, P. L., Y. J. Jeon, and C. J. Svenson. "Application of biotechnology to industrial sustainability." *Process Safety and Environmental Protection* 83, no. 6 (2005): 499–503.

[4]   Evans, Annette, Vladimir Strezov, and Tim J. Evans. "Assessment of sustainability indicators for renewable energy technologies." *Renewable and Sustainable Energy Reviews* 13, no. 5 (2009): 1082N1088.

[5]   REN21 Steering Committee. *Renewables 2014 Global Status Report.* Paris: Rewewable Energy Policy Network for the 21st Century, 2014. http://www. webcitation.org/6SKF06GAX.

[6]   Pimentel, David. "Ethanol fuels: energy balance, economics, and environmental impacts are negative." *Natural Resources Research* 12, no. 2 (2003): 127–134.

[7]   Iowa State University, "Corn for biofuel production." *American Cattlemen.* http://www.americancattlemen.com/articles/corn-biofuel-production.

[8]   Solomon, Barry D., Justin R. Barnes, and Kathleen E. Halvorsen. "Grain and cellulosic ethanol: History, economics, and energy policy." *Biomass and Bioenergy* 31, no. 6 (2007): 416–425.

[9]   Sun, Ye, and Jiayang Cheng. "Hydrolysis of lignocellulosic materials for ethanol production: A review." *Bioresource Technology* 83, no. 1 (2002): 1–11.

[10] Cardona, Carlos A., and Óscar J. Sánchez. "Fuel ethanol production: Process design trends and integration opportunities." *Bioresource Technology* 98, no. 12 (2007): 2415–2457.

[11] Yang, Bin, and Charles E. Wyman. "Pretreatment: The key to unlocking low-cost cellulosic ethanol." *Biofuels, Bioproducts and Biorefining* 2, no. 1 (2007): 26–40.

[12] Mouawad, Jad. "Exxon to invest millions to make fuel from algae." *The New York Times*, July 13, 2009.

[13] Shuler, Michael. *Bioprocess Engineering: Basic Concepts*, 2nd ed. Saddle River, NJ: Prentice Hall, 2002.

[14] "UC Berkeley Dept. of Chemical Engineering." http://cheme.Berkeley.Edu/.

[15] "MIT Course Catalog: Department of Chemical Engineering." http://web. Mit.Edu/catalog/degre. Engin.Chemi.Html.

CASE 8–2

Dr. Katheryn Rios, the chair of the Academics Committee at University of Ohio, writes a report to the president of the University Senate, Dr. Michael Coffman. The goal of the report is to explain the activities of the committee during the previous year. Dr. Coffman has asked the chair of each Senate committee for a report on the previous year's activities. Dr. Rios attempts to make her report readable by using clearly worded headings and concise descriptive information under each heading.

Note that she gives the main point in the first sentence and provides a succinct opening introduction/summary. She chunks the report itself into segments introduced by clearly worded headings. Dr. Coffman can read those segments that interest him.

This represents perhaps the most common type of routine report. The writer designs it to inform quickly and concisely.

CASE DOCUMENT 8–2

TO: Michael Coffman   DATE: June 9, 2017

FROM: Katheryn Rios

SUBJECT: Academics Committee and List of Concerns

The AC has not had significant business this year because of the ongoing financial stresses on the university. However, among the list of topics presented here, several need consideration by the full Senate. I call AC meetings when a meeting seems needed. Some of our business can be conducted by e-mail, and I have used that method, which works better for members who are out of town.

Should the University Switch to Plus/Minus Grading?

The AC prepared a memo for members to deliver to their college deans. We received little response and certainly no consensus for implementing this system. Liberal Arts expressed interest. In short, +/- grading would work well in some colleges but not in all. Our computing system could handle this feature, but our CIO did not want to assess the cost for programming unless more than one college wanted this change.

1|

Student Evaluations of Teaching

• Online or Paper?

No consensus emerged for moving to one system, even though the university Teaching Committee would prefer having to maintain only one system.

Faculty who want to have their students use the online system evaluation have trouble getting their students to respond. Paper evaluations seem to provide the most complete number of responses.

• What Will Happen to Student Teaching Evaluations?

Two years ago, the University Senate approved a streamlined group of questions that contained five questions required for all college teaching evaluations. Each college could choose an additional 15 questions. This method would allow comparison among colleges.

The AC then recommended that all evaluations require TWO questions: (1) Was this an effective teacher? and (2) Was this a good course? Still no action from the Council of Deans.

Extension of the Withdrawal Deadline from 50 to 60 Days

Request for this change came from the Student Senate. James Cornyn and I prepared an e-mail that went to advisors. Few responded. Thus, we phoned as many departments as we could. Results: Only departments which have few majors thought the deadline "might" be extended. The undergraduate dean opposed the idea, along with advisors from several large departments. Thus, this item was rejected.

Containing Textbook Costs

Helena Stone and I worked with Charlene Bangor on the bookstore contract. We never saw the final contract. Our concern: the 30% markup on books is too high, and that markup is added to the publisher's suggested price.

The AC approved an e-mail that was sent to the colleges for distribution to departments. It asked that faculty consider cost of books required and rationale for books required. The student senate reported that many required books are never used in classes. The library also has a website for posting class material.

The e-book, book rental options continue to grow. The AC just voted to send this e-mail every term to notify faculty of this option.

University Rule on Shortened Courses

The AC approved the rule I devised with help from the registrar's staff. Karan offered an amendment (see attachment, p. 4), which clarifies the approval process. I will send this to AC for their input/approval.

Students Connecting Through Service

This plan, which would be launched before our reaccreditation, was approved by AC this spring. We do not know what it means, other than it should broaden students' learning experiences. I sent an e-mail to the major service organizations, and four have responded positively thus far. The pre-med society and the dean want to work with the local physicians on the importance of innoculations. AC will have a report from the pre-med society at our next meeting.

Assessing Academic Advisors

The student senate has requested that academic advisors be evaluated. I responded to that request by stating that several questions would have to be clarified that deal with logistics: cost, number of advisors and which ones, procedure, method, and reports.

Recommendations for the Next Faculty Senate Discussion

1. **Review of Status of Preparation for Reaccreditation**

    What do faculty need to know about the coming reaccreditation visit?
    What is the status of recommendations from the Stephens report?
    What will happen with core curriculum after courses we approve are
      sent to the State?

2. **Discuss Way to Control/lessen Textbook Costs?**

    Do faculty consider costs when they select books?
    What is the cost of e-books versus paper copies?

3|

Elements of Formal Reports

Long reports should be developed as formal reports. The word *long* generally means ten pages or more, but as a writer you have to decide based on the report context. Writers should select elements to help readers find their way through the report. The longer the report and the more diverse the readership, the more elements will be required to help readers find what they need. Each element has a different purpose and often targets the needs of different readers. Reports are not novels: most reports will not be read cover to cover but selectively.

In general, formal reports may include any or all of the following elements. How these elements appear in practice differs widely and depends on the organization preparing the report. We will describe the general characteristics and content of each element and provide several examples. Most organizations develop report templates to fit the content routinely generated.

> Prefatory Elements
> > Letter of Transmittal or Memo of Transmittal
> > Title Page
> > Submission Page
> > Table of Contents
> > List of Illustrations
> > Glossary and List of Symbols
>
> Abstracts and Summaries
> > Informative Abstract
> > Descriptive Abstract
> > Executive Summary
>
> Discussion, or Body of the Report
> > Conclusion(s)
> > Recommendation(s)
> > References
>
> Appendices

Prefatory elements. Readers initially access a report from the prefatory elements. These show readers what your report will discuss and how it will approach the topic. Long, complex reports that will likely be read by many readers require prefatory elements. The letter/memo of transmittal, title page, table of contents, abstract, and/or summary are often your readers' first—and only—experience with the report. Each of these elements should help your readers grasp and accept the rhetorical purpose of the report. Never consider prefatory elements routine "paperwork" needing only perfunctory writing. No matter how well written and researched the report content, the effectiveness of the report begins with the effectiveness of the elements that state the content. Many readers may decide to read the body of the report if they find the prefatory elements compelling.

Letter/memo of transmittal. The letter/memo of transmittal is addressed to the individual who will initially receive the report. This person may not be the primary reader but the individual responsible for routing the report to appropriate readers who will digest and use the content. In consulting reports, the CEO of an organization that has solicited the consulting work is usually the person addressed in the letter of transmittal. Alternatively, the transmittal letter can address the person who has authorized and requested the information, analysis, work, or recommendations covered in the report.

Reports prepared for internal distribution will have a memo of transmittal, and reports prepared for external distribution will use a letter of transmittal. Both can be organized the same way.

The letter/memo of transmittal should include the following information at minimum:

- Statement of transmittal—subject and purpose of the report
- Reason for the report

In addition, transmittal documents may include the following items:

- Background material—the larger issue or problem addressed by the report
- Mention of earlier reports (or additional reports that may be needed)
- Information that may be of special interest or significance to the readers
- Specific conclusions/recommendations that might be of special interest to the person to whom the report is addressed
- Financial implications
- Acknowledgments—list of those who provided help in the project

The report in Case 8–1 includes a well-designed transmittal memo.

Title page. Title pages perform several functions. Basically, they provide critical identifying matter and may contain a number of identifying items to distinguish the report from others on a similar subject or from reports received in response to specific projects. Many organizations have a standard format for title pages. Whatever items you need—or your organization requires you to include—be sure to make the title page attractive. The following information often appears on title pages:

- Name of the company or individual(s) preparing the report, perhaps with his or her signature
- Name of the company or course for which the report was prepared
- Title and subtitle of the report
- Date of submission
- Code number of the report
- Contract numbers under which the work was performed
- Company or agency logo
- Proprietary and security notices
- Names of contact/responsible individuals
- Descriptive abstract

Submission page. Reports may use a submission page, which includes the list of contributors to the report and/or the names and signatures of the authorizing officer or project leader. Submission pages emphasize the point we make throughout this book: reports require accountability. Signatures on a submission page indicate that the authors stand behind the content. The submission page usually either precedes the title page or follows it. Alternatively, the title page and submission page may be combined.

Table of contents. The table of contents performs at least three major functions. First, it indicates the page on which each major topic begins and thus serves as a locating device for readers who may be searching for specific information. Second, it forecasts the extent and nature of the topical coverage and suggests the logic of the arrangement and the relationship among the report parts. The table of contents contains all major headings used in the report. For that reason, major headings, like the title of the report, should reflect the content of the material that follows. Skimming a table of contents should give readers a clear idea of the topics covered, the content presented under each heading, the amount of coverage devoted to that topic, the development of the report, and the progression of information. (See the table of contents in the report in Appendix C.) While you may wish to use "introduction," "conclusion," and "recommendations," avoid the term "discussion": it tells your reader nothing about how you will present the report content. Third, the table of contents should reflect the rhetorical purpose of your report—what you want your readers to know and why your report is important to them.

List of illustrations. If a report contains tables, graphics, drawings, photos—any type of visual—it's customary to use the page heading "Illustrations" or "Exhibits" to list all the visuals. These are usually divided into "tables" and "figures," but specific types may also be listed if you have an array of illustrations: for example, maps, financial statements, photographs, charts, and computer programs.

Glossary and list of symbols. Reports dealing with specialized subject matter often include abbreviations, acronyms, symbols, and terms not known to readers outside a specialized group. Readership should determine if a glossary or list of symbols is needed. Many reports prepared by government agencies for the general public provide glossaries to ensure that any reader can assess the report.

Glossaries sometimes appear as prefatory elements, but they can also appear in an appendix at the end of the report. When you first use a symbol or a term that you will include in the glossary, tell your reader where to find the list or the glossary. You may wish to place an asterisk (*) by a word or symbol that will be covered in the glossary. If you have only a few terms that you will need to define, you can define these as footnotes at the bottom of the page and use asterisks (* or **) to alert your readers.

Abstracts and summaries. Abstracts and summaries are the most important prefatory items in a report. The title page, table of contents, abstract, and summary may be the only parts of the report many recipients read. Thus, carefully plan these elements as each provides a slightly different perspective. While the title page and table of contents outline the report's content and direction, abstracts and summaries provide the essence of the report: topic, purpose, results, conclusions, and recommendations. Each item can stand alone or be designed in conjunction with other prefatory elements, but the wording of the table of contents should echo throughout an abstract or summary. These elements, when well designed, enable readers to quickly find a specific segment in the body of the report.

Often, the abstract follows or appears on the title page. Summaries may also follow the title page. While abstracts and summaries contain similar information, summaries usually provide more extensive information than abstracts and may be written for decision-makers whose needs differ from those of readers who want only the essence of a report. (See the table of contents and the summary in Appendix C.)

Abstracts fall into two categories, informative or descriptive. Abstracts, often accompanied by keywords, are prepared for use in online indexes and databases. Keywords allow your report, if it is stored in a database, to be retrieved. Thus, you should think carefully about the keywords that characterize the content of your report. What words will your readers most likely use in a database search? Abstracts with keywords may be separated from the report but linked to the full report. Based on the information in the abstract, readers can decide if they need to move to the full report.

Differences in abstracts have tended to disappear, and some abstracts have the characteristics of both traditional informative and descriptive styles. Organizations and journals usually have specific requirements for abstracts and may require, for example, a short descriptive abstract on the title page and an informative abstract after the table of contents.

Because the quantity of information that bombards readers increases, well-written abstracts have become critical to report access. The abstract should explain the purpose of the report, findings, conclusions, and recommendations—anything of significance. Because many readers now use abstract services, they may read only your abstract, or they may decide to retrieve/order your complete report if the abstract clearly shows the relevance of the content and key points.

A good way to plan an abstract is to create a file that contains the names of its main parts:

> Subject/Purpose
> Methods
> Findings or Results
> Conclusion(s)
> Recommendation(s)

Next, insert information under each heading. When you complete your abstract, you can remove the headings. Be sure to check the length of the abstract; publications rarely want abstracts longer than 100 to 150 words. This method allows you to track length easily. However, some publications want the headings included in the abstract. The example abstract (Figure 8–2) contains these items. Color helps you see each item.

Informative abstract. This type of abstract includes the research objectives, research methods used, and findings, including principal results and conclusions. Recommendations may also be included. Informative abstracts usually range from 50 to 500 words, depending on the length of the report and on the requirements

Chemical (Chlorpyrifos and Permethrin) Treatments Around Stacked Bales of Hay to Prevent Fire Ant Infestations

Keywords: chlorpyrifos, permethrin, Lorsban 4E, Astro Insecticide, fire ant, hay bale

Abstract: This research evaluated the efficacy of using a chemical barrier applied to the soil area under stacked bales of hay to prevent the red imported fire ant, *Solenopsis invicta* Buren (Hymenoptera: Formicidae), from infesting stacked hay. Specifically, we were interested in determining if we could protect "clean" hay bales stored in fire ant infested fields for up to several weeks. Chemicals selected as barrier treatments were Lorsban® 4E, active ingredient chlorpyrifos, which kills ants on contact, and Astro™ Insecticide, active ingredient the pyrethroid permethrin, which can also act as a repellent to ants. We established a series of 12ft × 12ft plots, with a 10ft buffer between plots along a fence row in a fire ant infested field. Plots were grouped into four blocks of three stacks each. Plots within blocks were randomly assigned to each treatment (four plots treated with Lorsban® 4E and four treated with Astro™ Insecticide, and four control plots). Treatments included spraying a 12ft × 12ft soil area with a 1-gal solution of each chemical and water formulation. After soil treatments, we placed four square-bales of hay, stacked two a side and interlocking in two layers, in the center of each plot. Stacked bales were sampled for fire ant infestation using 2.5 × 2.5cm olive oil–soaked index cards; one bait card was placed on each side of the top layer of hay in each stack. Results from ANOVA show a significant difference in mean infestation levels among treatments. Stacks of hay sitting in the chlorpyrifos plots had fewer ant infestations compared to the permethrin and control plots. Results after one week showed that only one stack in the permethrin and two in the control plots were infested with ants, while none in the chlorpyrifos plots were infested. Results show that after three weeks all four control stacks, three stacks in the permethrin treatment, and two stacks in the chlorpyrifos plots were infested. These results indicate that on a short-term basis, such as 1 to 7 days, chlorpyrifos may be an effective short-term treatment option for protecting stacked hay from fire ant infestations.

Ronald D. Weeks, Jr., Michael E. Heimer, and Bastiaan M. Drees, Chemical (Chlorpyrifos and Permethrin) Treatments Around Stacked Bales of Hay to Prevent Fire Ant Infestations, Texas Imported Fire Ant Research & Management Project, Red Imported Fire Ant Management Applied Research And Demonstration Reports, 2000–2002, Texas Cooperative Extension Service.

http://fireant.tamu.edu/research/arr/year/00-02/2000-2002ResDemHbk.htm #stackedbales.

● **FIGURE 8–2** Informative Abstract

of the disseminating organization and of the abstracting service. Informative abstracts help readers decide if they want or need to access the entire report for more thorough examination. They begin with a statement of the report's purpose, and the remaining sentences give major highlights and conclusions.

Descriptive abstract. This type of abstract states what topics the full report contains. Unlike informative abstracts, descriptive abstracts cannot serve as a substitute for the report itself. They begin with the report purpose and then explain content areas or topics covered in the report. Figure 8–3, a descriptive abstract, comes from a research report about the safety and reliability of automotive electronic control systems. See also the descriptive abstract on the title page of the sample report in Appendix C.

Executive summaries. Major reports often include executive summaries, which contain all of the items listed at the beginning of this chapter. However, unlike summaries for routine reports, executive summaries provide extensive development of each information segment to provide decision-makers the information they need without having to read or search the full report. An example of an executive summary can be seen in the report for a state study on shrub control on rangeland (Figure 8–4). Even though you know nothing about this project, the executive summary will provide you with extensive information. In many proposals, the summary often determines whether the proposal itself will be read.

Discussion, or body of the report. The main part of the report, the discussion, takes most of the writer's development time. The information to be reported constitutes the discussion. The discussion will explain in detail—information

---

This report summarizes the results of a study that assessed and compared six industry and government safety standards relevant to the safety and reliability of automotive electronic control systems. These standards include ISO 26262 (Road Vehicles - Functional Safety), MIL-STD-882E (Department of Defense Standard Practice, System Safety), DO-178C (Software Considerations in Airborne Systems and Equipment Certification), Federal Motor Vehicle Safety Standards, AUTOSAR (Automotive Open System Architecture), and MISRA C (Guidelines for the Use of the C Language in Critical Systems). The assessment was carried out along the following 11 dimensions: (1) type of standard, (2) definition of safety and hazard, (3) identification of safety requirements, (4) hazard and safety analysis methods, (5) management of safety requirements, (6) risk assessment approach, (7) design for safety approach, (8) software safety, (9) system lifecycle consideration, (10) human factors consideration, and (11) approach for review, audit, and certification. The observed strengths and limitations of the standards studied in this report could support the future development of a robust functional safety approach for automotive electronic control systems.

Source: US Department of Transportation, National Highway Traffic Safety Administration. *Assessment of Safety Standards for Automotive Electronic Control Systems,* Washington, DC: 2016. http:// www.ntis.gov.

• **FIGURE 8–3** Descriptive Abstract

EXECUTIVE SUMMARY

The disparity between the supply and demand for water in Texas has focused attention on brush control as a method for increasing the water supply for Texans. This review (1) summarizes how shrubs affect the water cycle at the tree, hillslope, small catchment, and landscape scales and (2) examines whether broadscale reductions in woody plant cover might increase the water supply. The greatest information need is at the landscape scale, but research at this scale is costly and difficult. Hence, most field investigations have focused on the tree or hillslope scales. However, extrapolating these data to larger scales is problematic because hydrological processes may change as the scale increases.

A review of the scientific literature has allowed the authors to make the following broad generalizations:

1. The relationship between shrub removal and increased water yields becomes stronger as annual rainfall increases.

2. The linkage between shrub removal and increased water yields is weaker in upland areas and stronger in areas adjacent to stream channels, where shrubs may be using shallow groundwater that is hydrologically connected to stream flow.

3. In upland areas, the linkage between removal of woody plants and increased water yield is stronger where water can move rapidly through the soil or parent material to recharge springs or shallow aquifers.

4. In areas where there is little subsurface water movement and woody plants are not accessing groundwater, shrub control is unlikely to significantly affect either groundwater recharge or stream flow.

The data suggest that water can be salvaged by controlling dense stands of saltcedar in riparian zones, dense stands of Ashe juniper in areas with rapid subsurface flow (e.g., where springs are present), and mesquite on soils that develop deep cracks when dry. However, in each case there is considerable uncertainty with respect to the amount of water that can be salvaged and the fate of the salvaged water.

Hydrologic models have predicted substantial increases in water yields following shrub control, but these results must be used with caution. There is a threefold variation in the predicted water yield increases from different models, and in some cases the predicted increases are at least three times larger than the measured values.

More research is needed to fully understand the extent to which shrub control can increase water yields. Key gaps include community-level evapotranspiration studies, isotope studies to determine the sources of water used by shrubs, groundwater–surface water connections, nested watershed studies, and landscape scale studies on saltcedar in various habitat types.

The hydrological condition of Texas watersheds and the partitioning of water within the hydrological cycle are determined by complex interactions between soils, vegetation, and climate. Brush encroachment is only one change that has affected the health of Texas watersheds—droughts, excessive grazing, water impoundments, urbanization, and water transfers also have altered the hydrological cycle. It would be valuable to shift the debate away from a focus on brush control and water yields to a broader assessment of best management practices for improving watershed health and sustainability.

• **FIGURE 8–4** Executive Summary

appropriate to the context, the readers, and the purpose of the report—why the report was done, its objectives, methods, findings, results, analysis of results, conclusions emerging from results, and recommendations for dealing with the results. The discussion is the heart of the report: without it, effective summaries and abstracts could not be written. All conclusions and recommendations derive from the discussion. Stated another way, the presentation of information in the discussion allows the report writer to draw conclusions and perhaps recommendations. In short, the discussion must support all conclusions and recommendations.

Ironically, however, the main discussion is the report segment read least. While most readers will look at the summary, abstracts, and table of contents, few will actually delve into the discussion. Nevertheless, the discussion becomes the source, foundation, and documentation for every statement written in the abstract and summary. Your conclusions need to evolve from the discussion. Recommendations need to evolve from the discussion and the conclusions. In a sense, then, the discussion develops as a narrative that comes to a conclusion. Many discussions end with a factual summary, a concise narrative of the report's main findings.

Many readers may need to read parts of the discussion carefully. They will read the introduction, check the table of contents, skip to relevant sections that interest them, and then move to the conclusion or factual summary that pulls together the main results or ideas.

Parts of the discussion. The main body of the report generally begins with an introduction (or introduction + summary) that forecasts what is to follow in the report. It directs the reader's mind to the subject and purpose. It tells the reader how to approach the content by explaining the scope of the report, the plan of development, and any additional information the reader may need. Examine the introduction from a report for the US Department of Transportation (Figure 8–5). Note that it presents the topic, the rationale, and the development plan for the report. Because of the length of the introduction, the writer partitions the parts of the introduction with headings. Also examine the introduction to the sample report in Appendix C.

Note that the introduction should always include the report subject, purpose, and plan of development. Some reports place the background and the scope in separate sections that follow the introduction, if these two items are extensive. What you include in the introduction depends on your readers. If your readers will expect your report, you can write a short introduction. However, if your report will be archived and read later by people who know little about the context,

---

1. Introduction

1.1 Background

The John A. Volpe National Transportation Systems Center (Volpe Center) provides technical support to the U.S. Department of Transportation's Federal Railroad Administration (FRA) Office of Research, Development and Technology on issues involving railroad safety including trespass prevention. The Volpe Center is developing a body of information to assist current and future researchers in their efforts to reduce rail trespassing, the leading cause of rail-related deaths in the U.S.

Railroads have experimented with alternative fencing strategies to deter illegal trespassing. High-security fencing—reinforced fencing used to prevent trespasser incursions or breaches whether from climbing, cutting, or other method of passage—is being used more frequently. Although this solution is more formidable than standard fencing, it tends to be more costly than standard chain link fencing or no fencing.

While cost constraints prevent many agencies from using high-security reinforced fencing, others have begun to use it in limited circumstances. As a result, there is a need for more industry best practices for installing high-security fencing solutions and identifying appropriate applications for them.

• **FIGURE 8–5** Report Introduction with Section Headings

At the 2012 Right-of-Way Fatality and Trespass Prevention Workshop[1], which was sponsored by FRA and the Federal Transit Administration (FTA), the attendees (mostly rail and transit safety experts) identified current research needs, including research into fencing design and the use of fences to mitigate rail trespassing. FRA tasked the Volpe Center to conduct this research with a focus on gathering information about the extent and use of high-security fencing.

1.2 Objectives

This report summarizes field observations that highlight the differences and similarities in railroad approaches to select and appropriate effective fencing strategies. As the market for high-security fencing solutions grows based on increased demand and costs become competitive in more balanced market conditions, it is vital to remember "safety first" when measuring safety concerns against associated economic realities.

1.3 Overall Approach

The FRA asked railroad industry representatives to provide information on current and proposed installations of high-security fences along railroad rights-of-way. Where feasible, site visits were made by Volpe Center researchers to visually inspect and photograph examples of fencing that are currently in use. In addition, most of the participating agencies were informally interviewed about issues such as policy implications, community acceptance, site choices, hazard analysis and cost concerns. From this information gathering exercise, a summarized list of best practices has been created and presented in this report.

1.4 Scope

The report's scope is by no means exhaustive and is limited to the specific transportation agencies and localities identified within it, which were chosen through informal outreach and its familiarity with fencing and rail right-of-way (ROW) security.

This research was also limited because some of the information is considered safety-sensitive or confidential by railroads or the public agencies that own or operate rail service. Much of the information that the team received was not available online and was made available to the authors for investigative purposes only.

1.5 Organization of the Report

Chapter 1 – Introduction: Addresses the objectives of this research, defines the intended scope, and presents an outline for organizing the information found within this report.

Chapter 2 – Physical Attributes: Focuses on the physical hardware and installation of high-security fencing designs used to limit access to the railroad ROW. Each agency's fencing solutions are examined; the agency's stated needs and the proposed/implemented solution are briefly discussed.

Chapter 3 – Location-Based Hazard Analysis: Describes some of the standardized methods that transportation agencies use to determine specific locations for high-security fencing.

Chapter 4 – Community Interactions: Highlights situations where with local communities and collaborating with local institutions led to improved high-security fencing solutions.

Chapter 5 - Organizational Prioritization: Briefly discusses why an official (or formal) organizational commitment to a uniform fencing policy plays an important role in implementing useful high-security fencing.

Chapter 6 – Conclusion summarizes best practices developed through interviews, site visits and observation.

---

[1] 2012 ROW Fatality Trespass Prevention Workshop, http://www.fra.dot.gov/conference/trespass2012/ cited on 9/18/2014.

• **FIGURE 8–5** *Continued*

Wind Power in the United States: Technology, Economic and Policy Issues

Introduction

Rising energy prices and concern over greenhouse gas emissions have focused congressional attention on energy alternatives, including wind power. Although wind power currently provides only a small fraction of US energy needs, it has grown more rapidly than any other electricity source. Wind energy already plays a significant role in several European nations, and countries like China and India continue to rapidly expand their capacity both to manufacture wind turbines and to integrate wind power into their electricity grids.

This report describes utility-scale wind power issues in the United States. We partition the report into the following sections:

- Background on wind energy
- Wind resources and technology
- Industry composition and trends
- Wind power economics
- Policy issues

● **FIGURE 8–6** Standard (and Readable) Report Introduction

you will need to provide a longer, more informative introduction (see Figure 8–6). Avoid long introductions: focus on the subject and purpose in terms that explain the relevance of the report to readers. But do anticipate your readers—who will read this report and when—as you plan the introduction. Without a proper introduction, readers will have difficulty following the main discussion.

Collecting and grouping information. As you gather information, try grouping your material and notes into specific categories. Label these categories, and then begin your report. You may want to develop your report according to major sections—introduction, followed by the topics or categories. Open and name the file and begin. For example:

- Introduction: state the purpose of your report—what you expect to accomplish, what you want your reader(s) to know. You can add information later. Stating your purpose at the beginning of the draft helps you stay focused.
- Category/topic 1: phrase describing the issue you want to present
- Category/topic 2: etc.
- Category/topic 3: etc.

Next, begin inserting information under each topic or subject category. Focus initially on inserting information that pertains to the topic heading.

This method helps employees who may be trying to work on a report, answer phone calls, answer/send e-mail, and deal with other routine business events throughout the workday. In short, you can arrange, insert material, save what you have, and add other material as time permits. You may also write notes to yourself or use different colors and fonts for text you may want to move, delete, or revise. How you design your report will depend on (1) the kind of report you are writing,

(2) your readers' informational needs, and (3) the purpose of your report. Introductions should be as short as possible while providing readers with the information they need to understand the report.

Strategy for presenting the discussion. The discussion should be planned around each topic mentioned in the plan of development. Note that the logic of the discussion should be evident from the headings as these are repeated in the table of contents.

Each paragraph in the discussion should begin with a topic sentence, followed by supporting sentences, data, and visuals.

Reports with standard arrangement patterns. Some kinds of reports have fairly standard arrangement patterns. Many government funding agencies require that all reports submitted follow a specific plan, and content may be inserted into a template and submitted online. Many organizations also have a standard plan for their policies—what sections to include and the order in which they appear. Policy and procedure manuals, detailing the rules that apply to the employees of a business entity, usually follow a standard pattern.

Often, however, as in Case 8–2, you will need to write a report in which you, as the writer, must decide how to present your research or the information you need to convey. Generally, you have two basic choices: topical arrangement and chronological arrangement.

Topical arrangement. In topical arrangement, the order in which you present your ideas should be logical and inclusive. For example, in a report on disease management of citrus fruit, the writer arranges the report by grouping information about specific citrus diseases. He describes, in parallel arrangement, the main diseases:

Introduction—Description of treatments for citrus diseases
Disease #1—Melanose
    Description
    Factors to be considered before the application of fungicides for melanose control
        Table I. Chemical controls for melanose
Disease #2—Greasy Spot
    Description
    Factors to be considered in the management of greasy spot
        Table II. Chemical controls for greasy spot
Disease #3—Foot Rot
    Description
    Factors to consider in managing the disease
        Table III. Chemical controls for foot rot
Disease #4—Citrus Nematode
    Description
    Sampling instructions to determine presence of citrus nematode
        Table IV. Citrus nematode counts considered low, medium, or high at specific times during the growing season
Conclusion: Factors to consider before applying nematicides
        Table V. Chemical controls for nematodes (summary)

Once the citrus researcher has arranged his information, he can begin inserting information beneath each topic heading.

Chronological arrangement. Some topics can be presented by time. You explain or present information sequentially, in the order in which it occurred. The following outline of a literature review, *Cultural Control of the Boll Weevil—A Four Season Approach—Texas Rolling Plains*, illustrates chronological arrangement. This technical report surveys and reviews existing research on how boll weevils can be controlled throughout the agricultural year. Note, too, that the segments use parallel development, and each segment ends with a summary of that segment. This approach allows the reader to choose where to begin and how much to read. For example, the reader may wish to read only the Summary and Introduction and the factual summary for each season:

> Summary
> Introduction
> Spring Cultural Control
>> Prepare the land for planting
>> Utilize delayed planting
>> Use uniform planting
>> Summary
> Summer Cultural Control
>> Shorten the growing season
>> Change the microclimate
>>> Row direction
>>> Bed shape
>>> Row spacing
>> Summary
> Fall Cultural Control
>> Utilize harvest-aid chemicals
>> Role of planting date
>> Terminate irrigations in August
>> Summary
> Winter Cultural Control
>> Eliminate the overwintering habitat
>> Modify the overwintering habitat
>> Avoid the overwintering habitat
>> Summary
> Conclusions
> Acknowledgments
> Supporting Research Studies

Persuasive arrangement and development. Many times reports argue for a specific point or position. When you argue, you need to persuade your readers. Understanding the likely objections of your readers will be critical to planning your report and presenting your arguments. Some reports may require writers to present conclusions and recommendations that readers will not welcome. Other reports may address readers without preconceived ideas. In each situation, the writer must design the report to anticipate the perspective of the readers and build a convincing case.

Case 8–3

William Allen is a student working on a project with his major professor in chemical engineering to discover more about the costs of fracking. William prepares a report that analyzes the water used and its monetary value in the specific case of hydraulic fracturing in the Eagle Ford Shale region of Texas.

CASE DOCUMENT 8–3

TO: Richard Harrington   DATE: November 12, 2017

FROM: William Allen

SUBJECT: **Water Use and Value Related to Hydraulic Fracturing: Texas Eagle Ford Shale**

Introduction
Hydraulic fracturing (also known as fracking) is a process that involves drilling vertically into shale formation (followed by drilling through the rock horizontally), then pumping a water and chemical mixture into the drill sites at high pressures. The high pressure causes the shale to crack (fracture) allowing the extraction of natural gas and oil. There is an abundance of shale in the United States and hydraulic fracturing has grown substantially over the past decade alone.[1]

The development of hydraulic fracturing technology has opened vast regions to production of oil and natural gas. There were more than 45,000 gas wells completed in the United States in 2012,[2] with almost 11,000 of those wells being completed in Texas.[3] With this boom in production has come a high demand for water in order to fracture the wells.

The high water demand for fracking has created controversy, especially in more arid parts of the United States, which struggle with the issue of continuing drought. This analysis targets Eagle Ford Shale in Texas, a region of limited water supply. I compare the value of water used in fracking and in alternative uses. The results will be largely based on the acre-foot unit of measurement which is the equivalent of 325,851 gallons.

Methodology
Water is a tricky commodity to value, in part because water, in its purest form (not including substances dissolved in the water such as salt and other minerals), is fairly uniform throughout the earth. Water value does differ, however, based on availability and location, as well as on the opinions of those who make use of water in their everyday activities. These groups include city water suppliers, farmers, hydraulic fracturing companies, and environmentalists. With the exception of values determined by environmentalists, water values are determined by the total

revenues generated from the use of the water. The determination of water values necessarily includes consideration of production costs such as the costs of distilling the water for drinking, pumping the water out of wells, and shipping the water to the appropriate sites. This analysis observes these differences among three basic water stakeholders: municipal and industrial (M&I) users, agricultural users, and fracking users.

For M&I use, the prices and quantities presented by the Texas Municipal League[4] served as an estimate of value while a residual value approach was applied for agricultural use. A residual value approach involves comparing irrigated (land that water is pumped onto in addition to rainfall) returns to net dryland (land that only uses rainfall; no additional water is added) returns with a charge to all inputs including land. The only difference is that there is an imputed value of water for irrigated land. In order to gain a strong idea of what these returns were, three crops were used to conduct this analysis, including cotton, sorghum, and bermuda (grass) pasture. Examples of both the M&I and agricultural approaches to determining the value of water are shown in Table 1.

I derived the value of water in fracking in a slightly different manner. This analysis assumes that a typical well is capable of being productive for 20 years.[5] Thus, I projected the revenues (based on the prices and quantities of oil and gas extracted) and costs of the well each year using an equation[6] that predicted the production cycle of a well. Then, I applied capital budgeting techniques to estimate the initial investment, annual costs related to production, and revenues generated from the use of the well. Next, I calculated the present value of the well by discounting the revenues and costs each year back to the present period using the projected economy discount rates.[6] Discounting values simply implies taking future values and converting those values back to a current stated value. The discount rate is the means by which this conversion process takes place. Discount rates take into account projected inflation and risk. By comparing the present value of revenue to the present value of costs, I estimated an approximation of returns to water. I then divided this present value to water by the water used in fracking to calculate approximate value per acre-foot, per acre-inch, and for every 1,000 gallons.

Results

First, I examined M&I and agriculture water value. I obtained water prices from the Texas Municipal League (mentioned in the methodology) and obtained the agriculture data from crop enterprise budgets presented by the Texas A&M AgriLife Extension Service.[7] For M&I, water rates (prices/costs) are reported for two water use levels, including residential (5,000 and 10,000 gallons per month) and commercial (50,000 and 200,000 gallons per month) customers, without reference to average typical quantities of water consumed by those users (see Table 1). Because of limited resources, I assumed that M&I companies have set prices that reflect what they believe the value of their water. Therefore, the prices from the

TABLE 1 The Value of Water per Acre-Foot for M&I and Agriculture Water Use in the Eagle Ford Shale Region

Municipal and Industrial

Classification	$ Per Thousand Gallons Low	High	$ Per Acre Foot Low	High
Residential				
5,000 Gallons	2.62	6.7	855	2,185
10,000 Gallons	2.17	5.05	708	1,647
Commercial				
50,000 Gallons	2.29	4.21	747	1,373
200,000 Gallons	2.31	4.05	753	1,321

Agriculture[a]

	Cotton Dry	Cotton Irrigated	Cotton Returns to Water	Sorghum Dry	Sorghum Irrigated	Sorghum Returns to Water	Bermuda Pasture Dry	Bermuda Pasture Irrigated	Bermuda Pasture Returns to Water
Yield	1,320 lbs.	2,272 lbs.	-	22 cwt	43 cwt	-	140 lbs.	600 lbs.	-
Total Revenue ($)	493.00	826.00	-	187.00	366.00	-	55.00	270.00	-
Land Charge ($)	123.00	207.00	84.00	62.00	121.00	59.00	25.00	100.00	75.00
Water Charge ($)	-	16.00	(16.00)	-	8.00	(8.00)	-	84.00	(84.00)
Irrigated (ac.in.)	-	14.00	-	-	14.00	-	-	12.00	-
Net Returns ($)	(77.00)	(28.00)	49.00	(49.00)	17.00	66.00	(26.00)	(102.00)	(76.00)
Net Returns to Irrigation ($)	-	-	117.00	-	-	117.00	-	-	(85.00)
Value (ac.in.) ($)	-	-	8.36	-	-	8.36	-	-	-
Value (ac.ft.) ($)	-	-	100.29	-	-	100.29	-	-	-

[a] The average farm in the Eagle Ford area has 2,037 acres.[8]

Texas Municipal league are deemed appropriate for this analysis. For agriculture, three crops being observed are shown from both a dryland and an irrigated perspective (see Table 1). Visual examples are shown in Figures 1 and 2.

The value for M&I water use has an estimated low value of $708 per acre-foot and a high value of $2,185 per acre-foot. In agriculture, estimated water value stays at slightly more than $100 per acre-foot for both cotton and sorghum. For bermuda pasture, no value is shown in the "Value per Acre-Foot" row because it

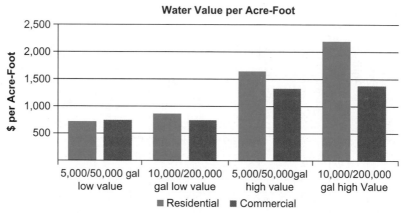

Figure 1 Water Value per Acre-Foot for M&I in the Texas Eagle Ford Region

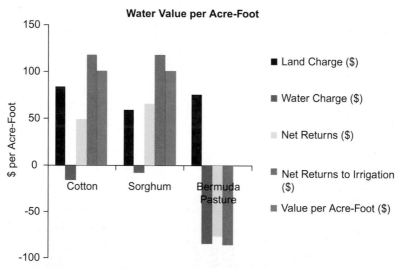

Figure 2 Water Value per Acre-Foot for Agriculture in the Texas Eagle Ford Region

assumed that ranchers will not choose to irrigate bermuda pasture if the returns to the irrigation are less than the returns to dryland methods.

Chesapeake Energy[9] and Dr. Stephen Holditch[10] provided the information for the initial investments in drilling, fracking, and operating wells as well as the elements (and their associated costs) contributing to the fracking process. They also provided most of the following facts aside from the calculations. Fracturing a typical well in the Eagle Ford Shale requires approximately 4.8 million gallons (14.73 acre-feet) of water. Drilling a well, including set-up costs, transportation, casing, contingencies, completion of the well, etc., totals at $6.3 million. A royalty payment worth 25% of the total revenue is included with the operating costs. Also included in the operating costs are the variable costs that account for the daily fracking activities. Shown in Table 2 are the present values of the revenues, operating costs (variable costs and 25% royalty payments), and the fixed costs (represented drilling costs previously discussed). I separated the revenues and variable into two components, oil and gas, because a typical well is assumed to produce both oil and gas. The fixed costs are lumped together because those represent the costs that are spread over both oil and gas extraction. The table is also separated into values per acre-foot, per acre-inch, and per 1,000 gallons. The values presented in the table represent the value per well. Therefore, the net returns are divided by the 14.73 acre feet needed to fracture the well.

TABLE 2 Net Value of Water used for Hydraulic Fracturing in the Eagle Ford Shale[a]

	$ per Well	
Returns to Gas	$ 6,058,814	
Returns to Oil	17,541,265	
Total Returns		$23,600,079
Operating Costs to Gas	$ 1,211,763	
Operating Costs to Oil	5,847,088	
Royalty Payments (25% of the Total Revenue)	5,900,020	
Total Variable Costs		$12,958,871
Drilling and Fracking Costs		$ 6,343,000
Net Returns per Well		$ 4,298,208
Net Returns per Acre Foot		$ 291,800
Net Returns per Acre Inch		$ 24,317
Net Returns per Thousand Gallons		$ 896

[a] This is the present value as of 2015.

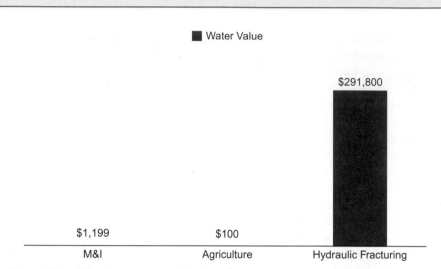

Figure 3 Water Value Comparison per Acre-Foot for M&I Users, Agriculture Users, and Hydraulic Fracturing Users

Based on the estimated oil and gas prices applied, the result is that water value in hydraulic fracturing is nearly $300,000 per acre-foot. Water, therefore, is far more valuable to the fracturing users than it is to agriculture and M&I users. Return on investment and risk management was not included in this analysis, creating the possibility that this result is high. However, sensitivity analysis would reveal that the value of water in fracking would not go below $35,000 and would not exceed $570,000 per acre-foot. Shown in Figure 3 is a comparison of the water values for the three water users discussed in this report. For M&I, the value shown is a simple average of the four values listed in Table 2.

Conclusions

We can see that the value of water used for hydraulic fracturing exceeds the value of water used in M&I and agriculture. This scenario is more than likely the case because the oil and gas industry is currently very lucrative despite attempts to move to alternative sources of fuel such as biofuels, hydrogen, electricity, etc. Public knowledge of the large difference in these water values could have considerable implications for how water is distributed in the near future.

To begin, economics assumes that a person is rational above all else. Realizing that water is far more valuable in hydraulic fracturing than it is in agriculture and in M&I, an economist would be led to believe that water owners would look to sell their water to fracking companies as soon as they could in order to make profit as soon as possible. This belief would stem from the fact that people need to make money in order to support themselves and their families. As such, could an understanding of this value cause a shift in water usage from, say, agriculture to hydraulic fracturing in Texas?

Now consider, however, the other side of the argument. Obviously the drought in Texas has been making agriculture and some M&I activity more difficult to partake in. Assuming that ownership of water falls largely in the hands of farmers, would money alone be enough of an incentive for Texas farmers to let go of their water? The point here is that farmers seem to understand better than most that water holds the key to life. Without water, they will be unable to grow crops and work the land from an irrigation standpoint. Would the understanding of this nonmonetary value of water be a force to cause farmers in Texas to keep their water at all costs? Would understanding the high value of water for fracking alter the balance of water used in fracking and water used for other purposes in Texas? The fact that this research indicates that water is more valuable in fracking opens the door to numerous questions that will need addressing in the coming years.

References

1   Sieminski, A. (2013). Status and outlook for shale gas and tight oil development in the U.S. Retrieved February 19, 2016, from the website: http://www. eia.gov/pressroom/presentations/sieminski_03062013.pdf.

2   Maugeri, L. (2013). The Shale Oil Boom: A U.S. Phenomenon: Study Forecasts Tripling of U.S. Shale Oil Output. Retrieved February 19, 2016, from the website: http://belfercenter.ksg.harvard.edu/publication/23191/shale_oil_boom. html.

3   Texas Railroad Commission, Texas Oil and Gas Divisions. (2013). Retrieved May 7, 2016, from the website: http://www.rrc.state.tx.us/data/drilling/ txdrillingstat.pdf.

4   Texas Municipal League. (2015). Municipal and Commercial Water Rates. Retrieved June 17, 2016, from the website: http://www.tml.org/surveys.asp.

5   Hodlitch, S. (2016). Director, Texas A&M University Energy Institute. Personal communication, November 28, 2016.

6   Swindell, G. S. (2012). Eagle Ford Shale—An Early Look at Ultimate Recovery. SPE Annual Technical Conference and Exhibition, San Antonio, Texas. Retrieved August 8, 2016, from the website: http://gswindell.com/sp158207.pdf.

7   Texas A&M AgriLife Extension Service. (2015). Crop and Livestock Enterprise Budgets. Retrieved November 12, 2016, from the website: http://agecoext. tamu.edu/resources/crop-livestock-budgets/by-district.html.

8   US Department of Agriculture. (2015). Agriculture Census. Retrieved June 27, 2016, from the website: http://www.agcensus.usda.gov/Publications/2007/ Online_Highlights/County_Profiles/Texas/.

9   Chesapeake Energy. (2012). Water use in Eagle Ford Deep Shale Exploration. Retrieved May 28, 2016, from the website: http://www.chk.com/Media/Educational-Library/Fact-Sheets/EagleFord/EagleFord_Water_Use_Fact_Sheet.pdf.

10 Hodlitch, S. (2016). Director, Texas A&M University Energy Institute. Personal communication, July 10, 2016.

Letter Reports

Reports can be prepared as letters. See Figure 8–7, prepared by an engineer at Bradshaw Engineering, who has been asked to explain the differences in corn and sugar ethanol to undergraduate engineering students at Texas A&M. Note the purpose statement and the plan of the report, in addition to the short summary statement. The effective document design allows her report to be read quickly.

BRADSHAW ENGINEERING, LLC
2104 RIVERSHIRE PLACE
HOUSTON, TX 77036

February 18, 2017

Dr. Jim Freeman
P.O. Box 3998
MSC Station
College Station, TX 77834

SUBJECT: **Advantages and Disadvantages of Ethanol from Brazilian Sugar**

Dear Jim:

Introduction

I was interested in your questions about Ethanol and Brazillian Sugar, which you introduced during your class last week. Mr. Bradshaw suggested that you might like to share the following information with your engineering interns.

With a growing demand for ethanol in the United States, producers are finding new, cheaper crops to convert to ethanol. Corn remains the major source of ethanol production in the United States. Recent discoveries find ethanol from sorghum in China and sugarcane in Brazil to decrease the overall demand for corn-based ethanol because of their advantages. This letter examines the advantages and disadvantages of Brazilian sugarcane-based ethanol from an economic, social, and environmental perspective.

Economic Impact

Brazilian Sugar is Cheaper and More Effective
For the past two years, prices for sugar have steadily decreased because of a global surplus in production.[1] Because of a large demand for ethanol, both in Brazil and globally, ethanol prices are much higher than raw sugar prices. This pushes Brazilian millers to focus more on producing sugar-based ethanol, helping them make a profit faster. A larger supply of ethanol from sugar creates an economic competition with corn ethanol.

Ethanol from Brazilian Sugar
Tasnim Mohamed   Page **1** of **4**

• **FIGURE 8–7**   Letter Report

If prices of corn ethanol remain higher, more consumers will prefer to use sugar ethanol. McKenna argues that the use of corn ethanol is inefficient. Even if all corn production was dedicated to ethanol production, he states, gasoline usage would only reduce by 15%.[2] The same reduction of gasoline usage could also be achieved with increasing fuel efficiency by 3.5 miles per gallon.

Increase in Food Prices

The increase of corn prices has greatly affected the food industry as well. A chain reaction began when prices of corn increased because many consumers depend on corn to make a variety of foods. Corn is often used to make animal feedstock, so if these prices increase, so will the price of meat. The prices of other common corn-based foods can increase as well, such as cornmeal, tortillas, corn bread, and many more. A drought at the end of 2012, for example, kept the price of corn high.[2] High corn prices cause many farmers to lose their businesses.

Using sugarcane-based ethanol can also have adverse effects on the price of raw sugar. High sugar prices create a ripple effect throughout the world economy on the cost of all products containing sugar, from candy and soft drinks to condiments and sauces. Rising sugar costs also raise prices at food service facilities, such as restaurants and bakeries.

Social Impact

Tension with the Indigenous Brazilian People

The mass production of sugarcane in Brazil currently presents a problem to its society. Most sugarcane plantations in Brazil are located on land reserved for the indigenous people, the Guarani. These people hold little power with the Brazilian government. Because of Brazil's increasing profit from exporting sugar-based ethanol, sugarcane millers are free to extend their plantations as they please, forcing the Guarani out of their homes and off their land.[3]

The Guarani have no other choice but to work on the plantations under horrible conditions and low wages. To decrease the negative social aspect of sugarcane production in Brazil, the landowners "need to be held accountable" and respect the rights of the local people.[4]

Jobs Created for the Uneducated

Despite tension with indigenous peoples in Brazil, sugarcane production has created a large workforce for the uneducated population. Table 1 lists a few demographics of people with various jobs in Brazil. Based on the table, the people working with sugar cane crops have the lowest education. Clearly, many people work in this industry relative to the number of people in the other job positions.

Table 1: List of Demographics for Various Jobs in Brazil4

Statistic	Sugar Cane Crops	Sugar	Ethanol	Food and Beverages	Fuels
People (x1000)	789.4	126.0	67.0	1507.0	104.7
Mean Age (years)	35.1	36.6	35.6	34.4	37.1
Mead Education (years)	2.9	6.5	7.3	7.1	8.9
Mean Income	446.6	821.3	849.9	575.0	1281.1

• **FIGURE 8–7** *Continued*

Environmental Impact

Ethanol in the Carbon Cycle
Both sources of ethanol have an equal environmental benefit. By converting any kind of biomass to fuel, the CO_2 absorbed by the plants (such as corn and sugarcane) is released again into the atmosphere. Because CO_2 is constantly removed and replaced from the atmosphere, burning bioethanol resembles the carbon cycle. Unlike bioethanol, burning gasoline simply adds CO_2 to the atmosphere without initially removing it.

Reduction of Greenhouse Gases
The production, distribution, and consumption of ethanol create nearly 20% less greenhouse gases than the same process used for gasoline.[5] However, if an increased demand for cropland leads to clearing out large forests or grasslands, ethanol would be useless. Forests and grasslands naturally absorb most of the atmospheric CO_2 and depleting this would greatly increase the amount of atmospheric oxygen even with the use of ethanol rather than gasoline.

Sincerely,

Tasnim Mohamed

Tasnim Mohamed, P.E.

● **FIGURE 8–7** *Continued*

Sources

1   Almeida, Isis. Brazil ethanol above sugar shows mills may favor biofuel.
    Bloomberg BusinessWeek. 2013. http://www.businessweek.com/news/2013-02-11/
    brazil-ethanol-topping-sugar-signals-millers-may-favor-biofuel

2   McDonald, Kay. Paying more for food? Blame the ethanol mandate. CNN. 2012. http://www.cnn.
    com/2012/08/20/opinion/mcdonald-corn-ethanol

3   Villarreal, Ryan. Deadly ethanol: Brazil's sugarcane farms take toll on
    indigenous people. *International Business Times*. 2012. http://www.ibtimes.com/
    deadly-ethanol-brazils-sugarcane-farms-take-toll-indigenous-people-912116

4   Smeets, Edward, Martin Junginger, Andre Faaij, Arnaldo Walter, and Paulo Dolzan.
    Sustainability of Brazilian Bio-ethanol. Report. *Copernicus Institute*. 2006. http://
    web.archive.org/web/20080528051851/http://www.bioenergytrade.org/downloads/
    sustainabilityofbrazilianbioethanol.pdf

5   Elmendorf, Douglas W. The impact of ethanol use on food prices and greenhouse-gas emissions.
    Report. *US Congress Congressional Budget Office*. 2009. http://www.cbo.gov/sites/default/files/
    cbofiles/ftpdocs/100xx/doc10057/04-08-ethanol.pdf

● **FIGURE 8–7** *Continued*

Writing Collaboratively

In the workplace, you may need to write reports, instructions, and letters on your own and as part of a team. Collaborative writing occurs in numerous ways that continue to evolve as technology evolves: you may draft a document or part of a document and then submit it, as an e-mail attachment, to several additional employees who will contribute information to your draft. Each writer may suggest changes to others' work by using a comment tool, such as that found in most word processing programs. Or the team leader may create a wiki on the organization's intranet and post material that can then be viewed, revised, and discussed by all team members.

Team members often cannot collaborate in "real time" because of other work responsibilities. Wikis and other forms of virtual collaboration allow teams to work more flexibly but could also raise questions about information security.

Warning: As we discussed in Chapter 1, the need for security of an organization's intellectual property cannot be underestimated. Hackers can destroy an organization by stealing intellectual property, trade secrets, contracts, and personnel information. The networking software students use in school cannot offer the security needed by business and research organizations. If used carelessly, social media can ruin lives, businesses, and reputations. Think carefully and cautiously before you use social media for any message.

The secret to effective collaboration is planning:

- Decide when, where, and how you will meet
- Decide who will do what
- Determine the deadlines and work schedule for each document segment
- Learn to work together. (Note: When people from different countries and cultures collaborate, the need for sensitivity to cultural differences is critical.)
- Assign one member the task of editing the document to make the style consistent throughout.

The team leader. Collaborative projects must have a team leader. This person may be the project manager or someone appointed. This person will convene meetings, virtual or on-site, and serve as facilitator to keep the project moving. In addition, the team leader

- ensures that all team members know their tasks
- launches the effort and provides everyone with the document requirements (purpose, audiences, rationale), format, timetable for completion, and methods of communicating about the document (see Figure 8–8)

Project Name	
Goals	
Team members and contact info	member mobile #
Responsibilities	member duties
Deliverables	
Timeline	
Meetings	
Initial Outline	

● **FIGURE 8–8** Collaborative Team Planning Sheet

- monitors development so that the report is completed as required
- ensures that the completed document has a single style throughout and contains all information and required segments

Requirements of team leaders. Team leaders, in either face-to-face or virtual settings, need to employ a few fundamental strategies:

- Maintain a positive approach
- Be aware of how you sound; be respectful
- Solicit input
- Encourage timely responses from everyone
- Keep records of all communications

After each draft, the team leader should encourage, as time permits, comments by each team member. The "comment" tool in doc or pdf files allows each team member to recommend changes and assess the effectiveness of the document.

Requirements of team members.

- Keep to the schedule as determined by the project leader. If you find you have an issue that threatens your keeping to the schedule, notify the team leader immediately.
- Be respectful and accept comments, recommendations, and suggestions from other team members and the project leader. Exhibit a good attitude, even when you feel stressed.
- Do your part of the project. Do not assume that if you get behind, another team member can or will do your work for you. The quality of the team project stems from the quality of the work performed by each member of the entire team.

Example Report for Study

An example report appears in Figure 8–9. Another appears in Appendix C.

MEMORANDUM

DATE: November 12, 2017

TO: Dr. Robin Pumphrey

FROM: Lamees Elnihum

SUBJECT: **Current and advancing issues in chemical engineering**

To assist the college's academic advisors in the preparation of majors, I am submitting the attached report that describes current and advancing issues in chemical engineering. The report provides relevant information about current topics of interest in chemical engineering and the directions the field is taking. Furthermore, critical issues are discussed as well as the problems most needing solutions (and any proposed solutions that have surfaced thus far).

Chemical engineering is an extremely broad branch of engineering that deals with many developing sciences and technologies.

I divided the discussion of foci in the chemical engineering field into six main sections:

- Chemotherapeutics
- Biofuels
- Food engineering
- Polymers
- Catalysis and reaction engineering
- Materials

The topics discussed cover a significant portion of the world of chemical engineering, but in no way is the list complete. I selected only a few of the more imperative issues for discussion from the vast number of areas present in chemical engineering. I then divided the discussion of critical issues in chemical engineering into three main sections:

- Energy, environment, and sustainability
- Nanotechnology, microelectronics and geometry
- Genetic engineering

The report concludes with an acknowledgement of the changes and innovations in chemical engineering, the future of chemical engineering, and the directions the field is taking.

• **FIGURE 8–9** A Formal Report, with Memo of Transmittal

Current and Advancing Issues in Chemical Engineering

Lamees Elnihum

November 12, 2017

Abstract:

This report has been prepared to discuss current and advancing issues in chemical engineering. Research methods made use of the online engineering database Compendex (Engineering Village 2). The report describes the history and scope of chemical engineering, current and advancing foci in the field, critical issues in chemical engineering and proposed solutions, as well as the future of chemical engineering. Selected topics of discussion of the growing field of chemical engineering include chemotherapeutics, biofuels, food engineering, catalysis and reaction engineering, materials, and polymers. Critical issues include the topic of energy, environment, and sustainability; nanotechnology, microelectronics, and geometry; and genetic engineering. Current and future chemical engineers will find themselves exposed to a wide variety of research areas and will have the unprecedented chance to explore firsthand the versatility of these issues of interest, as well as to improve upon the critical issues of today.

• **FIGURE 8–9** *Continued*

Table of Contents

● **FIGURE 8–9** *Continued*

Summary

Purpose of the report

This report is intended to note changes in chemical engineering and the direction the field is taking. It also covers the most critical issues in chemical engineering, and the problems most needing solutions, as well as any proposed solutions that have surfaced.

Research methods used

I used an online engineering database, Compendex (Engineering Village 2), to gather information about the field of chemical engineering.

Understanding chemical engineering

Chemical engineering, the broadest of the engineering disciples, applies natural sciences along with economics to develop efficient conversion processes of raw materials to desired products. Today, chemical engineering encompasses a wide variety of areas, from nanotechnology to mineral processing. Chemical engineers may work with process and product development and design, or they may work in areas such as production, research, environmental studies, and management. On an industrial scale, chemical engineers deal with the materials or production of almost every article manufactured in a process system.

Current and advancing foci in chemical engineering

Chemotherapeutics, biofuels, food engineering, catalysis and reaction engineering, materials, and polymers are all disciplines of chemical engineering that are intensely focused on in current research and technology developments. Chemotherapeutic engineering applies chemical engineering principles and devices for chemotherapy of cancer and other diseases. It aims to improve the current inefficiency and side effects of chemotherapy through development of controlled and targeted drug delivery systems.

Biofuels are recurring interests in chemical engineering as alternatives to petroleum-based transportation fuels. Chemical engineering focuses on the processes used to make these renewable fuels, as well as the combustion chemistry of the compounds that constitute biofuels.

Food engineering has become an increasingly prominent focus of chemical engineering. Chemical engineers in the food industry must focus on preserving food quality, as well as creating new food structures in large quantities and fast rates. Along with the food, the packaging materials are also focused on by chemical engineers.

In catalysis and reaction engineering, chemical engineers focus on chemical kinetics and its interaction in flowing materials. They also look at how catalysts influence the reaction rate in a chemical reactor. The challenge is to design the catalyst for process efficiency and stability. Nature-inspired chemical engineering in catalysis and chemical reactor design is a growing aspect of reaction engineering and shows great potential for innovation in this field.

Chemical engineers also research the making of new materials from inorganic compounds found in nature. Altering and modifying the molecular composition and structures of these compounds, all of which have various electronic, mechanical, and biological properties, can generate new materials that can do everything from bone regrowth to powering cell phones.

Polymers allow chemical engineers to build the molecular structure of polymers designed by material scientists. Chemical engineers aim to discover more efficient processes of accomplishing this goal, and, in so doing, focus on new catalysts, new macromolecular building blocks, genetic engineering of materials, and much more. Scattering and imaging methods help chemical engineers gain insight into polymer structure so that product characteristics can be measured.

Critical issues in chemical engineering and proposed solutions

Chemical engineers seek new sources of energy with reduced harmful effects on the environment and a focus on sustainability. "Green Engineering" aims to transform existing engineering practices to those

• **FIGURE 8–9** *Continued*

that promote sustainability and promote human health as well as protection of the biosphere. Chemical engineers are also working on alternative energy sources, waste minimization, and environmentally friendly ecological design.

Nanotechnology, microelectronics and geometry are another issue in chemical engineering, as the need for increased complexity is on the rise due to necessities for miniaturized systems. Fractals also aid in the advances in miniaturization of technological devices.

Finally, genetic engineering is another current issue in chemical engineering. Plants currently undergo manipulation of genes in order to acquire desired traits, and are dubbed "genetically modified". Chemical engineers also aid in gene therapy in fetuses with genetic diseases and disorders.

Future of chemical engineering

Chemical engineering, the broadest of the engineering disciples, has a strong foothold in many current issues of interest in the field of science today. The versatility chemical engineering offers allows chemical engineers to become involved in a wide variety of research areas. Chemical engineers of the future will have the unprecedented chance to work in the expanding areas of chemotherapeutics, biofuels, food engineering, and many of the other subsets of chemical engineering described above. They will also have the chance to improve upon the current issues of our day, including Green Engineering, nanotechnology and microelectronics, and genetic engineering.

Current and Advancing Issues in Chemical Engineering

Introduction

Chemical engineering, the broadest of the engineering disciples, has a world of research areas to explore. The purpose of this report is to highlight changes in the chemical engineering field and to note which directions the field is taking. This report will also focus on current foci in chemical engineering, as well as critical issues and their proposed solutions. The topic of this report is relevant as chemical engineering is a rapidly changing and expanding field that encompasses a vast amount of material in its research domain. This report will first cover the history and scope of chemical engineering, and then dive into current and advancing foci in chemical engineering. The report will then discuss current issues and proposed solutions in chemical engineering, and finally close with a discussion on the future of chemical engineering.

One reason for the large broadness in the field of chemical engineering research is due to its exposure to numerous types of processes. Figure 1 (below) shows how chemical engineering deals with processes ranging from nano-scale to mega-scale. Chemical engineers are active at the molecular level as well as at the level of complex, overall systems. This versatility allows chemical engineers the

choice to research from a wide variety of subsets within chemical engineering, many of which will be described in the report below.

Understanding Chemical Engineering

Definition of Chemical Engineering

Chemical engineering is the branch of engineering that applies chemistry, physics, and life sciences, along with mathematics and economics, to processes that convert raw chemicals and compounds into desired products.[2] Today, the field of chemical engineering covers a broad range of areas, from biotechnology and nanotechnology to mineral processing.[2]

History of Chemical Engineering

Chemical engineering grew out of the Industrial Revolution in the late 19th century due to a need for unit operations.[3] The development of chemical engineering, an integration of the chemistry and

• **FIGURE 8–9** *Continued*

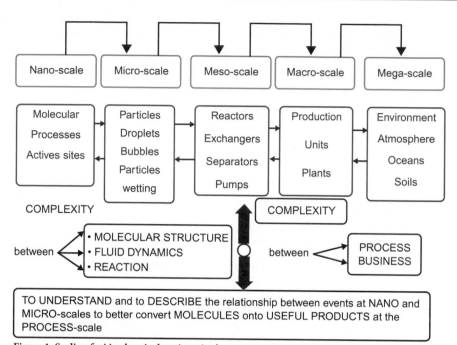

Figure 1: Scaling foci in chemical engineering[1]

mechanical engineering of the time, helped influenced the shift from batch processing to continuous processing in the production of industrial chemicals.[3] The desire for continuous processing, resulting from an unprecedented demand for bulk chemicals, created the need for an engineer conversant not only with machines, but also familiar with the chemical reactions and transport phenomena within the machines, as well as the influence the equipment had on how these processes operated on a large scale.[3] By 1910, the profession title "chemical engineer" was readily familiar in the United States.[3]

Scope of chemical engineering

Chemical engineering is the broadest field of engineering.[4] As such, chemical engineers are employed by almost all companies in the chemical process industry.[4] Chemical engineers work in numerous fields of science, including process and product development and design involving nuclear energy, materials science, food production, the development of new sources of energy, and medicine.[4] They also work in areas such as production, research, environmental studies, market analysis, data processing, sales, and management.[4] Chemicals engineers deal with the materials or production of almost every article manufactured on an industrial scale.[4]

Current and advancing foci in chemical engineering

Chemotherapeutics

Chemotherapy and other disease treatments provide a world of opportunities for chemical engineering in the field of human health care.[5] Chemotherapeutic engineering, a discipline of chemical engineering, applies chemical engineering principles, techniques and devices for chemotherapy of cancer and other diseases.[5] Currently, chemotherapy is one of the foremost treatments available for cancer and cardiovascular diseases that lead the nation in causes of deaths.[5] Chemotherapeutic

● **FIGURE 8–9** *Continued*

engineering aims to use chemical engineering techniques to improve current chemotherapy, which is far from satisfactory due to its limited efficiency and numerous (often life-threatening) side effects.[5] Chemical engineering in chemotherapy involves developing controlled and targeted drug delivery systems, especially emphasizing nanoparticles of biodegradable polymers and lipid bilayer vesicles.[5] One novel drug delivery system, known as liposomes-in-microspheres, is used as an example for possible combinations of existing polymer-based and lipid-based delivery systems.[5] Drug delivery systems rely on research of molecular interactions between the drug and the cell membrane.[5] The lipid monolayer at the air-water or oil-water interface and bilayer vesicles are used as models for the cell membrane.[5]

Biofuels

Biofuels, such as biodiesel and ethanol, are recurring interests in chemical engineering as alternatives to petroleum-based transportation fuels.[6] Biofuels offer long-term reliability as renewable fuel sources, as well as a decrease in harmful environmental impacts.[6] Chemical engineering focuses on the processes used to make such alternative fuels, as well as the compatibility of these substances with current fuel-delivery infrastructure and engine performance.[6] At the molecular level, chemical engineering works with the combustion chemistry of the compounds that constitute biofuels, as well as the formation of harmful or toxic emissions caused by decomposition and oxidation mechanisms.[6] Most importantly, chemical engineering seeks to explore the vastly diverse and complex chemical reaction networks of biofuel combustion, enabled by recent experimental investigations and combustion modeling that will aid in the selection of next-generation alternative fuels.[6]

Food Engineering

The past few decades have established chemical engineering as a crucial component of the food industry.[7] Engineers have faced the challenge of preserving food quality, as well as creating new food structures in large quantities and fast rates.[7] In preservation of food quality, active ingredients within the food must be kept within acceptable limits.[7] Regarding this fact, chemical engineering in the food industry requires collaboration with food and nutrition scientists, microbiologists, microelectronic engineers, etc.[7] Food preservation looks at not only the food itself, but also the packaging materials. Chemical engineering has a unique foothold in food engineering and the advances made in food engineering research and new technology developed.[7]

Polymers

The chemical variety of polymers allows material scientists to become architects at the molecular level.[10] Where material scientists are the architects, though, chemical engineers are the constructers.[10] Chemical engineers must build the molecular structures (as well as the internal micro and meso structures) of these polymer materials efficiently.[10] Currently, chemical engineers are consistently developing newer processes to accomplish this goal.[10] These developments involve focus on new catalysts, new macromolecular building blocks, reactive processing, self-assembly, manipulation of phase behavior, genetic engineering of materials, and much more.[10] One crucial aspect of polymer processing is the ability to measure product characteristics, thus making use of improving technologies such as scattering and imaging methods that allow insight into polymer structure.[10]

Catalysis and Reaction Engineering

Designed a chemical reactor requires a chemical engineer to consider how the chemical kinetics, often modified by catalysis, interacts with the transport phenomena in flowing materials.[8] Chemical kinetics, and how different catalysts influence the reaction rate, must be understood for chemical reactions that occur between two small molecules in a chemical reactor.[8] The challenge is designing the catalyst is to increase the efficiency and stability of the process.[8] One growing aspect of catalysis and reaction engineering is nature-inspired catalysis and reaction engineering, which looks at biology's use of robust

• **FIGURE 8–9** *Continued*

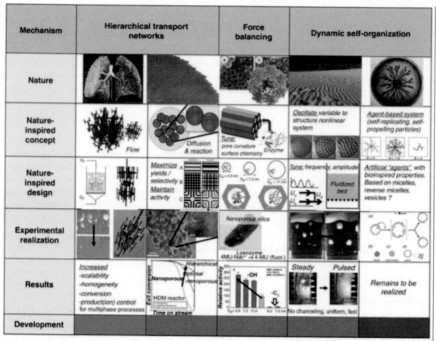

Figure 2: Catalysis and reaction engineering[8]

and highly efficient hierarchical networks to facilitate transport.[8] Nature-inspired chemical engineering shows great potential to innovate catalysis and reaction engineering by analyzing processes already functioning in nature.[8] Figure 2 (left) depicts a few of the processes seen in nature that chemical engineers seek to replicate in the laboratory.

Materials

New materials are made by modifying the molecular composition and structure of inorganic compounds found in nature.[9] These materials impact numerous novel technologies due to their varying electronic, mechanical, and biological properties.[9] With nature's molecules, chemical engineers create new materials that can do everything from bone regrowth to powering cell phones.[9]

Critical issues in chemical engineering and proposed solutions

Energy, Environment, and Sustainability

Chemical engineers are currently seeking new sources of energy with reduced harmful effects on the environment and a focus on sustainability.[11] The goal is to assure that a product, process, or production system is as sustainable as possible while having as little harmful effect on the environment as possible.[11]

Green Engineering

"Environmental," "ecological," and "clean" are only a few of the adjectives used to qualify this branch of engineering.[11] Green engineering aims to transform existing engineering practices to those that promote sustainability and promote human health as well as protection of the biosphere.[11] The promotion of sustainable development in chemical engineering is aided through a multitude of rules and guidelines defined by several institutions and individuals.[11] One of these strategic plans, known as the "12 Principles of Green Engineering", provides a framework for scientists and engineers to engage in when designing new materials, products, processes, and systems that are benign to human health and the environment.[11]

Air and Water Pollution Control

The Environmental Protection Agency defines "Green Chemistry" as the use of chemistry for pollution prevention at the molecular level.[11] The goal is to promote innovative chemical technologies that reduce the use or generate of harmful substances in the design, manufacture, and use of chemical products.[11] Pollution is minimized when green chemistry principles are incorporated into feedstock and reagent selection, solvent use, and overall synthetic design.[11]

Alternative Energy Sources

Chemical engineers look to use energy at high efficiency levels, which also helps to minimize the harmful effects of process systems unleashed on the environment.[11] Accepted renewable sources of energy include solar thermal energy, biofuels, hydroelectricity, wind power, and geothermal energy.[11] Regarding biofuels, chemical engineers are looking at sorghum-based ethanol fuel as an alternative to corn-based ethanol fuel.[11] Detailed planning and management of renewable energy use and resources using precise energy models can also aid chemical engineers in their goal to reach sustainability.[11]

Waste Minimization

Chemical engineers currently design processes with the knowledge that it is better to prevent waste than to treat or clean up waste after its creation.[11] The principle problems chemical engineers associate with the waste is that its only function is to be disposed of properly out of sight.[11] The main raw material of garbage companies is the waste itself, and the treatment method is burying everything underground and dumping harmful products everywhere.[11] For some chemical engineers, the goal is to seek the elimination of waste and instead associate so-called waste with a value.[11] "Universal recycling" aims to take each garbage item and convert it into something worthwhile.[11] Chemical engineers also strive to plan processes with minimized toxicity and harmful effects in chemical reactants and products.[11]

Ecological Design

Ecological design aims at designing minimally environmentally destructive processes.[11] The design respects wildlife, minimizes resource depletion, and maintains habitat quality, etc.[11]

Nanotechnology, Microelectronics and Geometry

Chemical engineers can design self-assembling systems with a modest degree of complexity, but the need for increased complexity is growing due to current necessities of miniaturized systems.[11] Scientists fabricate nanoparticle arrays based on the assembly of functionalized nanoparticles through self-assembly methods.[11] One of these methods is biomolecular nanolithography, in which nanoparticles self-assemble onto biopolymeric (DNA) scaffolds to form lines and more complex patterns.[11] A potential application of these methods is the generation of molecularly integrated nanocircuits as a greener, higher performance approach in the microelectronics industry.[11] Advances in miniaturization result from the combination of engineering with the mathematical field of geometry.[11] Fractals, or mathematically generate patterns, are reproducible algorithms or shapes that are produced by recursive subdivision.[11] Fractals can greatly aid in the design of smaller and more efficient technological devices.[11]

Genetic Engineering

Genetic engineering focuses on controlled manipulation of genes to change the genetic makeup of cells and produce novel organisms.[12] Plants that used to undergo selective breeding are now genetically modified to acquire traits including higher nutritional content, greater resistance to herbicide

• **FIGURE 8-9** *Continued*

and pesticide damage, increased resistance to disease, faster ripening, etc.[12] Genetic modification is a technique of changing the characteristics of a plant (or organism) to acquire certain desirable traits.[12] Chemical engineers help produce genetically modified foods that are produced using an artificial form of DNA called recombinant DNA.[12] rDNA with a positive trait is transferred into a plant lacking that trait to create the desired improvement.[12] Chemical engineers may also help with gene therapy, which is done when current technology detects genetic diseases and disorders in fetuses prior to their birth.[12]

Future of chemical engineering

Chemical engineering, the broadest of the engineering disciples, has a strong foothold in many current issues of interest in the field of science today. Chemical engineers of the future will have the unprecedented chance to work in a variety of expanding areas, including chemotherapeutics, biofuels, food engineering, catalysis and reaction engineering, materials, polymers, and much more. They will also have the chance to improve upon our experiences with Green Engineering in the field of energy, environment, and sustainability in chemical engineering, as well as upon the increasing need for research into nanotechnology and genetic engineering.

References

1. Four main objectives for the future of chemical and process engineering mainly concerned by the science and technologies of new materials production http://www.sciencedirect.com.lib-ezproxy.tamu.edu:2048/science/article/pii/S1385894704003870

2. Chemical Engineers and the Things They Do http://sciencenetlinks.com/lessons/chemical-engineers-and-the-things-they-do/

3. History of Chemical Engineering and Chemical Technology http://www.pafko.com/history/h_intro.html

4. Chemical Engineering http://portal.acs.org/portal/acs/corg/content?_nfpb=true&_pageLabel=PP_ARTICLEMAIN&node_id=1188&content_id=CTP_003380&use_sec=true&sec_url_var=region1&__uuid=10e583cd-9347-44bb-bd26-28aa6dc64eea

5. Chemotherapeutic engineering: Application and further development of chemical engineering principles for chemotherapy of cancer and other diseases http://www.sciencedirect.com.lib-ezproxy.tamu.edu:2048/science/article/pii/S0009250903002343

6. Biofuel Combustion Chemistry: From Ethanol to Biodiesel http://onlinelibrary.wiley.com.lib-ezproxy.tamu.edu:2048/doi/10.1002/anie.200905335/full

7. Food Engineering as an Advancing Branch of Chemical Engineering. http://web.ebscohost.com.lib-ezproxy.tamu.edu:2048/ehost/detail?sid=63dd1c18-e6dd-4dab-a97a-22923b7faddf%40sessionmgr4&vid=1&hid=9&bdata=JnNpdGU9ZWhvc3QtbGl2ZQ%3d%3d#db=a9h&AN=21517896

8. A nature-inspired approach to reactor and catalysis engineering http://www.sciencedirect.com.lib-ezproxy.tamu.edu:2048/science/article/pii/S2211339812000135

9. Materials http://web.mit.edu/cheme/research/areas/materials.html

10. Chemical engineering of polymers: Production of flexible, functional materials http://www.sciencedirect.com.lib-ezproxy.tamu.edu:2048/science/article/pii/0009250995002731

11. New trends for design towards sustainability in chemical engineering: Green engineering http://www.sciencedirect.com.lib-ezproxy.tamu.edu:2048/science/article/pii/S1385894707001192

12. Modification http://www.chemicalengineering.org/food/genetic.html

• **FIGURE 8–9** *Continued*

Report Checklist

Planning

☐ What is the purpose of your report? Have you stated it in one sentence?

☐ What is the scope of your report?

☐ Who are your readers? What are your readers' technical levels?

☐ What will your readers do with the information?

☐ What information will you need to write the report?

☐ How long should the report be?

☐ What format should you use for the report?

☐ What report elements will you need?

☐ What illustrations will you need to present information or data?

☐ How will you make your report accessible for all of your likely readers?

☐ What elements do you need to include in your introduction?

☐ What arrangement will you use in presenting your report?

Revision

☐ Does your report do the following:

- Introduce the subject and purpose?
- Present enough data in words and visuals to justify any conclusions drawn?
- Discuss and evaluate the data fairly?
- Summarize the data?
- Draw logical conclusions from the data?
- If necessary, present recommendations that are clearly based on the data and the conclusions?

☐ Are your data accurate?

☐ Do your illustrations immediately show what they are designed to show?

☐ Is your format suitable for your content, audience, and purpose?

☐ Have you properly documented all information sources?

EXERCISES

1. Visit the website of a major employer in your field. What kinds of reports do you find at this site? What are the elements and characteristics of the reports? How long are the reports? Who are the audiences for the reports? What are the purposes of the reports? Are the reports written by a single author or multiple authors? What do the reports indicate about this employer? What do the reports indicate about your field? How would you prepare to write a report like one of these?

2. Visit the website of the National Transportation Safety Board (www.ntsb.gov) to examine each of its six kinds of accident reports (aviation, hazardous materials, highway, marine, pipeline, and railroad). Describe the elements of the reports. How would you describe the writing style of the reports? What kinds of illustrations do you find integrated into the reports? Do the reports offer satisfactory descriptions of the accidents? Do the reports offer clear and convincing evidence of the causes of accidents? Do the reports make a persuasive case for changes in practices or materials to improve safety?

9

Proposals and Progress Reports

Proposals and progress reports, two common types of workplace documents, can be written collaboratively or individually depending on the size of the projects. Many projects, such as those funded by the National Science Foundation or the National Institutes of Health, begin with proposals and require regular progress reports during their course. This chapter explains development strategies for both types of documents.

Any work by an employee may require a progress report (or "status report," as it may be called). This report notifies the supervisor of all that the employee has accomplished. A project may begin with a verbal agreement between supervisor and employee, but the employee must typically submit a progress report either as an e-mail message or as a short memo report. Progress or status reports serve as a major source of documentation to show that employees have performed their assigned duties.

To help you understand how to design and write proposals and progress reports, we first discuss the development of proposals and use a student's research project proposal as an example. We also provide additional annotated examples of progress reports to illustrate how these occur in work settings.

The *proposal*, as its name implies, describes proposed work or research, the reasons for the work, the methods proposed to accomplish the work, the estimated time required, and the expected cost.

The *progress report*, as its name implies, describes and evaluates a project as work occurs. Thus, if an individual or an organization decides to begin a work project or research project, the individual or organization will report the progress on that project at intervals agreed upon when the proposal is accepted and the resulting agreement is being negotiated. Progress reports are especially important when a project requires several months or even several years to complete.

Employees may also need to report progress on the full range of projects or problems on which they are working. In situations like these, the employee writes a progress report to inform supervisors or other individuals about what has been accomplished in completing each job or solving each problem. The progress or status report thus becomes an official and even a legal record of work performed. Many organizations require employees to prepare annual status reports of their accomplishments for the past year and their objectives for the coming year.

 Quick Tip

Keep in mind that proposals and progress reports are persuasive documents. You write to convince your reader of the merit and integrity of your work. In a proposal, you write to persuade your reader that you have a good idea, a good method for implementing the idea, and the experience necessary to manage the implementation—that you can achieve what you propose. In a progress report, you write to persuade the reader either that you are making good progress on your project or that you realize you aren't making good progress but know why and are taking appropriate steps to fix the problem.

Proposals

The starting point of projects is often a proposal. A proposal offers to provide a service or a product to someone in exchange for money. Usually, when a business enterprise or government agency decides to have some sort of work or research done, it wants the best job for the best price. It will advertise the work it wants done and invite interested individuals or organizations to contact it. This document is usually called a *request for proposals* (RFP) or a *statement of work* (SOW). The soliciting organization may send selected individuals or institutions an RFP/SOW that includes complete specifications of the work desired.

Alternatively, the soliciting organization may first describe the needed work in general terms and invite potentially interested parties to prepare for this forthcoming opportunity. This type of announcement is usually called a *request for qualifications* (RFQ) or a *notice of intent* (NOI). The soliciting organization may also publish its RFQ or NOI—in newspapers and professional journals, on websites, or through industry-related e-mail distribution lists (see Figure 9–1). The responding parties will register their interest in the project with the soliciting organization and, as necessary, make a case for their qualifications. (This effort could include explaining past accomplishments, identifying companies for which they performed pertinent work, describing this work in detail, and providing a list of references.) The responses submitted to an RFQ or a notice of intent are called *letters of interest* or *intent*. Based on the letters of interest it receives, the soliciting organization will develop full descriptions of the work desired (RFPs/SOWs) for the groups it believes have the appropriate qualifications.

DE-FOA-0001708:NOTICE OF INTENT (NOI) FOR FUNDING OPPORTUNITY ANNOUNCEMENT (FOA) DE-FOA-0001628 "PRODUCTIVITY ENHANCED ALGAE AND TOOL-KITS (PEAK)"

This FOA will support an applied research focus on biological variables which contribute most greatly to the modeled minimum fuel selling price of algal biofuels. These biological variables include: biomass productivity; biomass composition (e.g., lipid, protein, and carbohydrate content); predation and pathogen resistance; halotolerance; heat and cold tolerance; and high-intensity light (i.e., direct sunlight) tolerance. Supporting advanced biology research approaches to enable high productivity in applied "field setting" cultivation conditions that are representative of future scaled-up cultivation conditions will accelerate progress towards EERE's technical milestones for algal biofuels and will fill a gap between bench-scale research and pre-pilot efforts. Improved and more environmentally robust strains are necessary for breakthrough reductions in algal biofuel costs. This FOA seeks to fund multidisciplinary biological innovation that will deliver strains, tools, data, and techniques to enhance algal biofuel potential and enable accelerated future innovation in algal biofuels and bioproducts.

It is anticipated that the FOA may include the following Areas of Interest:

Area of Interest One, Strain Improvement: Development of enhanced algal strains with increased areal productivity and biofuel intermediate yield. Strain improvement developments will include isolation, directed evolution, breeding, and/or genetic engineering of novel algal strains that can reproducibly out-perform the current best available strains in outdoor conditions, where "performance" is represented by productivity, robustness, and composition.

Area of Interest Two, Cultivation Improvement: Development of increased areal productivity and biofuel intermediate yield through enhanced management of ecological or abiotic contributions to cultivation biology. Cultivation biology development improvements will include leveraging natural or designed microbial assemblages of the cultivation ecosystem to boost performance and exclude pathogens, and understanding under what cultivation conditions should certain strains be employed. This Area of Interest is not focused on engineering a better cultivation system.

For both Areas of Interest, selected projects will support the development of at least one novel tool, technique (method), or dataset that upon completion of the project will enable developers to accelerate innovation in that topic area. The types of tools, techniques, and datasets of interest span biological, computational, and analytic strategies that will advance the state of technology and are complementary to the scope of improvement development work.

EERE envisions awarding multiple financial assistance awards in the form of cooperative agreements. The estimated period of performance for each award will be approximately 36 months.

This is a Notice of Intent (NOI) only. EERE may issue a FOA as described herein, may issue a FOA that is significantly different than the FOA described herein, or EERE may not issue a FOA at all.

DOCUMENTS
• Notice of Intent DE-FOA-0001708 for FOA DE-FOA-0001628 (Last Updated: 11/30/2016 03:16 PM ET)

CONTACT INFORMATION
• EERE-ExchangeSupport@hq.doe.gov
For technical problems with EERE Exchange system.

SUBMISSION DEADLINES
• Full Application Submission Deadline: TBD

• **FIGURE 9–1** Example Notice of Intent

In short, each aspect of the solicitation process, from the RFQ/NOI to the RFP/SOW, has an appropriate use, but one or more of them is necessary to initiate action on a project.

For example, academic institutions will sometimes support the research projects of their faculty by making funds available on a competitive basis. Faculty members will submit project proposals that explain the research they envision; how much time they will need to complete the project; any financial resources required for equipment, salaries, and release time from regular teaching duties; and the goals and benefits of the research. The competition for the available funds starts with a simple RFP (see Figure 9–2).

TO: All Faculty

FROM: Dr. Rolfe D. Auston, Vice President for Research

DATE: January 8, 2018

SUBJECT: 2019 Texas-Mexico Collaborative Research Grant Program

The Office of the Vice President for Research is pleased to issue a request for proposals for the Texas-Mexico Collaborative Research Grant Program. This program annually awards one-year grants of up to $50,000 to faculty members to advance inter-institutional cooperation in science, technology, and scholarly activities that have a direct application in industry or government through the complementary efforts of scientists and scholars from Texas and Mexican institutions.

Two main objectives of the Collaborative Research Grant Program are to provide seed funding to 1) support the completion of a 12-month inter-institutional project, and 2) support the development and submission of proposals for external funding of research from competitive granting agencies both domestic and international (e.g., NSF, NIH, DOE, World Bank, NATO, UNESCO, etc.) and industry.

Participating institutions in Texas and Mexico have agreed on several research priority areas as noted in the request for proposals. The research proposed must be linked to the private sector and have direct application to solving an industrial or governmental problem. All proposals must include research that directly improves security for the citizens of the region or explores issues relating to security challenges facing both countries.

A principal investigator (PI) is required from both a Texas institution and a Mexican institution. The PI from Texas must be a tenured or tenure-track faculty member. The PI from Mexico must be a scientist or scholar from any Mexican institution of higher education and research. Other investigators may include faculty from other US institutions or Mexican faculty, postdoctoral students, graduate students, or research staff. A letter of intent must be received by 5:00 p.m. on Monday, March 5, 2018, to be eligible to submit a full proposal. Full proposals must be submitted, routed electronically for appropriate signatures, and received by 5:00 p.m. on Monday, April 30, 2018. The request for applications is available on the Web at http://CRGP2019.edu.

For additional information, please contact Ms. Catharine J. Restivo (916-555-6093; crgp@texas.edu).

• **FIGURE 9–2** Request for Proposals

The context of proposal development. Because proposals are time-consuming to write—most require substantial research and analysis on the part of the proposing organization—individuals and organizations wishing to respond to an RFP study it carefully. They do not want to submit a proposal that is unlikely to be accepted. Thus, the proposer—whether a university professor seeking research funds or a highway construction firm seeking to win a contract from a county to upgrade its rural roads—will carefully approach the decision to prepare a proposal.

The individual or the company must first decide whether to respond to the proposal. Can we do the work requested? Can we show that we can do this work, based on what we have already done? Can we do it within the time limit given in the RFP? Businesses responding to RFPs must address economic issues. How much will our proposed approach cost? How much money can we make? Who else will be submitting proposals? What price will competitors likely quote for the same work? Can we compete? What other projects do we have under way? Could problems arise that would make us unable to complete the job on time and at the price we quote? Do we have personnel qualified to work on this project?

The RFP in Figure 9–2 would elicit other questions. Is my field applicable to the research opportunity described here? Can I develop a collaborative proposal by the deadline? What types of national security topics would be most likely to attract funding? The university research office often has a person who helps answer these questions.

Many business entities requesting proposals will hold a bidders' conference at which companies interested in submitting a proposal can ask questions about the project or seek clarification of the needs described in the RFP. Most RFPs require that proposals be submitted by a deadline and contain specific information. Proposals that miss the deadline or do not contain the information requested are ordinarily omitted from consideration. Therefore, once an organization decides to submit a proposal, staff members carefully study the RFP and identify the information requirements. Each information requirement is given to an individual or a group who will be responsible for furnishing necessary material and data.

Some proposals, such as university research proposals, may be written by one or two persons. In complex proposals, however, different sections may be written by individuals in different areas of the organization. An editor or proposal writer will then compile the final document. This writer/editor may be assisted by readers who help check the developing proposal to be sure that all requested information is included and that the information is correct. Many proposals have to conform to strict length requirements. Proposals that do not conform will be eliminated from consideration.

Once the proposal has been written and submitted, it becomes a legally binding document. The proposing company or individual is legally committed to do what is described in the proposal at the cost and within the time limit stated. For that reason, the proposing organization carefully checks all information for accuracy.

When a large number of bidders submit proposals in response to an RFP, the soliciting organization may select several finalists and allow each finalist to give an oral version of the proposal. During oral presentations, the soliciting organization asks questions. Representatives of the proposing groups have one more opportunity to argue for the value of what they propose, the merits of their institution, and the justification for the cost attached to the proposed work.

Students may be required to submit proposals for semester research projects or for other university programs, such as research opportunities with faculty. These proposals (two appear in this chapter) provide you with good practice in developing proposals you may write in your career: for example, government grants to fund your research, approvals to launch a work-related project, and actual bid proposals to help your business organization win work projects. Explaining to a potential employer during an interview that you have studied and written at least one proposal and a progress report can help you win a job offer.

Effective argument in proposal development. All writing is persuasive: it must convince the reader(s) that the writer has credibility and that the writer's ideas have merit. However, the success of a proposal hinges entirely on the effectiveness of its argument—how well the writer argues for a plan, an idea, a product, or a service and how well the writer convinces the reader that the proposing organization has all the best qualifications to do the work needed. In planning the proposal, the proposer must harmonize the soliciting company's needs with the proposer's capabilities. The writer must be acutely sensitive to what readers will be looking for but not propose action outside the capability of the proposing individual or organization. The proposing individual or organization has an ethical responsibility to explain accurately and specifically what work can be done and what work cannot be done so that no possibility exists of deceiving readers by making promises that cannot be fulfilled.

The following questions can help you analyze the effectiveness of the argument, from the solicitor's perspective, whether in a written or an oral proposal:

- What does the soliciting organization really want?
- What is the problem that needs to be solved?
- What approaches to the solution will be viewed most favorably?
- What approaches will be viewed unfavorably?
- What objections will our plan elicit?
- Can we accomplish the goals we propose?

To answer these questions, the proposer may be required to do research on the organization, its problems, its corporate culture, the perspective and attitudes stemming from its corporate culture, and its current financial status, goals, and problems. As the proposal develops, the writer should examine it from the intended reader's perspective.

- What are the weaknesses of the plan, as we—the writers—perceive them?
- How can we counter any weaknesses and the reader's potential objections?
- How can we make our plan appealing?
- How can we show that we understand their needs?

- How can we best present our capability to do this project?
- What are our/my strengths?
- From our own knowledge of our organization, what are our weaknesses—in personnel and in overall capability to complete this project as proposed?
- Do we need to modify our proposed plan to avoid misleading readers about our ability to perform certain tasks on time, as proposed, and at cost?
- Can we sell our idea without compromising the accuracy of what we can actually do?

As a proposal writer, you should consider each question and determine what evidence you will need to support the merits of your idea and the arguments needed to refute any objections. Every sentence in your proposal should argue for the merits of your plan and your or your organization's ability to complete it. Although the proposal must be designed as a sales document, you have an ethical obligation to present a plan that meets the soliciting organization's needs and requirements. (In considering the ethical issues that confront proposal writers, you will want to review Chapter 3, "Writing Ethically.")

Standard sections of proposals. Proposals generally include three main divisions: a summary, a main body, and attachments. The main body focuses on the three main parts of the proposal: the proposal's objectives (technical proposal), methods for achieving objectives (management proposal), and project cost (cost proposal). Proposals vary, but you will see the following segments embedded in some way:

> Project summary
> Project description (technical proposal)
>> Introduction
>> Rationale and significance
>> Plan of the work
>>> - Scope
>>> - Methods
>>> - Task breakdown
>>> - Problem analysis
>> Facilities and equipment
> Personnel (management proposal)
> Budget (cost proposal)
> Conclusion
> Appendices

Major business proposals are submitted in complete report format, which requires a letter of transmittal, a title page, a submission page (perhaps), a table of contents, and a summary—all items discussed in Chapter 8. Shorter proposals may be written in a memo or letter format. Many government proposals must be submitted online only. Whatever the format, the main elements will be required, although how they appear will vary with each proposal. In most RFPs, the soliciting organization explains what should be included in the proposal (either specific information to be included or major elements). Often, RFPs indicate the maximum number of pages allowed in a proposal. Writers should follow these instructions carefully to ensure that the proposal is not rejected during the initial screening process because it violates the specifications stipulated by the RFP.

Summary. The summary is by far the most important section of the proposal. Many proposal consultants believe that a project will be accepted or rejected based solely on the effectiveness of the summary. The summary creates your readers' first impression of what you are proposing. It should concisely describe the project, particularly how your work meets the requirements of the soliciting organization, your plan for doing the work, and your or your company's main qualifications. The summary should be a concise version of the detailed plan, but it should be written to convince readers that you understand what the soliciting firm needs and wants, that what you are proposing can be done as you describe, and that your approach is solid because you have the required knowledge and expertise. After reading the summary, readers should want to read more of your proposal.

Project description (technical proposal). The technical proposal describes what you or your company proposes to do. The description must be as specific as possible. The technical proposal has a number of elements.

Introduction. The proposal introduction should explain what you are proposing, why you are proposing this idea, and what you plan to accomplish. The introduction contains the same elements as any introduction. In short proposals, the summary and introduction can be combined.

Rationale and significance. Your success in writing a winning proposal will rest largely on your ability to convince your readers that you understand the project. In the section on rationale and significance, you need to make clear that you know the needs of the readers—as stated in the summary or introduction—and that you have designed your goals by analyzing and defining their needs. Although you will clearly try to sell your idea, you should recognize and answer any questions your readers may have as you argue the merits of your project. Convincing your readers that you fully understand what they are looking for is critical in establishing your credibility.

In short,

- You may want to define the problem to show that you understand it.

- You may want to explain the background of the problem and how it evolved by providing a historical review of the problem.

- You may want to explain why your project is necessary and what results can be expected from it.

- You may want to describe your solution and the benefits of your proposed solution.

Of greatest importance, however, is the *feasibility* of the work you propose. Is your proposed work doable? Is it suitable, appropriate, economical, and practicable? Have you given your readers an accurate view of what you can and will do?

Plan of the work. The section on the work plan is also critical, particularly to expert readers who will attempt to determine whether you understand the breadth of the work you are proposing. In this section, you will describe how you will go about achieving the goals you have stated. You will specify what you will do and in what order, explaining and perhaps justifying your approach as you believe necessary. A realistic approach is crucial in that a knowledgeable reader will sense immediately if your plan omits major steps. A flawed work plan can destroy your credibility as well as the merits of the goals or the solution you are proposing.

Scope. The work plan section may need to describe the scope of the proposed work. What will you do and not do? What topics will your study or your work cover and not cover? What are the limits of what you are proposing? What related topics will be outside the scope of your project? As the writer of the proposal, you have both an ethical and a legal obligation to make clear to your readers the limits of your responsibility. You must avoid creating expectations that you do not intend to meet.

Methods. A work plan may also require a statement of the methods you will use. If you are going to do on-site research, how will you do this research? If you plan to collect data, how will you do so? How will you analyze this data? How will you guarantee the validity of the analysis? If your research includes human subjects, how will you make sure that their participation is voluntary and their privacy is protected? If you are going to conduct surveys or interviews, how will you do so and what questions will you ask? If you plan to do historical research or a literature review of a topic, how will you approach such a review to ensure that your findings are representative of what is currently known about a subject area? What precautions will you take to verify that your research is conducted according to applicable ethical and legal standards? A precise, carefully detailed description of your work methods can add to your credibility as one who is competent to perform the proposed work.

Task breakdown. Almost all proposals require you to divide your work into specific tasks and to state the amount of time allotted to each task. This information may be given in a milestone chart, as illustrated in the methods section of the student research proposal (see Case 9–1). The task breakdown indicates how much time you plan to devote to each task. A realistic time schedule also becomes an effective argument. It suggests to readers that you understand how much time your project will take and that you are not promising miracles just to win approval of your proposal or business plan.

If a project must be completed by a deadline, the task breakdown and work schedule should indicate exactly how you plan to fit every job into the allotted time. However, do not make time commitments that will be impossible to meet. Readers who sense that your work plan is unrealistic will immediately question your credibility. Remember, too, that a proposal is a binding commitment. If you cannot do what you propose, what the soliciting organization requires within the required time, you can destroy your professional credibility and leave yourself subject to legal action.

Problem analysis. Few projects can be completed without problems. If you have carefully analyzed the problem or work you intend to do, you should anticipate where difficulties could arise. Problems that may be encountered can often be discussed in the rationale section. However, if you discover major obstacles that you believe will occur during the course of the project, you may wish to isolate and discuss these in a separate section. Many organizations that request work or solicit research proposals are aware of problems that may arise. Reviewers in these organizations look carefully at the problem analysis section, wherever it occurs, to see whether the proposer has anticipated these problems and explained the course of action that will be followed in dealing with them. Anticipating and designing solutions to problems can further build your credibility with readers, who will not be impressed if you fail to diagnose points in your work plan that could be troublesome and even hinder your completion of the project as proposed.

Facilities and equipment. The facilities section of the proposal is important if you need to convince the reader that your company has the physical capability to do the proposed work. Facilities descriptions are particularly crucial if a product must be built at a particular plant site or a service provided from a designated location. Even in research proposals, your readers may want to know what resources you will use. Sometimes existing facilities are not adequate for a particular job and your company will need to purchase specific equipment. The facilities section enables you to explain this purchase and how it will be included in the cost proposal.

Researchers may need to travel to visit special libraries or research sites. The amount of money needed for this travel will be part of the cost proposal. Thus, the nature of any extra research support, its importance, and its cost to the project should be explained here.

Personnel (management proposal). Any technical proposal or project is only as good as the management strategy that directs it. The management proposal should explain how you plan to manage the project: who will be in charge and what qualifications that person or team has for this kind of work. Management procedures should harmonize with the methods of pursuing the work described in the technical proposal.

Descriptions of your management philosophy and hierarchy should clearly reflect your company's overall management philosophy and culture. Readers should see the same kind of management applied to the proposed work as to the company and other projects it manages. Any testimony to or evidence of the effectiveness of the management approach will lend credibility to the technical proposal. Proposal reviewers must be convinced that you and your organization have a sound approach supported by good management of that approach.

In research proposals, the researcher who is soliciting funds will want to explain his or her expertise in the subject area proposed. This explanation may focus on educational background, previous projects successfully undertaken, published research on the topic, and general experience.

Budget (cost proposal). The cost proposal is usually the final item in the body of the proposal, even though cost may ultimately be the most crucial factor in industrial proposals. Cost is usually given last and appears as a budget for the length of the proposal period. The technical and management sections of the proposal—with their descriptions of methods, tasks, facilities, required travel, and personnel—should help justify the cost. They should have already explained the rationale for items that will produce the greatest cost. However, any items not previously discussed in the technical and management sections—such as administrative expenses, additional insurance benefit costs, and unexpected legal costs—should be explained. An itemized budget, often submitted as a separate document, includes items such as the proposing organization's liability for not meeting project deadlines, for cost overruns, and for unforeseen labor strikes and work stoppages. Many budget sections include standard statements such as descriptions of union contracts with labor costs, insurance benefit costs, non-strike costs, and statements of existing corporate liability for other projects—any existing arrangements that affect the cost of the proposed contract. Clearly, the budget should explain exactly how much the project will cost and how the cost is determined. The extensiveness of the budget depends on the magnitude of the project.

Conclusion. The proposal includes a final section that repeats what the proposal offers the potential client or the soliciting agency, why you or your company should be selected to perform the work, and the benefits that the project, when completed, will yield for the client. The conclusion presents the final restatement of your central argument.

Appendices. As in any report, the appendix section includes materials to support information you give in the main body of the proposal—in the technical, management, or cost proposals. For example, the appendix might include résumés of principal investigators, managers, or researchers. These résumés should highlight their qualifications as they pertain to the specific project.

CASE 9–1 **Research Proposal**

The undergraduate research office at a university solicits proposals from undergraduate students who wish to pursue research with a faculty member. Case Document 9–1 is a research proposal written by an undergraduate chemistry major, Tucker Folsom. The proposal includes sections required by the solicitation for these undergraduate research proposals. Readers of this proposal will be faculty from science and engineering. For that reason, the proposal is highly technical and is prepared to meet the strict length requirements given in the RFP.

CASE DOCUMENT 9–1

TO: Dr. Elizabeth Tebeaux   DATE: February 29, 2016

FROM: Tucker Folsom

SUBJECT: **Proposal for Research Paper—Application of Organometallic Chemistry to Medicine**

Rationale for the Topic and Approach
In addition to providing a source of knowledge to researchers, this report serves to increase nonscientists' understanding of how research affects lives.

It is important to always apply context to research, giving the public an understanding of the importance of continued research in the central sciences. Without a full understanding of the ways that advances in organometallic chemistry affect the lives of the public, interest is eventually lost. A loss of interest and understanding of relevance is one of the worst possible sets of circumstances for the future of science—when the general population becomes disinterested funding is lost.

Research Methods and Methodology
The American Chemical Society (ACS) Publication, Google Scholar, and LexisNexis are the main avenues used for research on this topic. ACS and Google Scholar provided most of journal papers used in this report, while LexisNexis provided supporting or background material.

Journal articles used in this report are selectively chosen from those published in the last 4 years as to keep the science reported current. Sources that date to later than 2012 were used for background understanding or for contextual purposes rather than for understanding of recent scientific advances.

Work Schedule

Week 1: February 28–March 5	Read sources, revise and expand outline, begin work on III
Week 2: March 6–12	Begin work on V and VII and finish III
Week 3: March 13–19 (Spring Break)	Update sources and finish V and VII
Week 4: March 20–26	Begin work on IV
Week 5: March 27–April 2	Begin work on VI and finish IV
Week 6: April 3–9	Finish first draft of report, begin revisions
Week 7: April 10–16	Final work on updating sources and citation of sources and figures, revise second draft
Week 8: April 17–23	Complete report by April 18

Note: Sections of the report are written out of order based on the author's experience and familiarity with sections V and VII.

Outline of Report

    I. Letter of transmittal
   II. Title page
  III. History of organometallic chemistry research and relation to medicine
      a) Context of organometallic chemistry within science
      b) Scope of impact of organometallic chemistry on medicine
          i. Figure depicting overall scope
         ii. Transition into specific categories to be discussed
      c) Introduction to major fields researched
          i. Cancer diagnosis and treatment
         ii. Luminescent complex research
       iii. Antimicrobial materials development
       iv. Photo CORM and cleating ligand research
  IV. Cancer and tumor diagnosis/treatment
      a) Opportunities with organometallic anticancer drugs
      b) Advances made with ruthenium-based metal complexes
          i. Explanation of properties of ruthenium
         ii. Applications of ruthenium with arene ligands
       iii. Complexes in testing process
      c) Tumor treatment
          i. Patents and medicines in use
   V. Luminescent complex research
      a) Discussion of author's research within field
      b) Importance of imaging in medical fields
      c) Introduction of radiometals and functionalized-lanthanide complexes
          i. Capabilities of radiometals
         ii. Capabilities of lanthanide complexes

VI. Antimicrobial materials development
 a) Chemistry and biomedical engineering
  i. Aging population and expansion of biomedical research
  ii. Biocompatibility problems
  iii. Solutions offered by organometallic chemistry
 b) Drug delivery and antimicrobial work
VII. Photo CORM and cleating ligand research
 a) Discussion of author's research on subject
 b) Use of carbon dioxide in bodily systems
 c) What is a photo CORM and how are they used?
  i. Photo CORM research
  ii. Cleating ligand use

Qualifications of the Author and Connection to Topic

The author is a second-semester freshman studying chemistry at Texas A&M University in College Station, TX. He has significant research experience in inorganic and organometallic chemistry with research groups at both Texas A&M (College Station) and The University of Texas at Austin. The author expects to attend graduate school to earn a PhD in chemistry. This report serves as an opportunity for the author to explore the field of organometallic chemistry and its applications to better understand possible opportunities in the future.

Working Bibliography

Abd-El-Aziz, Alaa S., Christian Agatemor, Nola Etkin, David P. Overy, Martin Lanteigne, Katherine Mcquillan, and Russell G. Kerr. "Antimicrobial Organometallic Dendrimers with Tunable Activity Against Multidrug-Resistant Bacteria." *Biomacromolecules* 16.11 (2015): 3694–703. Web.

Adini, A. R., M. Redlich, and R. Tenne. "Medical Applications of Inorganic Fullerene-like Nanoparticles." *Journal of Materials Chemistry* 21.39 (2011): 15121. Web.

Ariel Alberto, Roger. Combination of Intercalating Organometallic Complexes and Tumor Seeking Biomolecules for DNA Cleavage and Radiotherapy. Patent 6844425 B1. 2014.

Bazaka, Kateryna, and Mohan Jacob. "Implantable Devices: Issues and Challenges." *Electronics* 2.1 (2012): 1–34. Web.

Comby, Steve, Esther M. Surender, Oxana Kotova, Laura K. Truman, Jennifer K. Molloy, and Thorfinnur Gunnlaugsson. "Lanthanide-Functionalized Nanoparticles as MRI and Luminescent Probes for Sensing and/or Imaging Applications." *Inorganic Chemistry* 53.4 (2014): 1867–879. Web.

Eke, Uche B., and Tenimu A. Abubakar. "Bioorganometallic Compounds in Medicine: The Search for New Antibacterial Agents." *World Journal of Biomedicine and Pharmaceutical Sciences* 1 (2015): 22-27. Print.

Gasser, Gilles, and Nils Metzler-Nolte. "The Potential of Organometallic Complexes in Medicinal Chemistry." *Current Opinion in Chemical Biology* 16.1–2 (2012): 84–91. Print.

Guo, Zijian, and Peter J. Sadler. "Metals in Medicine." *Angewandte Chemie* 38 (1999): 1512-1531.

Hartinger, Christian G., Nils Metzler-Nolte, and Paul J. Dyson. "Challenges and Opportunities in the Development of Organometallic Anticancer Drugs." *Organometallics* 31.16 (2012): 5677–685. Web.

Iida, Joji, Elisabeth T. Bell-Loncella, Marc L. Purazo, Yifeng Lu, Jesse Dorchak, Rebecca Clancy, Julianna Slavik, Mary Lou Cutler, and Craig D. Shriver. "Inhibition of Cancer Cell Growth by Ruthenium Complexes." *Journal of Translational Medicine* 14.1 (2016): n.p. Web.

"In Memoriam: Nuclear Medicine Pioneer Alan Davison, PhD." Plus Media Solutions, 20 Nov. 2015. Web.

Küster, Tatiana, Nadine Lense, Fabienne Barna, Andrew Hemphill, Markus K. Kindermann, Joachim W. Heinicke, and Carsten A. Vock. "A New Promising Application for Highly Cytotoxic Metal Compounds: η6 -Areneruthenium(II) Phosphite Complexes for the Treatment of Alveolar Echinococcosis." *Journal of Medicinal Chemistry* 55.9 (2012): 4178–188. Web.

Motterlini, Roberto, and Leo E. Otterbein. "The Therapeutic Potential of Carbon Monoxide." *Nature Reviews Drug Discovery* 9.9 (2010): 728–43. Web.

Nazarov, Alexey A., Christian G. Hartinger, and Paul J. Dyson. "Opening the Lid on Piano-stool Complexes: An Account of Ruthenium(II)–Arene Complexes with Medicinal Applications." *Journal of Organometallic Chemistry* 751 (2014): 251–60. Web.

Niesel, Johanna, Antonio Pinto, Harmel W. Peindy N'dongo, Klaus Merz, Ingo Ott, Ronald Gust, and Ulrich Schatzschneider. "Photoinduced CO Release, Cellular Uptake and Cytotoxicity of a Tris(pyrazolyl)methane (tpm) Manganese Tricarbonyl Complex." *Chemical Communications* 15 (2008): 1798. Web.

Teo, Ruijie D., John Termini, and Harry B. Gray. "Lanthanides: Applications in Cancer Diagnosis and Therapy." *Journal of Medicinal Chemistry* 59.13 (2016): 6012-6024. Web.

Tenders Info, comp. *Bioinorganic Chemistry for the Design of New Medicines*. Rep. n.p.: n.p., 2011. *LexisNexis*. Web.

Ronconi, Luca, and Peter J. Sadler. "Using Coordination Chemistry to Design New Medicines." *Coordination Chemistry Reviews* 251.13–14 (2007): 1633–648. Web.

Yan, Yaw Kai, Michael Melchart, Abraha Habtemariam, and Peter J. Sadler. "Organometallic Chemistry, Biology and Medicine: Ruthenium Arene Anticancer Complexes." *Chemical Communications* 38 (2005): 4764. Web.

Yang, Xiaoping, Michael M. Oye, Richard A. Jones, and Shaoming Huang. "Anion Dependent Self-assembly of a Linear Hexanuclear Yb(iii) Salen Complex with Enhanced Near-infrared (NIR) Luminescence Properties." *Chemical Communications* 49.83 (2013): 9579. Web.

Yempally, Veeranna, Samuel J. Kyran, Rajesh K. Raju, Wai Yip Fan, Edward N. Brothers, Donald J. Darensbourg, and Ashfaq A. Bengali. "Thermal and Photochemical Reactivity of Manganese." *Inorganic Chemistry* 53.8 (2014): 4081-8. Web.

Zeglis, Brian M., Jacob L. Houghton, Michael J. Evans, Nerissa Viola-Villegas, and Jason S. Lewis. "Underscoring the Influence of Inorganic Chemistry on Nuclear Imaging with Radiometals." *Inorganic Chemistry* 53.4 (2014): 1880–899. Web.

Progress Reports

When a soliciting organization requests a proposal, it often states that a specific number of progress reports will be required, particularly if the project covers a long time period. As their name suggests, progress reports (sometimes known as "status reports") explain how work is progressing on a project. Their immediate purpose is to document the activities of your organization, but their long-range objective is to demonstrate your organization's competence in pursuing a task and completing it.

Status reports may also be prepared in paper format and submitted to supervisors. These progress reports help you or your work group provide evidence of your activities. They generally have three goals:

- To explain to the reader what has been accomplished and by whom, the status of the work performed, and problems that may have arisen that need attention

- To explain to your client how time and money have been spent, what work remains to be done, and how any problems encountered are being handled

- To enable the organization or individual doing the work to assess the work and plan future work

Structure of progress reports. Any work project, such as transportation or construction projects, requires regular progress reports. While these differ, they incorporate the same basic segments: goal of the project, work accomplished (for a given period), work remaining/planned (for the next period), problems encountered, and financial expenditures.

Structure by work performed. This is the standard structure for progress reports.

    Beginning
- Introduction/project description
- Summary

  Middle

• Work completed	or	• Task 1
Task 1		Work completed
Task 2		Work remaining
• Work remaining		• Task 2
Task 3		Work completed
Task 4		Work remaining
• Cost		• Cost

  End
- Overall appraisal of progress to date
- Conclusion and recommendations

In this general plan, you emphasize what has been done and what remains to be done and supply enough of an introduction to be sure that the reader knows what project is being discussed.

For progress reports that cover more than one period, the basic design can be expanded as follows:

Beginning
- Introduction
- Project description
- Summary of work to date
- Summary of work in this period

Middle
- Work accomplished by tasks (this period)
- Work remaining on specific tasks
- Work planned for the next reporting period
- Work planned for periods thereafter
- Cost to date
- Cost in this period

End
- Overall appraisal of work to date
- Conclusions and recommendations concerning problems

Structure by chronological order. If your project or research is broken into time periods, your progress report can be structured to emphasize the periods.

Beginning
- Introduction/project description
- Summary of work completed

Middle
- Work completed
   Period 1 (beginning and ending dates)
      Description
      Cost
   Period 2 (beginning and ending dates)
      Description
      Cost
- Work remaining
   Period 3 (or remaining periods)
      Description of work to be done
      Expected cost

End
- Evaluation of work in this period
- Conclusions and recommendations

Structure by main project goals. Many research projects are pursued by grouping specific tasks into major groups. Then the writer describes progress according to work done in each major group and perhaps the amount of time spent on that group of tasks. Alternatively, a researcher may decide to present a project by research goals—what will be accomplished during the project. Thus, progress reports will explain activities performed to achieve those goals. In the middle of the following plan, the left-hand column is organized by work completed and remaining and the right-hand column, by goals.

Beginning

- Introduction/project description
- Summary of progress to date

Middle

• Work completed	or	• Goal 1
Goal 1		Work completed
Goal 2		Work remaining
Goal 3, etc.		
• Work remaining		• Goal 2
Goal 1		Work completed
Goal 2		Work remaining
Goal 3, etc.		
• Cost		• Cost

End

- Evaluation of work to date
- Conclusions and recommendations

Online submission of progress reports. More and more organizations require that progress reports be composed and submitted online through interactive sites. For example, the National Science Foundation specifies four kinds of "project reports":

- Annual Reports (a progress report for NSF reviewers)
- Interim Reports (a progress report, only as necessary, for NSF reviewers)
- Final Reports (a report of results for NSF reviewers)
- Project Outcomes Reports (a report of results for public access)

Researchers submit their reports by logging in to the research.gov site and answering a series of questions (see Figure 9–3). The questions elicit all the key information of a standard progress report: that is, what you have accomplished, what problems you have encountered, what you did or plan to do to solve these problems, and what you intend to accomplish during the next reporting period.

Similarly, the US Federal Railroad Administration uses online quarterly progress reports to monitor the implementation of a new automatic safety system, Positive Train Control, across all railroads (see Figure 9–4). This kind of interactive online reporting makes it easier to compile consistent and reliable information, especially in the case of a multi-year project with scores of sources operating at multiple locations.

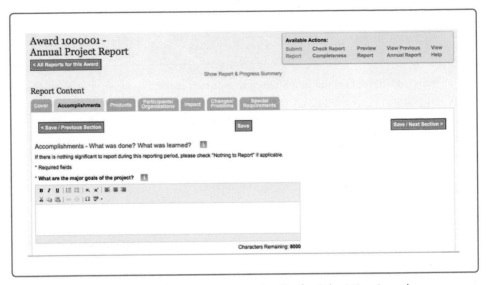

• **FIGURE 9–3** Sample Page from NSF's Interactive Site for Submitting Annual Progress Reports

Quarterly Progress Report – Positive Train Control Implementation

1. Summary

Category	Cumulative Quantity Completed to Date	Total Quantity Required for PTC Implementation
Locomotives Fully Equipped and PTC Operable	495	521
Installation/Track Segments Completed	8	12
Radio Towers Fully Installed and Equipped	102	120
Employees Trained	6,923	10,985
Route Miles in Testing or Revenue Service Demonstration	106.0	218.5
Route Miles in PTC Operation	603.0	897.5

Provide a narrative summary of overall PTC implementation progress during the applicable quarter:

During the 4rd quarter of 2016 Amtrak has worked with the commuter railroads operating on the NEC to address interoperability design issues. Amtrak and Septa are working on interfaces between the two railroads to allow Septa trains to begin to run with ACSES on the NEC by the end of the year.
PTC installation continues on the Michigan Line and all wayside signal and PTC installation will be completed by the end of 2016. Dispatching between Kalamazoo and Battle Creek has been changed over from NS to Amtrak and Amtrak is preparing to begin testing of both ITCS and I-ETMS on this portion of the Line. A Back Office Server (BOS) has been installed in Chicago to support I-ETMS on the State owned portion of the Michigan Line. This BOS will be migrated to the nation BOS in future. Chicago Union Terminal will be part of Michigan BOS for now.
ARINC has been chosen to provide a hosted BOS for Amtrak's Nation wide fleet. ARINC has begun work on the BOS and testing has been done between Amtrak and CSX simulating an Amtrak train initializing through the Amtrak BOS and being dispatched onto CSX territory. Amtrak performed 14I-ETMS major tests successfully.

• **FIGURE 9–4** Summary Page of a Quarterly Progress Report Submitted Through FRA's Interactive Site

CASE 9–2

Recipients of research funds are typically expected to submit a progress report. During his research project with his faculty member, Tucker must submit periodic progress reports. He submits the first progress report four weeks after the proposal for the project. The detail in this report of his progress suggests that his work follows the research methods he discussed in his proposal. Note that he also mentions a change in the audience he will address in his research report. Tucker knows that all substantive changes to the initial proposal must be covered in the progress reports.

CASE DOCUMENT 9–2

First Progress Report

TO: Dr. Elizabeth Tebeaux   **DATE:** April 4, 2016

FROM: Tucker Folsom

SUBJECT: **1st Progress Report for Research Paper—Application of Organometallic Chemistry to Medicine**

Summary:

This paper reports on specific studies in the field of organometallic chemistry that contribute to research or advances in the medical community. Organometallic chemistry is a specific category in chemistry focusing on molecules containing a carbon metal bond. I have conducted research in organometallic chemistry for several years at both The University of Texas at Austin and Texas A&M University. I chose to make area my topic as a way to introduce myself to new problems in my field. My objective is to create a report, accessible to those with an introductory knowledge of college chemistry, which catalogues and explains the applications of organometallic chemistry in medicine.

- Note: Since submitting the initial proposal for my report I have shifted my target audience to those with an introductory knowledge of college chemistry. I did this for several reasons: first, aside from my research experience, I myself only have an introductory knowledge of college chemistry; second, I realized my audience, being Dr. Tebeaux and other readers, will be more apt to read and understand my report when written in an approachable manner.

- I have already completed section I and am currently working on section III. I expect a significant increase in the rate at which the paper is written as I have already read many of my sources—cataloguing the information that will be included in each section of the paper.

- Research for this paper is going well, I am finding that the papers I curated as part of my original research are providing consistent and applicable data.
- Although writing for this paper is behind schedule, I initially planned for an earlier due date. Therefore, as I have more time than anticipated, I will be able to finish according to my revised work schedule.

Revised Work Schedule Going Forward:

April 3–9	All sources printed and read, section III
April 10–6	Finish sections II and V
April 17–23	Finish section IV and start VI
April 24–30	Editing
May 1–2	Editing and printing

Revised Outline of Report:

Changes made are *italicized*, additional changes are expected as the report is written.

I. History of organometallic chemistry research and relation to medicine
   a) Context of organometallic chemistry within science
      i. *What is organometallic chemistry?: Important concepts and terms*
         • *Ligands*
         • *Research*
         • *Organometallic chemistry vs. inorganic chemistry*
      ii. *Connection to other parts of science*
   b) Scope of impact of organometallic chemistry on medicine
      i. Organometallic compounds and medicine
         • Figure depicting overall scope
   c) Introduction to major fields researched
      i. Cancer diagnosis and treatment
      ii. Luminescent complex research
      iii. Antimicrobial materials development
      iv. Photo CORM and cleating ligand research
II. Cancer and tumor diagnosis/treatment
   a) Opportunities with organometallic anticancer drugs
      i. Advantages of metals in medicine
         • *Structural diversity of metal compounds*
         • *Replacement of organic compounds using metals*
            ◦ *Tamoxifen drug increased activity*
         • *Ligand exchange to bind with DNA*
         • *Catalytic properties*
      ii. *Drug/gene delivery*
         • *pH control*
         • *Nanoparticles*

        iii. *Radiation therapy*

   b) Advances made with ruthenium-based metal complexes

        i. Explanation of properties of ruthenium

        ii. Applications of ruthenium with arene ligands

        iii. Complexes in testing process

   c) Tumor treatment

        i. Patents and medicines in use

III. Luminescent complex research

   a) Discussion of author's research within field

   b) Importance of imaging in medical fields

        i. *Radio imaging vs. luminescent imaging*

   c) Introduction of radiometals and functionalized-lanthanide complexes

        i. Capabilities of radiometals

        ii. Capabilities lanthanide complexes

IV. Antimicrobial materials development

   a) Chemistry and biomedical engineering

        i. Aging population and expansion of biomedical research

        ii. Biocompatibility problems

        iii. Solutions offered by organometallic chemistry

   b) Drug delivery and antimicrobial work

V. Photo CORM and cleating ligand research

   a) Discussion of author's research on subject

   b) Use of carbon dioxide in bodily systems

   c) What is a photo CORM and how are they used?

        i. Photo CORM research

        ii. Cleating ligand use

VI. Conclusion

   a) *Looking forward*

   b) *Why scientific research is essential*

Research Consulted:

1) Lanthanides: Applications in Cancer Diagnosis and Therapy
   a. **Citation:** Teo, Ruijie D., John Termini, and Harry B. Gray. "Lanthanides: Applications in Cancer Diagnosis and Therapy." *Journal of Medicinal Chemistry* 59.13 (2016): 6012-6024. Web.
2) Anion Dependent Self-Assembly of a Linear Hexanuclear Yb Salen Complex with Enhanced Near-infrared Luminescence Properties
   a. **Citation:** Yang, Xiaoping, Michael M. Oye, Richard A. Jones, and Shaoming Huang. "Anion Dependent Self-assembly of a Linear Hexanuclear Yb(iii) Salen Complex with Enhanced Near-infrared (NIR) Luminescence Properties." *Chemical Communications* 49.83 (2013): 9579. Web.
3) Underscoring the Influence of Inorganic Chemistry on Nuclear Imaging with Radiometals

a. **Citation:** Zeglis, Brian M., Jacob L. Houghton, Michael J. Evans, Nerissa Viola-Villegas, and Jason S. Lewis. "Underscoring the Influence of Inorganic Chemistry on Nuclear Imaging with Radiometals." *Inorganic Chemistry* 53.4 (2014): 1880-899. Web.

4) Lanthanide-Functionalized Nanoparticles as MRI and Luminescent Probes for Sensing and/or Imaging Applications
   a. **Citation:** Comby, Steve, Esther M. Surender, Oxana Kotova, Laura K. Truman, Jennifer K. Molloy, and Thorfinnur Gunnlaugsson. "Lanthanide-Functionalized Nanoparticles as MRI and Luminescent Probes for Sensing and/or Imaging Applications." *Inorganic Chemistry* 53.4 (2014): 1867–879. Web.

5) In Vivo Anti-Tumor Activity of the Organometallic Ruthenium-Arene Complex RAPTA-C in Human Ovarian and Colorectal Carcinomas
   a. **Citation:** Weiss, Andrea, Robert H. Berndsen, Maxime Dubois, Cristina Müller, Roger Schibli, Arjan W. Griffioen, Paul J. Dyson, and Patrycja Nowak-Sliwinska. "In Vivo Anti-tumor Activity of the Organometallic Ruthenium(Ii)-Arene Complex [Ru(η 6 -p-cymene)Cl 2 (pta)] (RAPTA-C) in Human Ovarian and Colorectal Carcinomas." *Chemical Science* 5.12 (2014): 4742–748. Web.

6) The potential of organometallic complexes in medicinal chemistry
   a. **Citation:** Gasser, Gilles, and Nils Metzler-Nolte. "The Potential of Organometallic Complexes in Medicinal Chemistry." *Current Opinion in Chemical Biology* 16.1-2 (2012): 84–91. Print.

7) Opening the Lid on Piano-Stool Complexes: An Account of Ruthenium(II)-Arene Complexes with Medicinal Applications
   a. **Citation:** Nazarov, Alexey A., Christian G. Hartinger, and Paul J. Dyson. "Opening the Lid on Piano-stool Complexes: An Account of Ruthenium(II)–Arene Complexes with Medicinal Applications." *Journal of Organometallic Chemistry* 751 (2014): 251–60. Web.

8) Using Coordination Chemistry to Design New Medicines
   a. **Citation:** Ronconi, Luca, and Peter J. Sadler. "Using Coordination Chemistry to Design New Medicines." *Coordination Chemistry Reviews* 251.13–14 (2007): 1633–648. Web.

9) Metals in Medicine
   a. **Citation:** Guo, Zijian, and Peter J. Sadler. "Metals in Medicine." *Angewandte Chemie* 38 (1999): 1512-1531.

10) Bioinorganic Chemistry for the Design of New Medicines
   a. **Citation:** Tenders Info, comp. *Bioinorganic Chemistry for the Design of New Medicines.* Rep. n.p.: n.p., 2011. *LexisNexis.* Web.

11) The Therapeutic Potential of Carbon Monoxide
   a. **Citation:** Motterlini, Roberto, and Leo E. Otterbein. "The Therapeutic Potential of Carbon Monoxide." *Nature Reviews Drug Discovery* 9.9 (2010): 728–43. Web.

Working Bibliography

Abd-El-Aziz, Alaa S., Christian Agatemor, Nola Etkin, David P. Overy, Martin Lanteigne, Katherine Mcquillan, and Russell G. Kerr. "Antimicrobial Organometallic Dendrimers with Tunable Activity against Multidrug-Resistant Bacteria." *Biomacromolecules* 16.11 (2015): 3694–703. Web.

Adini, A. R., M. Redlich, and R. Tenne. "Medical Applications of Inorganic Fullerene-like Nanoparticles." *Journal of Materials Chemistry* 21.39 (2011): 15121. Web.

Ariel Alberto, Roger. Combination of Intercalating Organometallic Complexes and Tumor Seeking Biomolecules for DNA Cleavage and Radiotherapy. Patent 6844425 B1. 2014.

Bazaka, Kateryna, and Mohan Jacob. "Implantable Devices: Issues and Challenges." *Electronics* 2.1 (2012): 1–34. Web.

Comby, Steve, Esther M. Surender, Oxana Kotova, Laura K. Truman, Jennifer K. Molloy, and Thorfinnur Gunnlaugsson. "Lanthanide-Functionalized Nanoparticles as MRI and Luminescent Probes for Sensing and/or Imaging Applications." *Inorganic Chemistry* 53.4 (2014): 1867–879. Web.

Eke, Uche B., and Tenimu A. Abubakar. "Bioorganometallic Compounds in Medicine: The Search for New Antibacterial Agents." *World Journal of Biomedicine and Pharmaceutical Sciences* (2015): n.p. Print.

Gasser, Gilles, and Nils Metzler-Nolte. "The Potential of Organometallic Complexes in Medicinal Chemistry." *Current Opinion in Chemical Biology* 16.1–2 (2012): 84-91. Print.

Guo, Zijian, and Peter J. Sadler. "Metals in Medicine." *Angewandte Chemie* 38 (1999): 1512-1531.

Hartinger, Christian G., Nils Metzler-Nolte, and Paul J. Dyson. "Challenges and Opportunities in the Development of Organometallic Anticancer Drugs." *Organometallics* 31.16 (2012): 5677–685. Web.

Iida, Joji, Elisabeth T. Bell-Loncella, Marc L. Purazo, Yifeng Lu, Jesse Dorchak, Rebecca Clancy, Julianna Slavik, Mary Lou Cutler, and Craig D. Shriver. "Inhibition of Cancer Cell Growth by Ruthenium Complexes." *Journal of Translational Medicine* 14.1 (2016): n.p. Web.

"In Memoriam: Nuclear Medicine Pioneer Alan Davison, PhD." Plus Media Solutions, 20 Nov. 2015. Web.

Küster, Tatiana, Nadine Lense, Fabienne Barna, Andrew Hemphill, Markus K. Kindermann, Joachim W. Heinicke, and Carsten A. Vock. "A New Promising Application for Highly Cytotoxic Metal Compounds: η6 -Areneruthenium(II) Phosphite Complexes for the Treatment of Alveolar Echinococcosis." *Journal of Medicinal Chemistry* 55.9 (2012): 4178–188. Web.

Motterlini, Roberto, and Leo E. Otterbein. "The Therapeutic Potential of Carbon Monoxide." *Nature Reviews Drug Discovery Nat Rev Drug Discov* 9.9 (2010): 728–43. Web.

Nazarov, Alexey A., Christian G. Hartinger, and Paul J. Dyson. "Opening the Lid on Piano-stool Complexes: An Account of ruthenium(II)–Arene Complexes with Medicinal Applications." *Journal of Organometallic Chemistry* 751 (2014): 251–60. Web.

Niesel, Johanna, Antonio Pinto, Harmel W. Peindy N'dongo, Klaus Merz, Ingo Ott, Ronald Gust, and Ulrich Schatzschneider. "Photoinduced CO Release, Cellular Uptake and Cytotoxicity of a Tris(pyrazolyl)methane (tpm) Manganese Tricarbonyl Complex." *Chemical Communications* 15 (2008): 1798. Web.

Teo, Ruijie D., John Termini, and Harry B. Gray. "Lanthanides: Applications in Cancer Diagnosis and Therapy." *Journal of Medicinal Chemistry* 59.13 (2016): 6012-6024. Web.

Tenders Info, comp. *Bioinorganic Chemistry for the Design of New Medicines*. Rep. n.p.: n.p., 2011. *LexisNexis*. Web.

Ronconi, Luca, and Peter J. Sadler. "Using Coordination Chemistry to Design New Medicines." *Coordination Chemistry Reviews* 251.13–14 (2007): 1633–648. Web.

Weiss, Andrea, Robert H. Berndsen, Maxime Dubois, Cristina Müller, Roger Schibli, Arjan W. Griffioen, Paul J. Dyson, and Patrycja Nowak-Sliwinska. "In Vivo Anti-tumor Activity of the Organometallic Ruthenium(Ii)-Arene Complex [Ru(η 6 -p-cymene)Cl 2 (pta)] (RAPTA-C) in Human Ovarian and Colorectal Carcinomas." *Chemical Science* 5.12 (2014): 4742–748. Web.

Yan, Yaw Kai, Michael Melchart, Abraha Habtemariam, and Peter J. Sadler. "Organometallic Chemistry, Biology and Medicine: Ruthenium Arene Anticancer Complexes." *Chemical Communications* 38 (2005): 4764. Web.

Yang, Xiaoping, Michael M. Oye, Richard A. Jones, and Shaoming Huang. "Anion Dependent Self assembly of a Linear Hexanuclear Yb(iii) Salen Complex with Enhanced Near-infrared (NIR) Luminescence Properties." *Chemical Communications* 49.83 (2013): 9579. Web.

Yempally, Veeranna, Samuel J. Kyran, Rajesh K. Raju, Wai Yip Fan, Edward N. Brothers, Donald J. Darensbourg, and Ashfaq A. Bengali. "Thermal and Photochemical Reactivity of Manganese." *Inorganic Chemistry* 53.8 (2014): 4081-8. Web.

Zeglis, Brian M., Jacob L. Houghton, Michael J. Evans, Nerissa Viola-Villegas, and Jason S. Lewis. "Underscoring the Influence of Inorganic Chemistry on Nuclear Imaging with Radiometals." *Inorganic Chemistry* 53.4 (2014): 1880–899. Web.

CASE 9–3

Major projects usually require several progress reports, particularly to explain problems and their corresponding solutions or to record ongoing expenses related to the project. Tucker discovers he must narrow the scope of his research report and take more time to complete it. He explains and justifies the changes in his second progress report.

CASE DOCUMENT 9-3

Second Progress Report

TO: Dr. Elizabeth Tebeaux   **DATE:** April 13, 2018

FROM: Tucker Folsom

SUBJECT: **2nd Progress Report, Revised Topic, for Research Paper—Application of Organometallic Chemistry Research to Medical Imaging and Cancer Treatment**

Revised Topic: Application of Organometallic Chemistry Research to Medical Imaging and Cancer Treatment

Summary

The scope of my report has been narrowed from five sections to focus on only medical imaging and cancer treatment. I made this change to the scope of my report for several reasons:

1. The initial report was beyond my scope of ability this semester. While the topic is viable, the report would have reached 40+ pages and encompassed more than asked for in this report.
2. A massive amount of research has been done in these topics; if I was to write on each topic, there would be far too much material to process in order to produce a well-written report.
3. This assignment doesn't request such a lengthy report; the example reports provided discuss their topic thoroughly in 10–20 pages.

Being overly ambitious is the likely cause of the breadth of my original report. I love chemistry, and when given the ability to write and research about it I decided everything was too interesting to not cover. Having not written a paper of this size before or done this much research for a project, I made the mistake of planning too little and not specifying my topic enough.

I want to emphasize that I haven't lost interest in my topic or found it too difficult. My problem arises from the fact that I enjoy my topic too much and if I didn't cut the breadth of my report it would become far too large and take too much time from my other studies.

Rational for Specific Categories Chosen

I have chosen to focus on medical imaging and cancer treatment because I know the most and the least about these categories. I have done two summers of research in a lab focused on making materials for medical imaging and have consequently read a lot of scientific papers on the subject. Cancer treatment is an interesting and important field that I have no experience in, allowing me to learn a lot while conducting my research.

Revised Work Schedule

April 10–6	Sections I and II complete, Section III started
April 17–23	Section III and IV complete
April 24–30	Editing
May 1–2	Editing and printing

Revised Outline

I. History of organometallic chemistry research and relation to medicine
    a. Organometallic chemistry: important concepts and terms
    b. Scope of organometallic chemistry within medicine
    c. Introduction to major fields researched
        i. Medical imaging research
        ii. Cancer treatment
II. Imaging research
    a. Discussion of author's research
    b. Importance of imaging in medical fields
        i. Radio imaging vs. luminescent imaging
    c. Introduction of radiometals and functionalized-lanthanide complexes
        i. Radiometals
            1. Why are radiometal-based imaging agents used over existing nuclear imaging agents?
            2. Design, structure and research on RBIAs
        ii. Functionalized lanthanide complexes
            1. What applications do lanthanide compounds have?
            2. Lanthanides as MRI contrast agents
            3. Lanthanides NIR luminescence properties
        iii. Radiometal imaging in use
        iv. Lanthanide complexes in use
III. Cancer treatment
    a. Opportunities with organometallic anticancer drugs (potential of organometallic complexes in medicinal chemistry)
        i. Advantages of metals in medicine
        ii. Drug/gene delivery

       iii. Radiation therapy
       iv. A note on water solubility
    b. Advances made with ruthenium-based metal complexes
    c. Ferrocifen derivative compounds
  IV. Conclusion
    a. Looking forward
    b. Essential nature of scientific research

Research

A large amount of research originally curated for this report will not be used. Additional research is being conducted on medical imaging and cancer treatment. Because of the large change in research, a working bibliography is not listed here. The research used in this report will be reflected in the bibliography attached to the final report.

Summary

This paper reports on specific studies in the field of organometallic chemistry that contribute to research or advances in the medical community. Organometallic chemistry is a specific category in chemistry focusing on molecules containing a carbon metal bond. I have conducted research in organometallic chemistry for several years at both The University of Texas at Austin and Texas A&M University. I chose to make this area my topic as a way to introduce myself to new problems in my field. My objective is to create a report, accessible to those with an introductory knowledge of college chemistry, which catalogues and explains the applications of organometallic chemistry in medicine.

- Note: Since submitting the initial proposal for my report I have shifted my target audience to those with an introductory knowledge of college chemistry. I did this for several reasons: first, aside from my research experience I myself only have an introductory knowledge of college chemistry; second, I realized my audience, being Dr. Tebeaux and other readers, will be more apt to read and understand my report when written in an approachable manner.
- I have already completed section I and am currently working on section III. I expect a significant increase in the rate at which the paper is written as I have already read many of my sources—cataloguing the information that will be included in each section of the paper.
- Research for this paper is going well; I am finding that the papers I curated as part of my original research are providing consistent and applicable data.
- Although writing for this paper is behind schedule, I initially planned for an earlier due date. Therefore, as I have more time than anticipated, I will be able to finish according to my revised work schedule.

Style and Tone of Proposals and Progress Reports

The proposal and its related report documents serve as sales documents, but writers have an ethical commitment to present information about a project in a clear and accurate manner. Proposals, once accepted, become legally binding documents. Because contracts emerge from proposals, organizations must prepare to stand behind their proposals. Thus, the style should be authoritative, vigorous, and positive, suggesting the competence of the proposer. Writers must bolster any generalizations with detailed factual accomplishments. In progress reports, writers must acknowledge problems encountered during work on a project but should stress positive solutions to these problems. Neither the proposal nor the progress report should resort to vague, obfuscatory language.

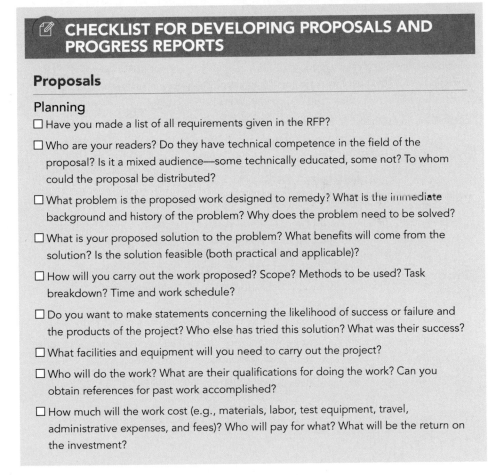

CHECKLIST FOR DEVELOPING PROPOSALS AND PROGRESS REPORTS

Proposals

Planning

☐ Have you made a list of all requirements given in the RFP?

☐ Who are your readers? Do they have technical competence in the field of the proposal? Is it a mixed audience—some technically educated, some not? To whom could the proposal be distributed?

☐ What problem is the proposed work designed to remedy? What is the immediate background and history of the problem? Why does the problem need to be solved?

☐ What is your proposed solution to the problem? What benefits will come from the solution? Is the solution feasible (both practical and applicable)?

☐ How will you carry out the work proposed? Scope? Methods to be used? Task breakdown? Time and work schedule?

☐ Do you want to make statements concerning the likelihood of success or failure and the products of the project? Who else has tried this solution? What was their success?

☐ What facilities and equipment will you need to carry out the project?

☐ Who will do the work? What are their qualifications for doing the work? Can you obtain references for past work accomplished?

☐ How much will the work cost (e.g., materials, labor, test equipment, travel, administrative expenses, and fees)? Who will pay for what? What will be the return on the investment?

☐ Will you need to include an appendix? Consider including biographical sketches, descriptions of earlier projects, and employment practices.

☐ Will the proposal be better presented in a report format or in a letter or memo format?

☐ How will you make your proposal accessible for all of your likely readers?

Revising

☐ Does your proposal have a well-planned design and layout? Does its appearance suggest the high quality of the work you propose to do? Do your readings both promote and inform?

☐ Does the project summary succinctly state the objectives and plan of the proposed work? Does it show how the proposed work is relevant to the readers' interest?

☐ Does the introduction make the subject and the purpose of the work clear? Does it briefly point out the importance of the proposed work?

☐ Have you defined the problem thoroughly?

☐ Is your solution well described? Have you made its benefits and feasibility clear?

☐ Will your readers be able to follow your plan of work easily? Have you protected yourself by making clear what you will do and what you will not do? Have you been careful not to promise more results than you can deliver? Have you carefully considered all the facilities and equipment you will need?

☐ Have you presented the qualifications of project personnel in an attractive but honest way? Have you asked permission from everyone you plan to use as a reference?

☐ Is your budget realistic? Will it be easy for the readers to follow and understand?

☐ Do all the items in the appendix lend credibility to the proposal?

☐ Have you included a few sentences that urge readers to accept the proposal?

☐ Have you satisfied the needs of your readers? Will they be able to comprehend your proposal? Do they have all the information they need to make a decision?

Progress Reports

Planning

☐ Do you have a clear description of your project available, perhaps in your proposal?

☐ Do you have all the project tasks clearly defined? Do all the tasks run in sequence, or do some run concurrently? In general, are the tasks going well or badly?

☐ What items need to be highlighted in your summary and appraisal?

☐ Are there any problems to be discussed?

☐ Can you suggest solutions for the problems?

☐ Is your work ahead of schedule, right on schedule, or behind schedule?

☐ Are costs running as expected?

☐ How will you make your progress report accessible for all of your likely readers?

Revising

☐ Does your report have an attractive appearance?

☐ Does the plan you have chosen show off your progress to its best advantage?

☐ Is your tone authoritative, with an accent on the positive?

☐ Have you supported your generalizations with facts?

☐ Does your approach seem fresh or tired?

☐ Do you have a good balance between work accomplished and work to be done?

☐ Can your summary and appraisal stand alone? Would they satisfy an executive reader?

EXERCISES

1. Your friend and colleague, Lydia Sasaki, has composed the following brief proposal and asks you to review it. What specific changes would you advise she make so that the proposal is as clear and persuasive as possible?

To: Chief William Ricco

From: Officer Lydia Sasaki

Subject: Surveillance Convention

Date: August 11, 2017

    I have received a letter of invitation from the Association of Certified Surveillance Professionals in reference to a convention being held by the association June 11–14, 2018, in Albuquerque. This convention will offer a comprehensive training program designed to thoroughly train police officers in all aspects of acoustic gunfire detection systems. Instructors at the convention consist of surveillance and security specialists from all over the world and the United States.

    I am respectfully requesting travel funds to attend the convention in order to represent our department. There are workshops at the convention that deal with the establishment of gunfire detection systems by police departments. There are also two preconvention courses that are being conducted specifically on acoustic surveillance. These are the AS Certification Course and the AS Maintenance Course, both on June 11. The remaining workshops and training sessions will be held June 12–14. The workshops include:

Equipment Selection	Urban Obstacles
Sensor Placement	Acoustic Gun Signatures
Legal Issues	Triangulation and Multilateration
Service and Maintenance	Acoustic-Based Patrolling
Community Relations	Video Surveillance Interaction
Critical Incident Evaluation	Incident Archiving

    As soon as we have the ability to acquire a gunfire detection system, it will be necessary for the department to obtain all of the facts regarding acoustic surveillance and to examine what other departments have done to implement and operate their systems. Attached to this memo is a copy of the letter I received from ACSP. In this letter are the details for registration, travel, and accommodations for individuals attending the convention.

    If you have any questions regarding the convention, please feel free to contact me at any time.

2. Visit www.grants.gov, locate a request for proposals (RFP) for a project related to your major, and answer the following questions:

- According to the RFP, what is the product or service the agency wants provided? Or what work does it want performed? Or what problem does it want solved?
- Does the RFP specify the categories of information the proposal must include? If so, what are those categories? If not, what would be appropriate categories?
- Does the RFP specify a minimum or maximum length for the proposal? If so, what is that length? If not, what length would you estimate as necessary?
- Does the RFP specify the organization of information? If so, what is that organization? If not, what would be an appropriate organization?
- Does the RFP specify the format for the proposal? If so, what is that format? If not, what format would you consider appropriate?
- Does the RFP identify the criteria by which proposals will be evaluated? If so, what are those criteria? If not, what are likely criteria? How might these criteria guide the writing, organization, and formatting of the proposal?

Write a memo summarizing your findings to distribute to the students in your class.

10

Instructions, Procedures, and Policies

Instructions versus Procedures

Procedures provide general guidelines for performing a task, while *instructions* provide specific, detailed steps. Procedures will be appropriate for some situations, while instructions may be necessary for readers who need step-by-step directions to perform a job task or process.

Instructions and procedures can appear in several formats: in letters and memoranda (see Chapter 7 for an example instructional memo), in reports, in technical papers or notes, in videos, in stand-alone documents to accompany mechanisms or processes, or in complete manuals. Complex procedures and instructions usually appear as manuals, which use many of the elements exemplified in formal reports. If instructions or procedures are short and simple, e-mail might be a suitable method of distribution, but variations in e-mail applications could produce awkwardly formatted, unreadable text. Ordinarily, prepare your instructions or procedures as an e-mail attachment (in a standard rtf, doc, docx, or pdf file) that can be printed or read on the receiver's computer or mobile screen.

Critical Role of Instructions and Procedures in the Workplace

Instructions and procedures often carry legal liability. Most new products and processes have instructions attached. Instructions may also appear on product websites. When a product fails because instructions lack clarity, completeness, and/or adequate warnings, the manufacturer can be held responsible for costs incurred by buyers who, in good faith, attempted to follow the instructions. Even worse, when a customer or an employee is killed or injured because instructions did not provide complete, readable information and safety warnings, the company can be liable for the injuries or death. Companies that provide incomplete or hard-to-follow instructions on products that must be installed or assembled by the purchasers can lose money on the product when purchasers return the product. Conversely, well-designed, clear, complete, accurate instructions instructions can improve the image of a product and its manufacturer and prove highly persuasive sales documents.

 Quick Tip

Many people resist reading instructions. They try to figure out for themselves how to operate a product or perform a task and will turn to the instructions only if all their efforts fail. When they do read the instructions, they want to understand everything immediately without having to read anything twice. A simple design, plain wording, and clear illustrations will be critical to encouraging readers to pay attention to your instructions or procedures.

1. Use concise headings and subheadings to describe and highlight each section.
2. Leave plenty of white space around headings.
3. Use numbers for every step in a chronological process; use bullets for lists of conditions, materials, and equipment.
4. Use white space to make items in lists easy to find and read.
5. Highlight safety information and warnings. Distinguish between *danger*, *warning*, and *caution*.
6. Keep illustrations as simple as possible.
7. Locate illustrations at the point where the reader needs them. Don't expect readers to locate an illustration that is several pages away from the instruction to which it pertains.
8. Label every illustration, and at the appropriate point in the related text, write "See Figure X."
9. Do not begin an instruction at the bottom of one page and complete it at the top of the next page. Insert a "page break" and move the entire instruction to the next page.

Planning Instructions and Procedures

Instructions and procedures should enable those who need them to perform the tasks covered. Successful instructions and procedures can be easily read, easily understood, and correctly followed by the intended users with minimal difficulty. Understanding your target readers and, in some cases, the context in which the documents will be used is paramount. If readers are confused or misled by statements, they can hold the company liable for injury or financial damage suffered. Instructions, particularly those that involve processes that could be hazardous, should explain the knowledge level required of the target audience.

Who are the readers? What do your readers know about the subject? These questions will determine how much information you provide and the type of language you use. The audience for instructions should be specified, especially if only the readers who conform to that audience should attempt to follow the instructions. Remember that many people resist reading instructions and will attempt to perform a task without doing so. When they do read the instructions, they want to understand them immediately without having to reread them. The format, choice of language, and use of visuals can be critical to encouraging readers to focus on the instructions or procedures.

What is your purpose? Successful instructions/procedures require that both you and your readers know your purpose. What should your readers know and be able to do after they read these instructions/procedures? Any information that isn't applicable should be omitted.

What is the context in which the instructions/procedures will be used? Knowing how documents will be read can help you design effective documents. Readers may be focusing on them while they are sitting at their desks. They may need to read the entire document before trying to perform the task, or they may perform the task as they go.

Do your readers need to read the instructions/procedures before they begin? If so, start with a brief explanation of how to read the instructions (e.g., "Read This First" or "How to Read These Instructions"). You can also include detailed instructions for first-time readers and then have a summary of steps for readers to use after they are familiar with the detailed instructions. Context-sensitive instructions can be found in numerous settings:

- Some sailboards have instructions for how to right the sailboard—if it turns over in the water—on the bottom of the boat where riders can see the instructions if the boat turns over.

- "Quick Start" instructions for many handheld devices can be found inside the top or lid of the device.

- Some products, such as bookcases, computers, and lamps, come with instructions composed only of pictures to show the steps in assembling the product. (If the product will be sold in other countries, "picture-only" instructions should be tested for interpretation.)

- Instructions used in machine shops and on manufacturing floors are often printed in large type on eye-saver yellow or green paper and laminated to resist grease. These instructions may be in metal loose-leaf notebooks so strong that pages cannot be torn out. Or they may appear on computer screens projected along the assembly line. Again, the readability, completeness, and accuracy of these screens can be critical to error avoidance and safety of employees.

- Basic operation instructions and troubleshooting information for large clocks can be found inside the clock mechanism access compartment. These are designed to accompany the booklet that contains detailed instructions.

Will the instructions be available online? If instructions will be available online, focus on readability of the material. Avoid "text-heavy" and lengthy instructions. Partition your instructions into distinct sections hyperlinked in a logical way. Or you may want to tell your readers immediately that they should print a copy of the instructions if they are too long and complex for online reading.

More and more organizations are making video instructions available to their clients and customers. The US Federal Trade Commission, for example, offers a wide array of video instructions, using film and animation, to offer training to businesses as well as individuals (see Figure 10–1). Videos prove especially effective in their ability to focus attention on key details, emphasize warnings and cautions, and display actions or operations with vivid efficiency. Videos will also motivate individuals to read more about the subject as well as find answers to related questions. Make sure to keep your video brief (1–3 minutes) and to include closed captioning for users with limited or no hearing. Also important is a transcript of the video instructions: this will allow users to review specific steps in the process, as necessary, without having to loop through the video again and again.

Planning is particularly critical for instructions or procedures. The document analysis checklist at the end of this chapter will help you determine the appropriate delivery medium, the information you will include, and the design strategies that will help you produce an effective document. Effective instructions require sensitive attention to readers' information needs, their current knowledge of the topic, and what you want them to know and do as a result of reading the instructions.

• **FIGURE 10–1** Clip of Video Instructions

Structure and Organization

First, analyze the audience, purpose, and context; then, begin to structure your procedures/instructions. Examine the following general guide for what elements to include, but remember that what you decide to include will depend on your topic, your audience, your purpose, and the reader's context. Comprehensive instruction/procedure manuals will usually contain most of these elements, while shorter instructions/procedures may require fewer. To illustrate instructions, we will focus on three examples.

Introduction. The introduction should familiarize readers with the task to be performed: specifically, what the instructions will allow readers to do and what skill level the person should have to perform the task successfully.

Theory governing the procedure or instruction. For some types of procedures, readers will perform tasks better if they have a general overview of the process. Knowledge of the process may help them understand when they have made an error and how to avoid errors. If you are going to give instructions for operating a mechanism, some explanation of the value and purpose of the mechanism can prepare readers for the process.

Warnings, cautions, hazards, and notes regarding safety or quality. Given the wide variety of technologically complicated and potentially dangerous products available today, you really cannot "overwarn" readers. And a failure to warn could have costly legal implications for your organization. Any level of hazard needs to be described—what will happen and why—if a particular action is performed. Identify the hazard, followed by the reason or the result, and be sure that all warnings (i.e., risk of fatal or serious injury), cautions (i.e., risk of moderate or minor injury), and notes (i.e., risk of damage) are clearly visible. Hazard messages given at the beginning of the instructions will alert readers to possible problems they may encounter later. Hazard messages should also be repeated if they are associated with a particular step or direction. The American National Standards Institute (www.ansi.org) and the International Organization for Standardization (www.iso.org) publish guidelines for the design of hazard messages. A review of research on hazard messages is available from the US Occupational Safety and Health Administration (www osha.gov).

Conditions under which the task should be performed. Some instructions, particularly laboratory instructions, may require you to describe the physical conditions required to perform a task: room size, temperature, time required for the entire process and for individual steps, specific safety processes. If time is a constraint, readers need to know how much time will be needed before they begin the task.

Name of each step. For each step, give the following: purpose of the step; warnings, cautions, and notes; any conditions necessary for performing the step; time required to perform the step; and the list of materials needed for the step.

- Place instructions in chronological order.
- Number each step.
- Limit each instruction to one action.
- After you have written the instruction, state in a separate sentence the reason for the instruction but only if you believe that the reader may not follow the instruction without an explanation. Include hazard messages whenever necessary. And be sure to explain the reason for a hazard message if the risk involved is not apparent to your readers.

CASE 10-1   The Grignard Reaction

Tasnim Mohamed, a graduate student who supervises a biochemistry lab, decides that the laboratory instructions for the Grignard reaction, presented as linear text in the lab manual, need clarification. She believes that labs should focus on teaching students chemical concepts rather than having them try to read confusing process discussion. She decides to revise these before assigning the lab. A portion of the original instructions appears in Case Document 10–1A. She will expect students to read the original instructions but also to have the revision (Case Document 10–1B) available when they actually begin the experiment. Her first step in revising is to do a prewriting analysis.

Prewriting Analysis for Revising the Grignard Reaction

1. Identify your reader(s).

Texas A&M students in the second semester of organic chemistry (CHEM238).

2. Describe your reader profile: background, attitude toward following instructions.

Most of the students are sophomores, so they may have had lab experience in another class before taking this lab. Generally, all students have already taken the first organic chemistry lab (CHEM237), so they have some idea of what to expect. They already know common lab safety rules such as wearing goggles and appropriate clothing, tying back long hair, no eating or drinking, no horseplay, etc.

Students in organic chemistry lab do not usually do everything correctly to obtain high yield or even the expected results. Most hope to finish the lab as quickly as possible in order to leave early, so they do not take the time to read huge paragraphs of instructions.

3. How much does your reader(s) know about the subject?

The students in the lab are not experts in the subject, so they will need some background information to understand what happens throughout the experiment. The course has a prerequisite of CHEM228, which is the Organic Chemistry II lecture course. Usually, the lecture material aligns with the laboratory experiments, but sometimes the professors teach at a different pace. The lecture gives students some background information, but they will probably still need help assessing what happens when something in the lab goes wrong.

4. Will you have to define any terms that your reader(s) may not understand?

Students should already know most vocabulary words that align with the organic chemistry material. A new technique specific for this lab alone will be the only thing I may have to define.

5. What do you want your reader(s) to be able to do after reading the instructions or procedures?

The reader should be able to fully understand the purpose of the lab, methods that will be used in the lab, and how to conduct the lab in a timely fashion. They will know where to obtain all the material and exactly what equipment to use for a specific setup. They will also know if they need to begin another task while they wait for something else to be completed so that they can safely and efficiently complete their experiment. The reader should know that actually understanding the lab is more important than the result.

6. What is the situation that led to the need for these instructions?
Organic chemistry lab is designed to teach students several lab techniques aligned with material learned through the lecture course. Some techniques include purification, spectroscopy, classification tests, computer modeling, and reaction kinetics. This specific set of instructions guides students through the Grignard reaction and allows them to practice their distillation and infrared spectroscopy techniques.

6a. If you are revising an existing set of instruction, explain why the original needs revision.
The Grignard reaction was the hardest lab of this semester. Students are often not able to perform the complete experiment because they are unable to prepare the Grignard reagent. If the lab instructions are revised, students may stand a better chance of performing the experiment correctly. The format of the procedure and the lab manual as a whole deter the students from actually reading the entire lab instructions, which hinders their success. They also find it difficult to reread bulky paragraphs of instructions while conducting the experiment, causing them to miss some vital steps.

7. How will your reader(s) use these instructions? Will they need to be read every time, or will they be used only for reference?
Readers should be able to read the instructions before lab, understand the lab, and then perform the experiment on their own. However, they can have the instructions available to them while conducting the experiment for a reference guide. Students should not have to read everything over again during the lab.

8. What kinds of pictorials, if needed, will you use?
I will include graphics of some experimental setups that are used during the lab. I will also need to include pictures of the equipment that the setup will need in case the student does not know some of the terms.

9. What kinds of formatting and page design strategies will you use to make the instructions easy to read?
I will make sure to use bold headings to separate main topics and subtopics. I will put each picture or graphic beside the text to which it corresponds. I will also make sure to leave plenty of white space to help the eye find what it needs on the page.

10. What content segments will you include?
I will include the following sections:
- Purpose
- Summary of Methods
- Materials
- Reagents and Solvents
- Methods
- Experimental Procedure
- Safety Precautions
- Discussion for In-Lab Notebook
- Discussion for Post-Lab Report

CASE DOCUMENT 10–1A: ORIGINAL, GRIGNARD REACTION

THE GRIGNARD REACTION

Purpose: To prepare and use the Grignard reagent as a method of studying the utility of organometallic reagents in synthesis and to identify starring materials by using retrosynthetic analysis

Reading: McMurry: See assignment for this lab session on the website.
Review *Separation and Purification Techniques:* "Distillation" (Simple)
Review *Separation and Purification Techniques:* IR, "Extraction"

INTRODUCTION

Organometallic compounds are defined as organic compounds which have a carbon-metal bond. These are valuable tools for use in organic synthesis. The reactivity of the organometallic compound is highly dependent upon the nature, ionic or covalent, of the carbon-metal bond. Organosodium and organopotassium reagents are highly ionic and the carbon acts like a powerful base. These reagents react explosively with water and are pyrophoric (burst imo flame upon exposure to air). Organomagnesium (Grignard reagents) and organolithium reagents are sufficiently ionic in nature to be strong carbon nucleophiles, strong bases, and yet covalent enough to be soluble in many organic solvents. These two reagents are widely used in organic synthesis. In recent years the use of other organometallic compounds, particulacly organocuprates and organomercury compounds, have been widely used in research. However, the very low toxicity of the Grignard reagents makes them a reagent of choice.

Organomagnesium reagents are perhaps the easiest to prepare and use of all the organometallic reagents. These reagents were discovered in 1901 (Victor Grignard won the Nobel Prize in 1912 for this discovery; the reagents are named after him.) Grignard reagents are prepared by the reaction of alkyl or aryl halides with magnesium. The exact mechanism involved in the formation of Grignard reagents is unknown, and despite their wide use, there is some doubt as to their exact structure. Grignard reagents are conventionally written as RMgX, where X is chlorine, bromine, or iodine, although there is some evidence that in solution a di-organomagnesium structure exists in equilibrium with the magnesium halide. Generally speaking, any organic halide can be used in the preparation of a Grignard reagent. There are a few exceptions to this, most notably are aryl chlorides. Aryl chlorides, where the chlorine attaches to the aromatic ring, do not react with magnesium. Methyl iodide (todornethane), a liquid cancer suspect agent, is used instead of either of the poisonous gases methyl bromide or methyl chloride to prepare methyl Grignard. Likewise, ethyl bromide (Qachrymator) is commonly used in place of the gas ethyl chloride to prepare ethyl Grignard. In the preparation of the Grignard reagents, the alkyl halide used can affect the rate of formation with $I > Br > Cl$ (organomagnesium fluorides have not been prepared). Once the reagent is formed, the halide ion has no apparent effect on either the rate or the yield of the subsequent reaction.

The use of an ether solvent is a key step in the synthesis of Grignard reagents. The electron pairs on the oxygen of the ether coordinate with the magnesium of the Grignard reagent and form a soluble complex. Most frequently diethyl ether or tetrahydrofuran (THF) is used as the solvent for both the synthesis and the subsequent reaction of the Grignard reagent. Grignard reagents have been made and used in the absence of ether; however in both situations (synthesis and use) the yields were "disappointingly poor."

> **Safety!** The use of ether is the only serious safety concern associated with the Grignard reaction. Ethers are highly flammable. Exercising a few simple safety precautions dramatically reduces the hazard of fire. Unfortunately that is not the only hazard associated with ethers. Ethers react readily with atomspheric oxygen to form peroxides. Organic peroxides can undergo explosive decomposition and the risk of an explosion is directly proportional to the concentration of the peroxide. This is why it is important *not* to distill to dryness (the distillation pot goes dry). Ethers are usually checked with special peroxide detection strips prior to their use.

Grignard reagents are most often used in reactions with aldehydes, ketones, or esters. The products of these reactions following work up are alcohols. Reaction with formaldehyde produces a primary alcohol that has a carbon chain one carbon longer than the beginning Grignard reagent. Reactions with other aldehydes give secondary alcohols, and reactions with ketones or esters yield tertiary alcohols. Notice that esters react with *two* moles of Grignard. After the first reaction, the intermediate loses alkoxide to form a ketone. The ketone then reacts with a second mole of Grignard to form a tertiary alcohol. Grignard reagents have also been used in reactions with nitriles, R-C/N to yield ketones, with epoxides to yield alcohols, and with anhydrous carbon dioxide, CO_2, to yield carboxylic acids (*Note:* Grignard reagents will also react with oxygen in air. The reaction with air is so slow that it is not normally necessary to run the reaction under an inert atmosphere. Atmospheric moisture is a real problem and will be discussed in detail.)

$$RX \ + \ Mg \ \xrightarrow[\text{or THF}]{\text{ether}} \ RMgX$$

$CH_3CH_2-O-CH_2CH_3$
diethyl ether
"ether"

tetrahydrofuran
THF

soluble Grignard complex

CASE DOCUMENT 10–1B: REVISED GRIGNARD

Tasnim Mohamed
CHEM 301
Spring 2018

Lab 18: The Grignard Reaction

Purpose

You will learn how to prepare and use a Grignard reagent to study the utility of organometallic reagents in synthesis and to identify starting materials by using retrosynthetic analysis.

Introduction

Organometallic compounds are organic compounds with carbon-metal bonds. These compounds are valuable tools for use in organic synthesis. Their reactivity is highly dependent upon the nature, ionic or covalent, of the carbon-metal bond. Organomagnesium (Grignard reagents) are widely used in organic synthesis due to their low toxicity.

Grignard reagents are conveniently written as RMgX, where X is chlorine, bromine, or iodine. When preparing Grignard reagents, the halide used can affect the rate of formation (fastest: I > Br > Cl: slowest). However, once the reagent is formed, the halide ion has no effect on the rate or yield of the reaction.

The use of an ether solvent is the key step of Grignard reagent synthesis. The election pairs on the oxygen of the ether coordinate with the magnesium of the Grignard reagents and form a soluble complex, as shown in the figure below. Most frequently, diethyl ether or tetrahydrofuran (THF) is used as the solvent for the synthesis and subsequent reaction of the Grignard reagent.

$$RX \ + \ Mg \ \xrightarrow[\text{or THF}]{\text{ether}} \ RMgX$$

$$CH_3CH_2-O-CH_2CH_3$$
diethyl ether
"ether"

tetrahydrofuran
THF

soluble Grignard complex

Grignard reagents are most often used in reactions with aldehydes, ketones, and esters to form alcohol products. They can also react with nitriles to yield ketones, epoxides to yield alcohols, and anhydrous carbon dioxide to yield carboxylic acids. Grignard reagents may be viewed as a strong base. They react readily with acids to form alkanes. Even weak acids such as water, alcohols, phenols, and terminal acetylenes will protonate Grignard reagents to give the corresponding alkane. If the synthesis does not call for the formation of an alkane, this side reaction only lowers the yield of the desired product. Also, if the starting material contains an unprotected alcohol, phenol, or terminal acetylene, you must use two equivalents of the Grignard reagent. One equivalent will be destroyed by the proton source.

Warning!

THF is flammable and volatile. Work close to the hood. If flames occur, smother with a watch glass or beaker, and do not use a fire extinguisher.

Diethyl ether is even more flammable and volatile than THF, has a low affinity for water, and reacts readily with oxygen to form peroxides, which can undergo explosive decomposition.

Materials

- ☐ 25mL round-bottom flask
- ☐ Heat gun
- ☐ Magnetic stir bar and stir plate
- ☐ Small beaker
- ☐ Weighing scale
- ☐ Weighing paper
- ☐ Claisen head
- ☐ Air condenser
- ☐ Micro-scale distillation glassware
- ☐ Sonicator
- ☐ Large test tube
- ☐ Plastic pipette
- ☐ Pasteur pipettes

Reagents and Solvents

Tetrahydrofuran (THF)
Magnesium turnings
Ammonium chloride
Distilled water
10% Sodium bicarbonate
Saturated sodium chloride
Magnesium sulfate

Alkyl Bromides:
1-Bromopropane
1-Bromobutane
1-Bromopentane

Ketones:
Acetone
2-Butanone
2-Pentanone
3-Pentanone

Methods to Achieve Objectives

Simple Distillation
Extraction
Infrared Spectroscopy

Experimental Procedure

This experiment will take two lab periods. In the first period, you will prepare the Grignard reagent and run the reaction with the ketone. The crude product will be allowed to dry until the second week. In the second week, you will purify and identify the product.

Week 1

Step 1: Obtain your alkyl halide and ketone.

- The alkyl halides and the ketones will be at the front of the lab room. Alkyl halide vials are labeled "#H" and ketone vials are labeled "#K." The # on the vial you obtained should be your workstation number. Record the unknown codes in your lab notebook.
- The alkyl halide vials contain 0.025 moles of one of the three alkyl bromides (1-bromopropane, 1-bromobutane, or 1-bromopentane) and the ketone vials contain 0.025 moles of one of the four ketones (acetone, 2-butanone, 2-pentanone, or 3-pentanone). Each of the compounds is dissolved in THF.

Step 2: Examine glassware.

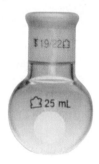

- Your 25mL round-bottom flask should contain no cracks. If you find a crack, return the flask to the stockroom and obtain a new one. If the glassware appears dry, you can begin with **Step 3**.
- If the glassware is not dry, dry it with a heat gun. You can speed up the drying process by rinsing glassware with acetone, which evaporates faster than water. All glassware must be *completely* dry before proceeding with the next step.

Experimental Set-up

Step 3: Add a stir bar to the flask and place it in a small beaker.

Step 4: Weigh 0.61g (0.025mol) of magnesium turnings on a piece of weighing paper and add to the flask.

Step 5: Add 4mL of THF to the flask.

Caution!

THF is flammable. Handle it carefully under the hood at your work station. In case of fire, DO NOT use a fire extinguisher. Smother flames with a watch glass or beaker.

Step 6: Attach Claisen head and air condenser to the flask.

- All ground-glass joints should be greased. If not, notify your TA. Use a Keck clamp to secure all connections.
- Refer to the figure below to set up the glassware.

Step 7: Use a plastic pipette to add 1mL of your alkyl bromide solution to the flask through the top of the air condenser.
- The diagram below shows an example of the reaction taking place in this step.

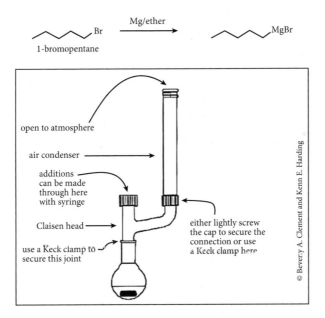

Step 8: Swirl the flask.

Your reaction may begin immediately, and the solution will become very warm and cloudy. In this rare case, you may proceed to Step 10.

Step 9: Place the flask with the attached condenser in an ultrasound bath.

- The bath is located underneath one of the main hoods in the back of the lab room.
- Look for agitation of the liquid and magnesium particles. The more agitation, the faster the reaction will initiate.

- After a couple minutes of agitation, the THF solution should appear somewhat cloudy. Warm the flask if it is not already warm.
- Hold the glassware by the condenser to immerse your flask so that the liquid inside is at or below the level of the liquid in the tank.

> **Ultrasound Bath Sonication Tips**
> – Make sure the bath is turned on. If it is on, you should hear a constant clicking noise. Turn the time dial if it is off. DO NOT turn on the heat on the ultrasound bath.
> – Sonication is greater near the corners of the bath.
> – Four reaction set-ups can be placed in the bath at the same time.

Step 10: Clamp the glassware so that the flask is centered over a magnetic stir plate underneath the hood at your work station.

- Make sure the mixture is stirring evenly.

Step 11: You can use the same plastic pipette from **Step 7** to add the rest of the alkyl bromide, slowly and in small portions, through the air condenser.

- The reaction should warm the solution to the boiling point of THF (~66°C) but not so much to the point where the solution boils over the top of the condenser.
- If the reaction does become too hot, moderate the temperature with an ice bath, without letting the temperature drop below room temperature (~25°C).
- If the reaction does not become warm and the magnesium turnings disappear, return the flask to the sonication bath.

Step 12: Stir the mixture for 15 minutes after adding all the alkyl bromide.

Addition of Ketone to Grignard Reagent

Step 13: Use the same pipette to add the ketone solution to the flask while continuously stirring.

- Add ketone at a rate to keep a steady, gentle reflux.

Step 14: Stir the mixture for 15 minutes after adding all the ketone.

- The reaction flask now contains the magnesium salt of the alcohol product.
- The diagram below shows an example of the reaction taking place in this step.

Hydrolysis

Step 15: Add 6g of ammonium chloride and 20mL of distilled water in a beaker with a stir bar.

Step 16: Stir salt solution vigorously and add contents of the reaction flask.

Step 17: Transfer the solution in the beaker to a large test tube.
- The diagram below shows an example of the reaction taking place in this step.

> **Why ammonium chloride?**
> The slightly acidic ammonium chloride solution is more effective in hydro-lyzing the magnesium alkoxide than water alone but keeps the acidity of the solution lower than the addition of excess HCl.

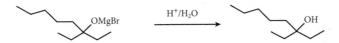

Step 18: Allow phases to separate and remove the lower *aqueous* phase.

- This phase will later be discarded. However, keep all extractions of the aqueous phase in a waste beaker at your work station. Never discard anything until the end.

Step 19: Wash the organic layer with 5mL of water and remove the lower aqueous phase.

Step 20: Wash the organic layer with 5mL of 10% sodium bicarbonate and remove the lower aqueous phase.

Step 21: Wash the organic layer with 5mL of saturated sodium chloride and remove the lower aqueous phase.

Step 22: Transfer the organic layer to a sample vial.

Step 23: Add magnesium sulfate to the vial to remove any water remaining.

- Cap the vial.
- Label it with your name and section number.
- Store it in your section's storage drawer until next week.

> **Why saturated sodium chloride?**
> The saturated salt solution wash helps remove the water from the organic layer. Some water is dissolved in the organic layer because of the presence of THF.

Week 1 Clean-Up Procedures
- Rinse vials that contained your starting alkyl halide and ketone with acetone in the acetone wash bottle in the sink.

- Return the vials to the rack where you obtained them.
- Dispose of plastic pipette by placing it in the labeled plastic beaker in one of the main hoods in the back of the lab.
- Dispose of used Pasteur pipettes in the broken-glass box after rinsing out any organics.
- Dispose of unreacted magnesium metal in your beaker in the solid waste container provided.
- Aqueous washes may be disposed of in the sink.

Week 2
Purification

Step 1: Decant the dry THF solution into a 25mL round-bottom flask with a few boiling chips.

Step 2: Set up the micro-scale distillation apparatus.

- All ground-glass joints should be greased. If not, notify your TA. Use a Keck clamp to secure all connections.
- Refer to the diagram on the right to set up the glassware.

Step 3: Distill THF with the steam bath.

- THF should distill around 55°C–65°C.
- Periodically remove the THF from the collection well of the distillation head with a Pasteur pipette.
- Dispose of the collected THF in the appropriate waste bottle in the main hood.
- After most of the THF has distilled (3–4mL remains in the round-bottom flask), the temperature will begin to fall.

Step 4: Remove distillation set-up from the steam bath and allow product to cool.

- You can speed up this process by placing the flask in a bath of *room-temperature* water, NOT an ice bath.

Step 5: Transfer the liquid in the flask to a conical reaction vial.

Step 6: Attach the vial to the micro-scale distillation head.

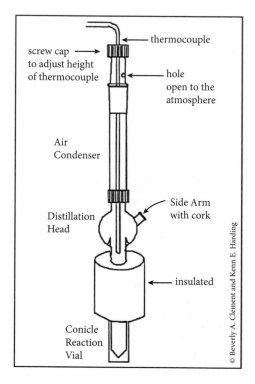

thermocouple

screw cap to adjust height of thermocouple

hole open to the atmosphere

Air Condenser

Distillation Head

Side Arm with cork

insulated

Conicle Reaction Vial

© Beverly A. Clement and Kenn E. Harding

Step 7: Distill the tertiary alcohol product using the heating mantle.

- Depending on your starting materials, your product may be expected to boil somewhere between 122°C and 215°C (see Table 1).
- Discard the initial distillate that collects in the collection well.
- Periodically transfer the product from the collection well of the distillation head with a Pasteur pipette into a pre-weighed, capped vial.
- At some point when you have already collected some product, and at a constant boiling point, collect a drop of the product from the collection well with a Pasteur pipette into a separate vial to later obtain an IR spectrum.
- Distill until most of the liquid has distilled or until temperature starts to drop.
- DO NOT distill until no liquid remains!

TABLE 1: Boiling Points for Possible Products				
Compound Name	b.p. (°C)		Compound Name	b.p. (°C)
2-methyl-2-pentanol	122		3-methyl-3-heptanol	163
2-methyl-2-hexanol	142		4-methyl-4-octanol	183
3-methyl-3-hexanol	142		3-ethyl-3-heptanol	184
3-ethyl-3-hexanol	160		3-methyl-3-octanol	185
4-methyl-4-heptanol	161		3-ethyl-3-octanol	199
2-methyl-2-heptanol	162		4-methyl-4-nonanol	212

Step 8: Turn off heating mantle and remove it from the distillation set-up.

Step 9: Obtain an IR spectrum of the product while allowing the glassware to cool.

- Identify your product by comparing your spectrum to the spectra of the products in Table 1.
- After identifying your product, determine your starting alkyl bromide and ketone.

Step 10: Weigh the vial to determine the yield.

- Calculate the percent yield of your product.

Caution!
DO NOT put the hot conical vial into water. Thermal shock will cause the vial to crack instantly.

Week 2 Clean-Up Procedures

- Rinse glassware with acetone in the acetone wash bottle in the sink.
- Dispose of THF in the labeled container in one of the main hoods at the back of the lab room.
- Dispose of the magnesium sulfate used to dry the extract in the appropriate waste container hood.
- Aqueous washes may be disposed of in the sink.

Discussion Items for the In-Lab Notebook

- Boiling point of alcohol product
- Identity of alcohol product (attach your IR spectrum)
- Theoretical yield
- Product yield
- Percent yield

Discussion Items for the Post-Lab Report

- Which starting alkyl halide and ketone would give the alcohol product you obtained?
- Write a Grignard synthesis scheme using different starting materials to obtain your same product.

CASE 10-2 Job Instructions

These procedures were written by Alicia Torres, an employee who had just completed several years as a runner/floater at an animal hospital. Before she left, she decided to write a set of instructions for the next person who would take her place. Job procedures have become increasingly important in many organizations. These help the organization in writing job descriptions, training new employees, and evaluating employee performance.

CASE DOCUMENT 10–2

How to Be a Runner at Kingsland Animal Clinic

Congratulations! You have the job! You will now need to learn how to perform your job. These instructions will tell you what you will need to do here at the clinic. Keep these instructions close by until you know all the tasks you will need to perform.

Your First Day

You will be training most of your first day. Read these instructions before you arrive, then observe, listen, and practice. Make notes on the instructions to help you remember what to do.

As you learn the job, you will see other ways that you can help. Always ask questions. The better all employees work as a team, the better service and care the clinic can provide. Every person employed at the clinic has an important job. The runner's tasks, as you will see, prevent gaps in service and care and improve efficiency.

What the Runner/Floater Does

As the runner/floater, you will experience all aspects of a working veterinary clinic: reception, technical, kennel cleaning, pharmacy, and any other job that the clinic does. If someone needs your help, respond immediately. The runner/floater position exists to make everyone's job easier and more efficient.

IMPORTANT: Realize that other employees do not always know when they need help, or they do not want to bother you by asking for it. Learn to know when someone needs you. Take initiative. If someone is having trouble, offer to help. Communicate openly with the receptionists, technicians, and vets to find where you are needed at any given time. Don't be afraid to ask questions about where and how you can help.

CAUTION: Since you will work with animals, remember that sick animals are often frightened. Remain alert and aware of your surroundings. Treat all animals with caution. Just because an owner tells you that her chihuahua is a "sweetie" doesn't mean that it won't try to bite you when you take its temperature.

Your Responsibilities—Discussed in These Instructions

- Knowing the Phone Room
- Helping the Receptionists
- Helping the Technicians
- Completing End-of-Day Duties
- Working on Other Tasks When You Have Time

The Phone Room

If you are assigned to the phone room, when one of the receptionists is at lunch, or if no one is in the phone room, **answer the phone**. A receptionist will teach you to work the phone.

- Greet callers with "This is [insert your name] at Kingsland Animal Clinic. How can I help you?"
- Always have a notepad ready to write down owner's name, animal's name, and any other important information.

Most common calls you will receive:

1. **Make an Appointment.** Use the Cornerstone program on the computer in the phone room to make an appointment. A receptionist will show you how to work Cornerstone. Practice using the program so that you don't make mistakes.
2. **Make a Boarding Reservation.** The boarding book should always remain in the phone room, on one of the desks. Write the owner's last name under the requested dates **in pencil**. If the reservation changes, you can easily erase.
3. **Ask about the Status of a Pet.** If the owner dropped off a pet for any reason, the owner may call to find out how the animal is doing. Put the owner on hold, run to the back, and ask the techs: "Does anyone know if [pet name] is ready to go?" or "[pet name]'s owner wants to know how he or she is doing. Does anyone know?" When you find out, run to the phone and relay the information.
4. **Ask about Prices.** The price book, like the boarding book, is located in the phone room. This book lists prices for common procedures, surgeries, details, etc. You may put the client on hold for a moment as you find the information.
5. **Ask a Pet Health Question.** Transfer the caller to the phone in the tech station. Put the caller on hold and first make sure a tech can take the call, then transfer the caller. If no tech is free, get the caller's information and

tell them a tech will call shortly. Pull the file, and put it in the Drop-Off box with a note attached. Describe question clearly.

6. **Refill Medication.** Simply transfer the caller to the pharmacy station. Check first to make sure someone is available in pharmacy.
7. **Other.** Many calls do not fit into these categories. If you do not know the answer, put the caller on hold and find someone who does know the answer. Any employee will be glad to answer your questions.

NOTE: Make sure to "roll the phones over" when the clinic closes so that no more calls come in.

Helping the Receptionists

The receptionists should never leave the front desk vacant. Your job: make sure they do not leave unless someone else takes their places. You must constantly run between the reception area and the back of the clinic to relay information from the receptionists to the vets and to retrieve animals. This requirement describes the "runner" part of your job. You will keep the receptionists from leaving by doing the following tasks for them:

- When a client drops off a pet, take the pet into the back.

  1. Make a cage card for the animal. Include name, date, and reason for the visit.
  2. Write the animal's name on one of our temporary collars.
  3. Put these in your pocket.
  4. Walk to the front and gently take the animal from the owner. Leave the dog's collar with the owner. CAUTION: Be gentle in dealing with an unknown animal.
  5. Take dogs to the kennel in the back. Take cats to the cat room. Take exotics to the exotics room.
  6. Put the temporary collar on the animal, and tape the cage card on the cage.
  7. Take the animal's folder, which the receptionist will have given you, and place it in the Drop-Off box.

- When a pet is ready to go home, bring it up to its owner.

  1. Ask the receptionist which animal needs to go home.
  2. Ask the owner for leash and collar or carrier.
  3. Find the pet.
  4. Take it to its owner.

- When a client arrives for an appointment, the receptionist will prepare the file and hand it to you. Take it to the back and place it in the appropriate appointment slot. Tell the techs that Dr. [vet's name]'s appointment has arrived or that a walk-in has arrived.

- If a receptionist needs to leave the front for some reason, ask if you can help so that he or she doesn't have to leave. You can help when bathroom breaks are needed.

Helping the Technicians

If the phone room is quiet and the receptionists seem to have things under control, you can head to the treatment area. Here you will find the techs, who can almost always use a hand. How you can help:

1. Restraining an animal. Usually two people are needed to deal with an animal. If you see a tech trying to take an animal's temperature or trim nails with no help, offer to hold the animal. Special techniques are needed to restrain the animal for a blood draw: the tech will be glad to show you.
2. Taking temperatures. You can help hold the animal, particularly the head if the animal decides to bite.
3. Trimming nails. If you do not know how to trim nails by looking for "the quick," ask a tech to teach you.
4. Weighing an animal.
5. Drawing blood. If you do not know how and would like to learn, the techs will teach you. This process requires two people.
6. Fecal tests. A tech will teach you how to attain a fecal sample and set up the fecal test.
7. Giving vaccines. Once you learn how to use needles, the techs and vets will let you draw and administer some vaccines, such as *Bordetella* and leptospirosis vaccines. You can also learn to give Adequan shots.
8. Pharmacy work. Once you have worked at the clinic for several months, you will be allowed to fill prescriptions. The tech will print a label, which will tell you how many and what kinds of pills to place in the pill bottles. Apply the label to the bottle and initial it.
9. Anything else the tech asks you to do. NOTE: You will not be asked or allowed to perform any task for which you have not been trained. Take every opportunity to learn.

End-of-the-Day Duties

End-of-day clean-up duties depend on how busy the clinic has been that day. Whenever business begins to slow down toward closing time, you can begin working on the following tasks:

- **Take out the trash.** Try to get someone to help you carry the bag to the dumpster.

1. Every trash can in the clinic must be emptied at the end of EVERY DAY.
2. Retrieve one large black trash bag from under the sink in the cat room.

3. Begin dumping the contents of every small trash can into the large bag. You do not need to change the trash bags in the smaller trash cans unless they are particularly dirty.
4. Make sure you get every trash can—the ICU, employee lounge, and every treatment room. The cans in the cat room and the kennel will be cleaned by the kennel kids.
5. Take the now full trash bag out to the dumpster.
6. Replace any of the small bags that you had to throw away.

- **Vacuum**

  1. The vacuum is located in the freezer room, off the exotics room.
  2. Insert one end of the vacuum into the various outlets in the hallway.
  3. Begin vacuuming the entire clinic (aside from the cat room and kennel).

Other Tasks to Complete When You Have Time

When you have time, such as a lull in activity or no appointments scheduled, other tasks must be completed.

1. Filing. The receptionists will show and help you with filing.
2. Restock the boxes of flea/tick/heartworm prevention on display in the reception area. (New boxes are stored in the stockroom. One of the vets must unlock that room for you.)
3. Restock the shelves of shampoos, ear cleaners, collars, ointments, etc. Most of these items are stored in the pharmacy area.
4. Clean the lobby. If at any point during the day the lobby looks dirty, sweep it. The broom is stored in the filing hall.
5. Clean up any messes the techs have made in the back (blood, nail trimmings, animal hair, etc.) by spraying the treatment tables with Trifectant, found in spray bottles.
6. Clean thermometers with alcohol swabs from the metal tins on the treatment tables.
7. Rinse the fecal loops and cups in the sink and place them in the Trifectant bath along with the urine collection trays.
8. Clean any dirty cages in ICU, isolation, or the cat room. For each cage:

   - Remove the mat from the cage and take it to the nearest bathtub.
   - Rinse off the mat and spray it with Trifectant. Leave it while you complete the next tasks.
   - Remove the newspaper from the cage and throw it away.
   - Spray the cage down with Trifectant.
   - Wipe the Trifectant with paper towels.
   - Line the cage with fresh newspaper found in every room of the clinic.
   - Go back to the mat—rinse it, dry it, and put it back in the cage on top of the newspaper.

9. Make sure all the animals have water, unless they will have surgery that day, and that all cats have litter boxes.

10. Check to see if the kennel kids need help cleaning cages, feeding, or walking dogs in the kennel.

Final Words—If you don't understand, always ask!

| CASE 10-3 | **Instructional Letter** |

Jonathan Varner, president of a professional organization, writes the incoming vice president to explain how elections for new officers should be conducted. The letter can be used as a reference by Dr. Dawson as she plans and executes the elections.

CASE DOCUMENT 10–3

May 12, 2017

Dr. Gabrielle Dawson
Department of Geography
Long Beach State University
Long Beach, CA

Dear Gabrielle,

As we discussed at AAGG in April, the VP of AAGG is responsible for conducting elections. From my experience in conducting the last election, I recommend that you begin this process immediately. The more persistent you are, the more efficient the process. Do as much as you can before faculty leave campus for the summer.

Timetable

In general, the election process begins during the late spring–early summer. Ballots are developed and mailed by mid-September. Ballots should be returned and tabulated by mid-January and the results reported to the vice president. As vice president, you will inform me and then inform each person whose name appeared on the ballot about the results of the election.

Election Procedures

The following procedures should help you conduct the elections. Ultimately, when the AAGG procedures manual is complete, these will appear in the manual. Until that time, however, each president will be responsible for informing the vice president of election procedures.

Materials Needed

- Election File—contains ballots of previous elections and names of nominating committees

- AAGG Directory—names, phone numbers, and e-mail addresses of all members
- Annual Meeting Agendas—protocols for announcing newly elected officers

Step 1: Creating a Nominating Committee

1. At the annual meeting in April, the president asks the general membership to consider individuals who should be asked to run for the offices of vice president and member-at-large. Members are invited to send the names of individuals to the president.
2. The president also organizes a nominating committee. At AAGG, I asked three individuals—Steve Jones, Pam Quintana, and Casey Morgan—to serve as the nominating committee. All agreed to serve. Casey Morgan agreed to chair the committee.

Note: Former officers of the organization are good choices for the nominating committee. They know what is involved in executing the duties of each office. I attempted to select a nominating committee of individuals from different regions in the United States. With that point in mind, you could select four members of the committee, but more than four makes the committee unwieldy.

3. The nominating committee must recommend a minimum of six nominees for three member-at-large positions. Eight nominees will be fine, too. Three vice presidential nominees should be chosen. The chair of the nominating committee should ask each candidate for a one-paragraph (150-word) biographical profile.

Step 2: Selecting Nominees for Offices

1. In selecting candidates, you will want to consider several factors:

   - Interest in AAGG and its Activities. Look for people who have shown interest in the organization: attendance at regional and national meetings, people who engage in conversation on the listserv, people who make solid suggestions.
   - Interest in Geography and/or Geoscience. AAGG needs individuals who are effective spokespersons for both as fields of teaching and research, i.e., individuals who are effective and prolific researchers; people who are in charge of programs; people who are regular presenters at meetings.
   - Record of Dependability. VERY IMPORTANT! Try to determine how reliable an individual is. Stay in touch with the nominating committee. Once you get a list of individuals you are considering, please post the list on the Executive Committee listserv. If any of the potential nominees has a track record of nonperformance, this is the time to determine that! Many people are "all talk" but slow on performance.

Note: Choose the VP very carefully. This is the person who will succeed you as president of AAGG. Because the president and vice president have so much responsibility and because they will need to work as a team, be sure you choose someone with whom you can work and someone who will continue to strengthen the organization. AAGG has very strong individuals in its thirty-year history. Each group of officers must be committed to developing AAGG further.

Step 3: Developing the Ballot

1. Be sure to let Casey Morgan know when you want the final list of names. I would suggest that you ask for the list by the second week in September.
2. Contact each individual whose name was submitted by the nominating committee. Be sure that each person has agreed to be nominated and understands what is involved.
3. Develop the ballot and a letter that will be sent along with the ballot.
4. Send the ballot and the letter to Dan Jameson. He will make copies. Try to get this material sent to Dan no later than the third week in September.

Step 4: Monitoring the Election Process

1. Once he has the materials, Dan will duplicate and mail the ballots to the membership. He knows the procedure: each ballot has a self-addressed, stamped return envelope, to ensure better response.
2. Dan will want all ballots returned by early January. He will count the ballots and report the results to you. As soon as you have the results, please call—rather than e-mail—each candidate and report the results.

Step 5: Preparing for the Annual Meeting

1. Plan to introduce each of the officers at the meeting. After my farewell speech, I will hand the reins of the organization to you.
2. To help prepare for the meeting, discuss your agenda as president with the incoming VP. I would also suggest that you plan an informal gathering of all the new officers BEFORE the Executive Committee meeting. At the EC, as you are aware, we always have several hours of business to conduct. Meeting with the incoming officers will give you planning time with them.
3. Be prepared to tell the general membership at the annual meeting your goals for the next two years. The newly elected VP and members-at large should be allotted time to express their concerns. Usually we allot fifteen minutes for this segment of the annual meeting.

Final Thoughts

The election process has been developed to attempt to involve a variety of individuals. The nominating committee should seek members who want to become more involved in AAGG. In no sense does the election process seek to exclude individuals other than those who are not interested in geography and/or geoscience and the development of AAGG as an organization.

As you work with the elections, keep notes on ways that the EC can improve the election process. Then add these to the agenda for the annual meeting.

As always, if you have any questions, please call me.

Sincerely,

Jonathan

Online Instructions

Online instructions need to follow the same readability considerations as those used in letters, reports, and stand-alone instructions. Otherwise, the instructions will create confusion for many users.

CASE 10-4

A university explained how to download class rosters from the faculty e-mail system and insert these into an Excel spreadsheet (see Case Document 10–4A). Note how the application of principles discussed in this chapter improves the instructions (see Case Document 10–4B). The revision diminished the number of calls to the computing help desk from faculty who were confused by the original instructions.

CASE DOCUMENT 10–4A

Original Online Instructions

How exactly do I get my roster from NEO, put it into an Excel spreadsheet, and make a comma separated value (.csv) file for uploading?

1. Log into your neo account, go to the **Class Roster** tool, and go to the appropriate semester. Click on **Download Class Roster**, and you'll see your class roster as a .csv file. Copy and paste this file into a blank Excel spreadsheet and save. There's no way right now to actually download it.

2. To get rid of the quotes, go to **Edit**, then **Replace**. You want to find every "and replace it with a blank, so in the **Find what** box type "and put nothing into the **Replace with** box. Click on **Replace All**, and the quotes disappear.

3. Under **File**, select **Save As** option, and save the file as **Formatted Text (space delimited) (.prn)**. Choose **OK** and **Yes** to the questions that pop up. Close the file, and answer **no** to the popup box. You don't want to save the changes.

4. Open the .prn file. The **Text Import Wizard** will pop up. Leave the file type as **delimited** and click **next**. Change the delimiter from **tab** to **comma** and click **next**. Holding the ctrl key down, click on the last four columns, and choose **Do not import column (skip)**, then choose **Finish**.

5. Under **File**, select the **Save As** option. Change the file name to something that does not include the .prn, and save the file as **CSV (Comma delimited) (.csv)**. Always answer **yes** to the question, "Do you want to keep the workbook in this format?" and when you close the file, answer **no** to "Do you want to save the changes?" Voila! This is the file the administrator needs.

6. To upload your students into your course, you only need the UIN numbers as a .csv file. So just delete the name columns and any rows above your data and save. Always answer **yes** to the question, "Do you want to keep the workbook in this format?" and when you close the file, answer **no** to "Do you want to save the changes?" Upload your file according to the directions in CPR. **[top]**

CASE DOCUMENT 10–4B

Revised Online Instructions

How to Download Your Roster

1. Click on **download class roster**. Note that your class roster now appears as a comma separated value (.csv) file.
2. Without closing the roster on Neo, create an Excel spreadsheet.
3. Go back to your roster. Copy the names.
4. Paste the roster into the Excel spreadsheet just as you copied it.

Get Rid of the Quotes (" ")

1. Go to **Edit** on the toolbar, then to **Replace**. You want to find every " " and replace it with a blank.
2. In the **Find what** box, type " ", and put nothing into the **Replace with** box.
3. Click on **Replace All**, and the quotes disappear.

Prepare to Format the Information in the Excel Spreadsheet

1. Under **File**, select **Save As** option, and save the file as **Formatted Text (space delimited) (.prn)**.
2. Choose **OK** and then **Yes** to the questions that pop up.
3. Close the file and answer **no** to the popup box.
4. Open the **.prn** file. The **Text Import Wizard** will pop up.

Select the Correct Boxes on the Wizard

1. Leave the file type as **delimited**. Then click **next**.
2. Change the delimiter from **tab** to **comma**. Then click **next**. You will see the opening rows of your roster appear in neat columns.
3. Under **File**, select the **Save As** option.
4. Save the file as an Excel file. Be sure to save with a name descriptive of the contents.

With your file in an Excel spreadsheet, you can now create a grade or attendance sheet.

Testing Your Instructions

Because users of products and services typically resist reading instructions, their interpretations of your instructions could easily surprise you. Test your instructions on a group of representative users to determine the accuracy and consistency with which users decipher the intended meaning and take the appropriate action. The more complicated the instructions or the more dangerous the product or service, the more important is this usability testing. Your objective is to make sure that users will be able to operate the product or service as safely, correctly, and efficiently as possible.

To conduct the usability test, you would give your representative users a copy of your instructions and access to the product in question and ask each user to complete a designated activity (e.g., for a word processing application: insert a table, do a grammar check, change autocorrect settings). You would observe each user as he or she reads your instructions and tries to operate the program. You would assess the time it takes to complete each activity, the points of confusion, the time to recover from confusion, and the strategies adopted to recover from confusion.

You could thereafter interview the representative users, either individually or jointly, about their experience with your instructions: What was helpful? What caused confusion? What would make the instructions easier to use? How effective were the illustrations? How could the wording be simplified? How evident were the warnings and cautions?

The results of this usability test will guide you as you revise your instructions to create a more productive and satisfying user experience.

✎ Checklist For Developing Instructions/Procedures

Because instructions require careful analysis of readers from many perspectives—the context in which the instructions will be read and used—following a checklist can be helpful in ensuring that you have considered critical issues needed for the instructions.

☐ Who are your readers? Describe them in terms of their knowledge of the subject: educational level, technical level, responsibilities in the organization.

☐ What do you want them to be able to do as a result of these instructions/procedures?

☐ What is the situation that has led to the need for these instructions/procedures to be written?

☐ How will these instructions/procedures be used? Will readers need to read all of them before they begin the task? In what context will readers be using these instructions?

☐ What problems could readers encounter in attempting to use these instructions/procedures?

☐ What types of problems in safety and/or quality control do you need to emphasize? What warnings or notes will you need to include?

☐ What topics do you want to be sure to include/exclude?

☐ What format will you use: online, online to be printed, paper, manual, poster, video?

☐ What types of illustrations will you need to include?

☐ What is the basic outline of your instructions? Does this outline meet the needs of your readers? Will it achieve your purpose?

☐ How will you make your instructions accessible for all of your likely readers?

☐ How will you test the usability of your instructions?

 EXERCISES

1. You are employed by a small manufacturer of all-organic ready-to-eat breakfast cereals. The company would like to develop its sales overseas, especially in China. Because ready-to-eat cereals would be relatively new to the majority of the population in China, your package will have to include instructions on how to prepare and eat a serving of cereal. Compose step-by-step instructions, including rough illustrations, to fit on the side of the package.

2. Find a set of instructions you have tried to use with limited success: for example, instructions for posting to a blog or wiki, for checking the security settings on your Internet browser, for operating a new mobile device, or for assembling the several components of a home entertainment system. Develop a slide presentation for your class explaining the deficiencies in the original instructions and the revisions you perceive as necessary.

3. Working individually or collaboratively, revise the document on portable heater safety habits shown in Figure 10–2. Start by conducting a brief usability test on the original document with at least five potential users. Which of the seven safety habits are most easily understood and remembered? Which are least understood and remembered? After you have reviewed your test results, ask yourself two questions:

   • What changes, if any, must you make to the text?
   • What changes, if any, must you make to the illustrations?

4. Working individually or collaboratively, adapt the original or your revised version of Figure 10–2 as a brief video of instructions about portable heater safety. Be sure to include closed captioning and to create a transcript of your video.

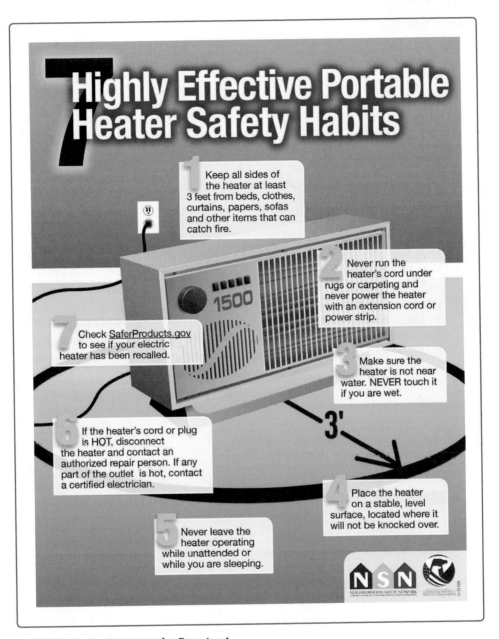

• **FIGURE 10–2** Document for Exercise 4

11

Oral Reports

Your ability to give oral reports can be as critical to your success as your ability to write reports. In today's work environment, knowing how to speak effectively and use presentation slides has become an increasingly valuable asset, whether you work in an academic or a nonacademic environment. In applying for a job or admission to a graduate program, for example, you may need to give an oral presentation on a specific topic or on a research project you have conducted. As an employee, you may need to convert a written document into an oral presentation or vice versa. Whether preparing a live presentation, a video, a screencast, or a podcast, you will benefit from knowing and then practicing the basic strategies of developing and delivering effective oral reports.

 Quick Tips

- In the introduction, interest the audience in the subject, and number the key points you will cover (e.g., "I will cover the following five points.").

- As you discuss your subject, help your audience pay attention and track your progress in your presentation by numbering each point as you cover it (e.g., "Now let's proceed to my second point.").

- When you reach the conclusion, announce it. You want your audience to pay extra attention. Audiences know that you will now summarize what you've said, and they can catch anything they missed earlier. Once you've said "in conclusion," you have about a minute before you must stop talking. Audiences grow hostile to speakers who promise to stop but then don't. In the remaining minute of your presentation, fix in the audience's mind the one or two ideas you want them to take away from your presentation.

- Keep your presentation as short as possible. Nobody wants to listen to a long presentation.

Understanding the Speaking–Writing Relationship

Effective oral and written presentations share the following requirements:

- Analyze your audience. Know what they will expect.
- Articulate your purpose clearly.
- Develop sufficient and appropriate supporting material.
- Understand the context in which your presentation will be received.
- Organize the material so that your audience can follow your points easily.
- Choose a style—level of language and formality—that is suitable to your role as well as your audience and purpose.
- Select the presentation format—text, text and visuals, text and sound and/or animations—that will enhance your audience's understanding of your message.

Analyzing the Audience

Analyzing your listening audience follows the same principles discussed in Chapter 2:

- How much does my audience know about the subject?
- How much do they know about me?
- What do they expect from me?
- How much interest will they have in what I say?
- What is their attitude toward me?
- What is their attitude toward my subject?
- What is their age group?
- What positions do they occupy in the organization?
- What is their educational background?
- What is their cultural/ethnic background?
- What is their economic background?
- What are their political and religious views?
- What kinds of biases will they likely have toward me and my topic?

To be an effective speaker, you must know your audience, establish an effective relationship with them by being sincere and knowledgeable about the subject, and conform to their expectations about appearance, demeanor, choice of language, and attitude toward them and the topic. If you are speaking to people from other countries, you should research the culture of that country as their expectations might differ from yours. If your audience includes individuals with limited hearing or vision, you will need to make appropriate accommodations in your presentation (e.g., closed captioning of videos, oral descriptions of photographs).

If you will be visible to your audience in a live or video presentation, be aware that hand gestures you use routinely may have different meanings in other cultures. Also, the clothing you choose to wear should be selected with the culture of the audience in mind: you want your audience to pay attention to what you are saying instead of what you are wearing.

Determining the Goal of Your Presentation

Oral presentations, like written reports, must be designed for a specific purpose. If you alert audiences at the beginning of your presentation to your reason for speaking, they will better understand their reason for listening. State your goal in one sentence. Announce your purpose early in the presentation to prepare your audience for the key ideas to come.

Keep in mind that oral presentations can enhance your reputation within an organization. Therefore, consider every speaking opportunity as a chance to sell not only your ideas but also your competence and your value to the organization.

Choosing and Shaping Content

Preparing the oral presentation often requires careful research:

- Determine what information you will need.
- Choose information that will appeal to your audience—particularly their attitudes and perspectives toward the topic.
- Consider a variety of information types: statistics, testimony, cases, illustrations, history, and particularly narratives that help convey the goal you have for your presentation. Be sure that every item you include pertains to the goal of your presentation.

Analyzing the Context

Analysis of context and analysis of audience are often inseparable:

- What broader concerns underlie the need for the presentation?
- What immediate issues underlie the presentation?
- How does your presentation relate to these issues?
- What will be happening in the organization when you make your presentation?
- How does your presentation fit into the organizational situation?
- If you are one of several speakers, what kinds of presentations will other speakers make?
- In what surroundings will you make the presentation?

Choosing the Organization

Like written documents, oral presentations must address your audience's needs and perspective:

- Does your audience have interest in what you will say?
- What key ideas do you want to convey?
- Based on your purpose and the needs and expectations of the audience, in what order should you present these ideas?

Answers to these questions will help you decide how to go about organizing your presentation. Generally, oral presentations have distinct parts—an introduction, a body, and a conclusion. The introduction should clearly tell the audience what the presentation will cover so that the audience is prepared for what is to come. The body should develop each point stated in the introduction, and the conclusion should reiterate the ideas presented and reinforce the purpose of the presentation.

As you do your research and collect information for your presentation, begin categorizing and organizing your ideas:

    I. Idea 1
    II. Idea 2
    III. Idea 3, etc.

Once you have identified the major divisions, start to subdivide each idea with its supporting explanations and evidence. Order the subdivisions so that the information proceeds in a logical sequence (e.g., most important to least important, similarities and differences, chronological order):

    I. Idea 1
       A, B, C . . .
    II. Idea 2
       A, B, C . . .
    III. Idea 3
       A, B, C . . .

Introduction. Be sure that you state your goal near the beginning. Even if you use some type of opening statement or anecdote to engage your audience, state the goal of your presentation next. Then state how you will proceed in your presentation—the key issues you will discuss. The main ideas should be announced and numbered here.

Body. Here you must explain each of your key ideas in order. Also try to include at least one example or anecdote for each point to make your ideas more vivid and memorable. The more you make your presentation easy for your listeners to follow and understand, the more they (and you) will profit from your presentation.

Conclusion. At a minimum, restate the key issues you want your audience to remember, but be concise. Nothing is worse than a conclusion that drags on and doesn't really conclude. Tell your audience exactly what you want them to remember from your presentation. Try to find a concluding narrative or statement that will have a lasting impact on your audience. The conclusion should leave the audience with a positive feeling about you and your ideas.

Choosing an Appropriate Speaking Style

How do you sound when you speak? You may have brilliant ideas and accurate and supporting statistics, but how you speak can make or break the effectiveness of your presentation. Avoid sounding condescending, arrogant, disrespectful, overexcited, glib, angry, or insincere. Use a conversational style with short sentences and plain language. Avoid speaking too rapidly. Remember that your organizational role and your relationship to your audience will dictate the degree of formality of your presentation.

- Do they know you?
- Is your rank in the organization above or below them?
- Are you speaking to an audience of individuals from all levels within the organization?
- What behaviors and interactions does the organization usually expect from those giving oral presentations?
- Is the audience composed of people who understand American English? How well do they understand American English?

If you are speaking before a group composed largely of people from other countries, determine beforehand their fluency in American English. If they are not comfortable with American English, speak more slowly, avoid idiomatic expressions (e.g., it was a piece of cake, it cost an arm and a leg), choose simple words, use short sentences, and limit each sentence to one idea.

Designing the Slides to Enhance Your Purpose and Your Meaning

Illustrations will make your ideas more persuasive and more professional. And the guidelines for using visuals in oral presentations mirror those for written documents: everything must exemplify simplicity and clarity and fit the needs of the audience.

Formal presentations should use PowerPoint or similar presentation software to help listeners follow your ideas. PowerPoint allows you to give your listeners the outline of your presentation and to insert supporting pictures, graphs, tables, drawings, photographs, diagrams, and flow charts as well as sound and video. But because these materials will be seen while the audience is listening to you, you must make sure that all visuals are as straightforward and as easy to read as possible. The slides in Figure 11–1 exemplify the following guidelines:

- Limit the information on any single slide.
- Choose a standard type font that is familiar and easy to read in various sizes (e.g., Helvetica, Arial, Times, Garamond). Skip the unconventional options as these could distract from your information and prove a challenge to read.
- Limit the type fonts you use to two per slide (e.g., Helvetica for headings and Times for text)
- Use consistent sizes of type (e.g., 36 points for headings, 24 points for subheadings, 20 points for text)
- Avoid using all capital letters.
- Use a background that does not overpower the text or illustrations. Dark type on a plain light background is usually the easiest to read.
- Don't crowd the information on your slide: blank space will make your slides easier to read and your ideas easier to understand.

● **FIGURE 11–1** Slides from a Presentation Prepared for the Public by the US Fire Administration

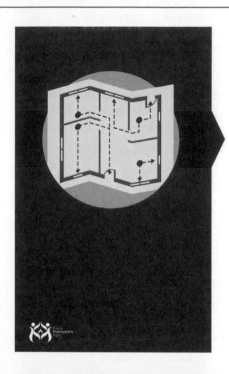

Make an escape plan.

- Know 2 ways out of every room.
- Have a meeting place outside your home.
- Know how to call 9-1-1 from outside to report a fire.
- Practice your escape plan with everyone who lives in your home at least twice a year.

Cooking is the number one cause of home fires.

- Stay in the kitchen when you are frying, grilling, broiling, or boiling food.
- If you leave the kitchen, turn the burner off.
- Keep things that can burn away from your cooking area.
- Turn pot handles toward the back of the stove so they won't get bumped.

• **FIGURE 11–1** *Continued*

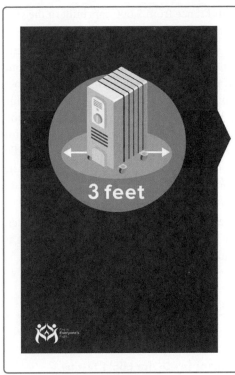

Heating is the second leading cause of home fires.

- Keep anything that can burn at least three feet away from fireplaces, wood stoves, portable heaters, and radiators.
- When you leave a room or go to bed, turn heaters off or unplug them.

• **FIGURE 11–1** *Continued*

Other issues to consider as you design your slide presentation include:

- With the exception of the title slide, avoid illustrations that are decorative instead of genuinely informative. While a picture related to the subject might be appropriate on the title slide to create interest before you start your presentation, illustrations during the presentation must contribute essential information or reinforce your message. With decorative illustrations, you risk distracting your audience instead of focusing their attention on your key points.

- Avoid slides with complicated illustrations, as in Figures 11–2 and 11–3. Viewers won't have the time or the ability to study tables and figures in detail. The message of your illustrations should be immediately evident. Choose illustrations (e.g., line graphs, bar graphs) that will be familiar to your audience and keep the design simple and consistent (see Figures 11–4 and 11–5)

- Do not read your slides to your audience (unless you have listeners with limited or no vision). Let the audience read each slide as you make it available, then begin to comment. Don't look at your slides as you speak. Look at your audience.

- If you have listeners with limited or no vision, be sure to identify and briefly describe each illustration (e.g., "Here is a line graph that displays the rapid growth of the automobile industry in France from 5,000 cars built in the year 1900 to 3,350,000 in the year 2000.")

- If your presentation will include a video or audio clip, keep it brief—to only about 10% of the entire presentation. You want the clip to support your presentation but not to dominate it, and you want your audience to remember all of your key points instead of only the information covered in the clip. Prepare your audience for the clip by introducing it with a brief description and noting about how long it is. Make sure the clip is closed-captioned for the hearing impaired. Also practice your presentation to verify that the audio or video clip (and the closed captions) will operate as you anticipate. If the technology fails, summarize the information covered in the clip and offer to make a copy of the clip available following the presentation.

- Avoid overwhelming your audience with a rapid fire of innumerable slides. You will keep your viewers attentive if your keep your presentation as short as possible, using only two or three slides per minute.

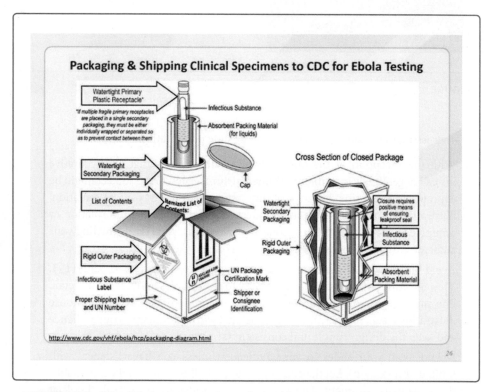

- **FIGURE 11–2**   This slide from a presentation for healthcare workers about the Ebola virus includes too much information for viewers to skim through and make sense of quickly. This level of descriptive detail is better suited to a handout that your audience could later give more time and attention.

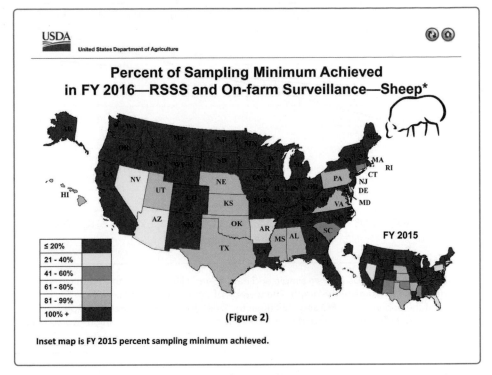

• FIGURE 11–3   This illustration from a slide presentation asks viewers to do a quick deciphering of a lot of information. Viewers must examine the legend on the left and the smaller map on the right and interpret their relationship to the larger map in the center. This would be a more effective illustration in a report, where adjacent paragraphs could make the meaning and implications of the illustration easier for readers to recognize.

- If your presentation has no specified time limit, estimate for your audience about how long you will speak and number your slides (e.g., 7 of 25, 8 of 25) to alert the audience to your progress through the presentation.

- If you are unfamiliar to your audience, do not miss the opportunity of this presentation to introduce yourself. Make sure your name is on the title slide and in a footer on every slide. And make sure your contact information is displayed on the final slide during the question-and-answer period.

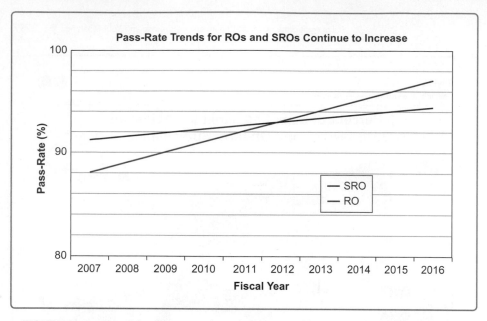

• **FIGURE 11–4**   This slide offers a simple and easy-to-read line graph with a heading that states clearly the message of the graph. The viewers of this presentation likely know the meaning of the abbreviations RO and SRO (Reactor Operator and Senior Reactor Operator), but labeling each data line with the job title instead of using a legend of abbreviations would make the graph more striking and efficient.

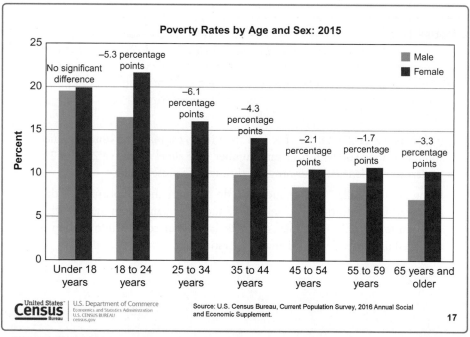

• **FIGURE 11–5**   This slide displays a series of paired column graphs, each clearly labeled, in left-to-right order by chronological age. (Note that bar graphs are more appropriate for reporting numerical information at the same point in time.) The graph nicely avoids the stereotypical coloring of pink for females and blue for males. The meaning of this illustration is easy to decipher but might be more readily evident if the figure title summarized the key message: Poverty Rates Were Significantly Higher for Adult Women in Every Age Group in 2015. Certainly, the speaker would make this point, but seeing it as well as hearing it would reinforce this message for the audience.

Planning Your Presentation—Questions You Need to Ask

Determine the right approach to your presentation by answering the following questions:

Audience.

- Who is my audience?
- What do I know about my audience—background, knowledge, position in the organization, attitudes toward me and my subject, abilities/disabilities?

Purpose.

- What is my purpose in giving this oral presentation?
- Is there (should there be) a long-range purpose?
- What is the situation that led to this presentation?
- Given my audience's background and attitudes, do I need to reshape my purpose to make my presentation more acceptable to my audience?

Context.

- What kind of presentation will I give: live and in person, live by video, live screencast, recorded video, recorded screencast, podcast (audio only)?
- What do I know about the location where (or from which) I will give my presentation? How will this location influence my presentation?
- What events have occurred or may transpire in the organization (theirs or mine) that may affect how my audience perceives my presentation?

Content.

- What ideas do I want to include and not include?
- Based on the audience and the context, what difficulties do I need to anticipate in choosing what to discuss?
- Can any ideas be misconstrued and prove harmful to me or my organization?

Illustrations.

- What kinds of illustrations will I need for the ideas I will present?
- Where should I use these illustrations in my presentation?

Style.

- What image—of myself and my organization—do I want to project?
- What level of language do I need to use, based on my audience's background and knowledge of my subject?

- What approach will my audience expect from me?
- How formal or informal should I be in addressing my audience?

Speaking to Multicultural Audiences

As organizations become more international, you may find that you need to give presentations to groups in other countries. Because you will want any audience to respond positively to your presentation, you will need to know how people from other cultures are likely to interpret the words you choose and your manner of speaking as well as how you interact with your listeners. If you are visible to this audience, your appearance itself will be a factor to consider. The illustrations you use may also have to be changed as symbols in one culture may have an entirely different meaning in another culture.

In your library and online, you will find a list of resources and books that can help you understand the perspectives of international audiences. As you consider your audience and the ideas you want to present to this audience, remember that your understanding of the cultural profile of your listeners is every bit as important as your correctly discerning their knowledge of your topic and their interest level.

Designing Each Segment

In reading, if you find a sentence or paragraph difficult to understand, you can stop and review the passage as many times as necessary. In listening to live presentations, however, rarely are you able to interrupt the speaker and ask him or her to repeat something you did not understand. You must hope that either something said later will clarify the confusing point or that you will be allowed to ask questions after the presentation. Even if a presentation is recorded, rewinding the recording and listening again to a confusing passage can be awkward and time consuming.

In designing your presentation, therefore, keep in mind the listening limitations of your audience. Listening is challenging, and audiences will tire even if a presentation is thoroughly interesting. For that reason, look for ways to keep your message as concise as possible. Don't omit information your audience needs, but look for ways to eliminate nonessential material.

Choose an interesting title. Grab your audience's attention by developing a title that, at the very least, reflects the topic of your presentation but does so in an interesting way. The title of an oral presentation should prepare your audience for the ideas you will present.

Develop your presentation around three main divisions. Helping your audience follow your message requires that you build into your structure a certain amount of redundancy. You must repeat your main points. In the introduction, you "tell them what you are going to tell them"; in the body, you "tell them"; and in the conclusion, you "tell them what you told them." This kind of deliberate repetition helps your audience follow and remember the main points you are making. To design your presentation with planned repetition, you must know your purpose precisely and decide exactly what you want your audience to know.

Focus the introduction. During the introduction focus your audience's attention on your central claim and the way you plan to present this claim. Unless the introduction is effective and interests the audience, you will have difficulty keeping your audience's attention. The effective introduction thus tells your audience how to listen, what to expect, and the path you will follow in presenting your message. You may also wish to introduce your topic with an attention-getting device: a startling fact, a relevant anecdote, a rhetorical question, or a statement designed to arouse your audience's interest. Again, the device you choose will depend on the audience, the occasion, and the purpose of the presentation.

Or if your audience is unreceptive or unfamiliar with the subject, you may want to include background material to help them grasp and process your main points:

1. Acknowledge that you perceive the problem that your audience has with you or your topic.
2. Establish a common ground with the audience—your points of agreement.
3. Attempt to refute (if you can do so efficiently) any erroneous assumptions that you believe the audience may have about you or the subject.
4. Ask the audience to allow you the opportunity to present your information as objectively as possible.

Organize the body. In the introduction state the main issues or topics you plan to present. Thus, in designing the body of the presentation, develop what you want to say about each of these main points or ideas. You may want to present your ideas in a logical sequence (e.g., first to last, top to bottom, cause and effect, before and after). This method will help your audience follow your ideas if you are giving an informative speech, an analytical speech, or a persuasive speech. Always demarcate and announce each point in the body as you come to it so that your audience knows when you have completed one point and begun another (e.g., "After drug and alcohol addiction, the next most important factor is . . ." or "Moving down from the cornice of the building, let us now examine . . .").

Fortify the conclusion. The conclusion reinforces the main ideas you wish your audience to retain. How you design the conclusion will depend on your initial purpose. A strong conclusion is nearly as important as a strong introduction because both the beginning and the ending will be the parts that audiences are most likely to remember.

Choosing an Effective Delivery Style

Avoid speaking in a "written" style of long sentences riddled with innumerable phrases and clauses. If you concentrate on getting your point across by having a conversation with the audience, you will likely use a natural, conversational style. Many suggestions for clarity in writing also apply to clarity in speaking:

- Avoid long, complicated sentences. Long sentences are often more challenging to decipher by listening than they are by reading.
- Avoid abstract, polysyllabic words. Instead, use concrete language that your audience can visualize.
- Avoid jargon, unless you are sure that your audience will be familiar with the specialized vocabulary.
- Use active-voice sentences with easy-to-identify subjects and verbs.

Techniques to Enhance Audience Comprehension

Look for ways to help your audience follow your ideas.

1. **Identify sections:** Be sure you clearly demarcate the beginning and end of each point and section of your presentation:
   - Announce each main topic as you come to it so that your audience knows when you have completed one topic and are beginning the next one.
   - Allow a slight pause to occur after you have completed your introduction, and then announce your first topic.
   - After completing your final topic in the body of your presentation, allow a slight pause before you begin your conclusion.
2. **Enunciate:** Speak slowly, vigorously, and enthusiastically. Be sure you pronounce your words carefully, particularly if you are addressing a large group or a multicultural group.
3. **Practice:** Rehearse your presentation until you feel comfortable. Try walking around, speaking each segment, and then speaking aloud the entire presentation. Rephrase ideas that are difficult for you to say—these will likely be hard for your audience to follow. Be sure to time your presentation so that it does not exceed the time limit. Keep your presentation as short as possible.
4. **Record:** If possible, record or videotape your speech. Listen to what you have said as objectively as possible. As you listen, consider the main issues of audience, purpose, organization, context, content, and style. Listen for tone, attitude, and clarity. Is the tone you project appropriate for your audience and your purpose? Is each sentence easy to understand? Are you speaking too rapidly? Are the major divisions in your presentation easy to hear? Are any sentences difficult to understand?

5. **Keep it conversational:** Do not memorize your presentation. You don't want it to sound like something you are reciting to your listeners. You want it to seem like your turn in a conversation you are having with your listeners.

6. **Use gestures and eye contact:** If you are visible to your audience, use natural gestures to emphasize your points. Move your body deliberately to aid you in announcing major transition points. In short, avoid standing or sitting transfixed before your audience. Also maintain eye contact with your audience. Doing so helps you keep your listeners involved in what you are saying. If you look at the ceiling, the floor, or the corners of the room, your audience may sense a lack of self-confidence or credibility. By looking at your audience you can often sense their reaction to what you are saying and adjust your presentation if necessary.

7. **Save handouts for the end:** Try not to provide the audience handout material before you begin. Doing so encourages your audience to read rather than listen. If you must provide written material, be sure the material is coordinated with your presentation so that you keep your audience's attention on what you are saying as you are saying it.

8. **Explain your slides:** If you use slides, tell the audience what they will see, show them the slide, give them time to digest what they are seeing, and then comment on the slide. Do not begin talking about another topic while a slide depicting a past topic is still showing.

9. **Encourage questions:** Keep in mind that the more interactive your presentation, the more engaged your audience will be. When you are planning your presentation, determine how you will handle questions. Are you willing to be interrupted by questions while you are speaking? Or must the audience save their questions till you finish speaking? Will you allow your audience to ask questions or make comments through social media while you are speaking? Will you address their online messages during or following your presentation?

10. **Answer questions concisely:** Always prepare for questions your audience might ask and decide how you will answer each one. Again, unless you have analyzed your audience and the reason for your presentation, you will not be able to anticipate questions that will likely arise.

If the question-and-answer period is live, keep it moving briskly. Answer each question as concisely as possible, and then move to the next question. If you are asked a difficult or complicated question, reword it or divide it into several parts and answer each part. Be sure to restate questions if people in the back of the room are unable to hear questions asked by those near the front. Also be sure to read aloud the questions or comments submitted through social media for those members of the audience without access. If you can't answer a question then and there, offer to find the answer and follow up with a text or e-mail message to the questioner. If a question is hostile or highly specific, kindly offer to discuss it with the questioner following the presentation. Remember to thank the audience for their questions as well as their time and attention.

If you are answering questions through social media, keep your answers brief and appreciative. If necessary, direct the questioners to pertinent resources or offer to explain in more detail through e-mail.

Designing and Presenting the Scripted Presentation

Papers presented at professional meetings are frequently read from written manuscripts if the material to be delivered is complex. These papers may then be published in the official proceedings of the professional society. However, presentations may also need to be read for other reasons:

1. A presentation that discusses company policy, a sensitive issue, or a topic that must be approved by someone in the organization before the presentation. In situations like these, the presentation is carefully written and read from the approved, written manuscript to ensure accuracy.

2. A presentation that will be circulated or filed as documentation. In a situation like this, a spokesperson may read a carefully prepared statement, particularly if a possibility exists that material may be misconstrued by those in the audience.

3. A presentation that will be the script for a video, screencast, or podcast.

4. Inexperienced speakers who must deal with a difficult problem may be more comfortable reading from a prepared manuscript. With the manuscript in front of you on the lectern, you don't have to worry about losing your train of thought or forgetting important details.

Scripted presentations can be effective if the speaker plans and writes the presentation carefully and then practices the oral reading enough to make it lively and engaging.

Structuring the scripted presentation. The structure of the scripted presentation is the same as that of the extemporaneous oral report. The presentation still has three main parts: the introduction, the body, and the conclusion. Each section should be structured like the extemporaneous presentation. However, you will need to write each section completely. If you know that the script will be published, you may wish to write it like an article for publication or a report and use headings and subheadings to reveal the content and organization of the presentation.

Writing the script. After you have designed the content of your paper and made final revisions in your ideas, you will need to give close attention to your sentences and paragraphs since you will be reading these directly from the page.

1. Be sure that you separate each section from the other sections. This means that each section should have an overview that clearly announces that the section is beginning. Each paragraph should also begin with a topic sentence that summarizes the content of the paragraph. In short, in writing a script to be

read, you are making a concerted effort to accentuate every device for revealing organization since your audience cannot stop and rehear what you have just said.

2. Limit each section and each paragraph within sections to one idea. Watch length so that your audience will not lose track of the key idea you are presenting.

3. Avoid excessive detail.

4. Use numbering to help your audience follow your key points and to know when one point has ended and the next point is beginning.

5. Avoid long sentences. Long sentences are more difficult to understand when listening than they are when reading.

6. Prune every sentence of unnecessary words to make it as clear and concise as possible.

7. Use active voice so that your sentences will preserve the natural quality of spoken language.

8. Type your presentation in a large type—12-point type or larger. Triple space and leave wide margins on each side of the page.

9. With a marker, draw a "break" line after the introduction, between each main point in the body, and before the conclusion. Underline or highlight important phrases or sentences throughout the presentation.

10. Consider using illustrations that will clarify or emphasize your key points.

Practicing the presentation. You will need to practice your scripted presentation until it no longer sounds like you are just reading words but instead are delivering ideas in a lively conversational style.

1. Read each sentence aloud. Rewrite sentences that are difficult for you to finish in a single breath or words that are difficult for you to pronounce.

2. As you practice reading the presentation, also practice looking directly at your audience and speaking important phrases or sentences to your audience.

3. Use overviews and topic sentences to announce each major topic as you come to it. To further alert your audience to the beginning of a new point, pause briefly, look at your audience, then read your overview statement or topic sentence. If possible, try to speak these to your audience instead of reading them.

4. As you practice reading your presentation, continue to listen for any sentences or words that might cause problems. Revise sentences and paragraphs that do not sound organized, logical, and clear.

5. As you read, speak slowly and enunciate clearly and distinctly.

6. Once you can read each sentence with ease and without haste, time your presentation to be sure that you do not exceed your allotted time. Audiences grow hostile to speakers who ignore time limits.

7. Read your speech into a recorder. Leave time between recording and listening so that you can gain some objectivity. As you listen, check once more for sentences that are hard to follow. Listen for the clear breaks between major sections and major points.

🗒 Checklist For Preparing Oral Reports

Audience

☐ Who is your audience?

☐ What do you know about your audience—background, knowledge, position in the organization, attitudes toward you and your subject? What is the relationship between you and your audience?

☐ What is the attitude of your audience toward you and your presentation likely to be?

☐ Is your audience from a culture markedly different from yours? What adjustments to your presentation will any such differences require?

☐ Does your audience include individuals with limited hearing or vision? What accommodations will you need to make for these individuals?

Purpose

☐ What is your purpose in giving this oral presentation? Is there (should there be) a long-range purpose?

☐ What is the situation that led to this presentation?

☐ Given your audience's background and attitudes, do you need to adjust your purpose to make your presentation more suitable for your audience?

Context

☐ Where and when will you be speaking?

☐ What events will be transpiring in the organization (theirs or yours) that may affect how your audience perceives what you say?

☐ What equipment, applications, and materials are available to you?
  - Slideware?
  - Audio?
  - Video?
  - Internet connection?
  - Chalkboard?
  - Flipchart and easel?
  - Handouts?

Content

☐ What ideas do you want to include and exclude?

☐ Based on the audience and the context, what difficulties do you need to anticipate in choosing content?

☐ Can any ideas be misconstrued and prove harmful to you or your organization?

☐ Do you have a good opening that will interest your audience and create a friendly atmosphere?

☐ Have you limited your major points to fit within your allotted time?

☐ Does your talk contain sufficient examples, analogies, narratives, and data to support your generalizations? Have you repeated key points? Can you relate your subject matter to some vital interest of your audience?

Illustrations

☐ What kinds of illustrations will you need to clarify or reinforce the ideas you will discuss?

☐ Where should you use these illustrations in your presentation?

☐ Are these illustrations immediately readable and understandable?

☐ Do these illustrations successfully focus the listeners' attention and augment and clarify your message?

Style

☐ What kind of tone do you want to use in addressing your audience? What kind of image—of yourself and your organization—do you want to project?

☐ What level of language do you need to use, based on your audience's background and knowledge of your subject?

☐ What approach will your audience expect from you? How formal should you be?

☐ Which delivery technique will be more appropriate? Extemporaneous? Manuscript? If you are speaking extemporaneously, have you prepared a speech outline to guide you? If you will be reading from a manuscript, have you introduced a conversational tone into your talk? Is your typed manuscript easy to read from?

☐ Do you have a good ending ready, perhaps a summary of key points or an anecdote that supports your purpose? If you began with a story, do you want to go back to it now?

☐ Have you rehearsed your talk several times?

Questions and Answers

☐ What questions might the audience ask? Are you prepared to answer these questions?

☐ How will you keep your answers concise and direct?

☐ Do you have a specific time limit for questions? How will you keep track of your time?

☐ Will your audience have access to social media to ask questions and make comments during your presentation?

EXERCISES

1. Examine Figure 11–6, a script for a presentation about ethanol that a college agricultural engineering student will give to a high school Future Farmers of America class. The speaker will read and glance at the script as she does not feel confident in speaking to this group. She will also give the FFA teacher a copy of the script. Examine the script in terms of the advice given above for delivering a presentation that will be read. How could this speaker improve the style of the script? What handouts could the speaker prepare to accompany the presentation?

2. Prepare a five-minute/ten-slide presentation explaining to new majors in your field the job opportunities available to graduates of the program.

3. Prepare a slide presentation of your major semester project, including a copy of your slides to distribute to your audience as a pdf file for their mobile devices.

4. Your friend at the CDC (Centers for Disease Control and Prevention) is preparing a new version of a slide presentation on the flu vaccine for a meeting of the Advisory Committee on Immunization Practices. He shares with you two slides from the previous version (see Figure 11–7) and asks your opinion of their design. What advice about each of the slides would you offer your friend?

Ethanol in Fuel: Benefits and Problems

Introduction

Good morning! Mr. Miller has asked me to speak to you this afternoon about the advantages and disadvantages of using gasoline that contains ethanol in engines of farm equipment. I will talk for about five minutes and then I'll be happy to answer any questions you might have.

The abundance of corn and the price of gasoline have made ethanol seem like the obvious solution to the energy crisis. However, when we look deeper into the issue, we find that ethanol has many flaws that possibly lower its value as a fuel alternative. Although ethanol does have benefits, the negative effects, from my perspective, outweigh the benefits.

• **FIGURE 11–6** Script for Exercise 1

Ethanol at a Glance—The Main Facts

- Gasoline that contains ethanol is named for the percentage of ethanol in the mixture (i.e., E10 contains 10% ethanol and E15 contains 15% ethanol).
- Fermented plants such as corn and grain produce ethanol.
- Ethanol oxygenates gasoline, which causes the gasoline to combust cleaner and more completely.
- Those of you who live on farms which have corn as the main crop see ethanol as a financial windfall.

Keep in mind that there are two sides to the ethanol story.

Benefits of Ethanol

First, Ethanol burns cleaner than gasoline. As a result, ethanol added to gasoline reduces harmful exhaust emissions. **Second**, ethanol supplements gasoline, which is often imported. Thus, ethanol advocates see it as a way to reduce dependence on foreign oil. **Third**, advocates see ethanol as a renewable resource, created from plants. Thus, many people see it as a solution to the energy crisis. After ethanol is created, the remains of the corn are used for animal feed. Ethanol can be recycled into the agriculture industry. **Fourth**, agriculture benefits from ethanol production because of the stability that the need for grains brings to the industry. Over the past few years, the price of corn has increased. This increase has a strong correlation to the increased use of ethanol in gasoline.

However, ethanol also creates problems:

Ethanol's Problems

Reduced Fuel Efficiency

Yes, ethanol can be used as an alternative to gasoline, but the efficiency of ethanol as compared to gasoline is lower. Studies have shown that E10 and E15 produce 3–5% less mileage than the equivalent volume of pure gasoline.

Corrosive Properties of Ethanol

Ethanol is a solvent, which means that it can wear away plastics, rubbers, and other materials. This becomes a problem when the worn parts are inside an engine. These parts must be replaced regularly to keep the engine running at its full potential. Repairs can cost hundreds, even thousands, of dollars. The residue from the dissolved parts builds up on filters and valves and further reduces engine efficiency.

• **FIGURE 11–6** *Continues*

Seals, hoses, and even fuel tanks can weaken, causing interruptions in the fuel flow and, in worst-case scenarios, leaks of the highly flammable fuel.

Tendency to Attract Water

Alcohols have a tendency to attract and retain water, and ethanol is no exception. When you store an ethanol–gasoline mix for long periods of time, the ethanol will attract all of the water in the tank. As a result, water will cause the mixture to separate. If the separated mixture is then used in the engine, the engine will not run properly, and damage could occur.

So the price of engine repair also has to be considered when you figure the cost of ethanol.

Conclusion

I believe this information should raise concerns about the use of ethanol in mechanized farm equipment. Although the increased use of corn, to distill ethanol, has provided growth and stability in the agriculture industry, the harmful effects of ethanol could outweigh its positive effects. While ethanol reduces the release of greenhouse gases into the air, the loss of fuel efficiency requires more fuel to be burned. The result: release of more greenhouse gases. This factor counteracts any positive effects of ethanol.

Agriculture equipment companies are looking at ways to slow damage of ethanol to engine parts. E10 causes the lowest amount of damage, but many states have already shifted to E20, which has proven extremely corrosive.

Okay, who has questions?

• **FIGURE 11–6** *Continued*

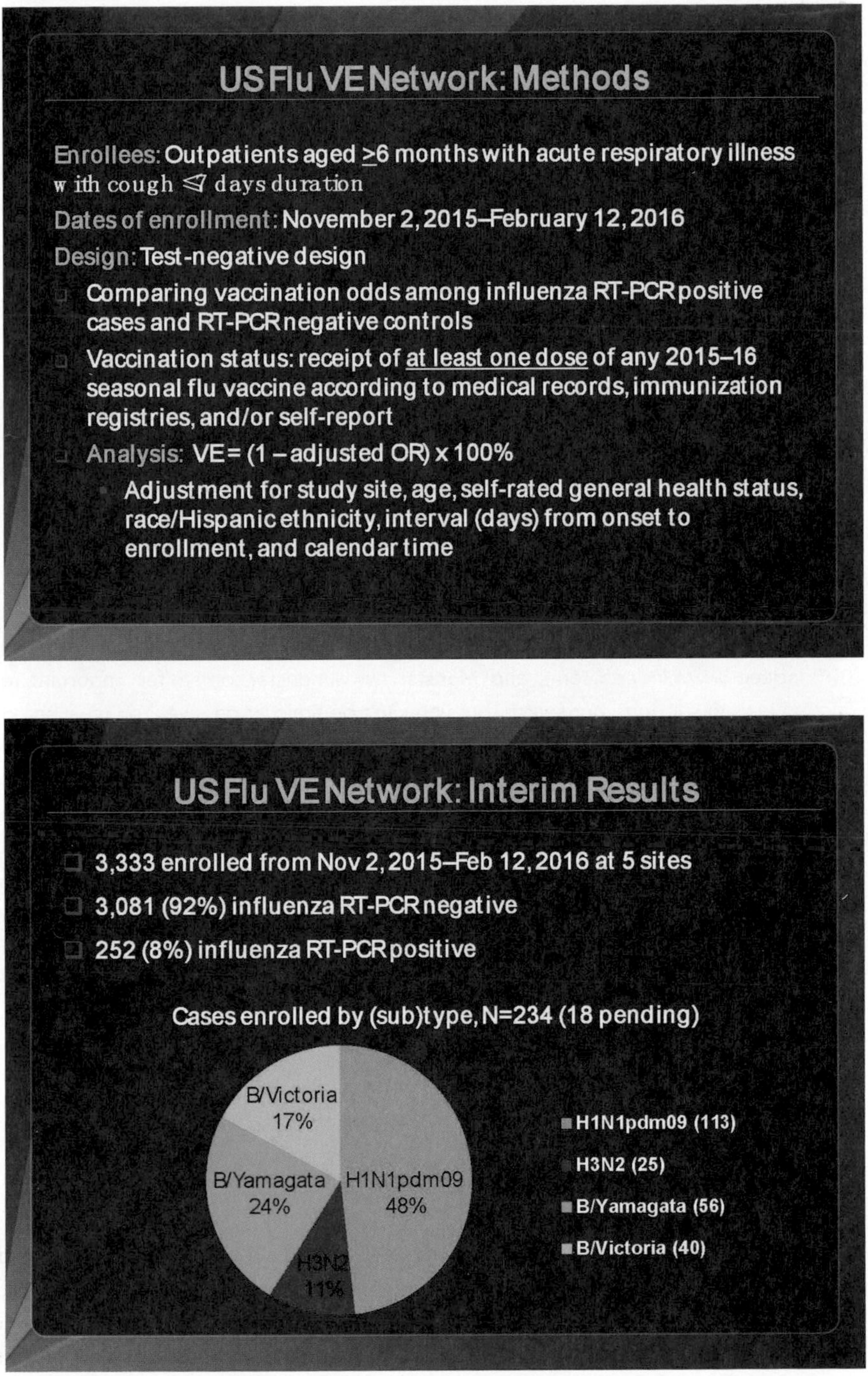

US Flu VE Network: Methods

Enrollees: Outpatients aged ≥6 months with acute respiratory illness with cough ≤7 days duration

Dates of enrollment: November 2, 2015–February 12, 2016

Design: Test-negative design

- Comparing vaccination odds among influenza RT-PCR positive cases and RT-PCR negative controls
- Vaccination status: receipt of <u>at least one dose</u> of any 2015–16 seasonal flu vaccine according to medical records, immunization registries, and/or self-report
- Analysis: VE = (1 – adjusted OR) x 100%
  - Adjustment for study site, age, self-rated general health status, race/Hispanic ethnicity, interval (days) from onset to enrollment, and calendar time

US Flu VE Network: Interim Results

- 3,333 enrolled from Nov 2, 2015–Feb 12, 2016 at 5 sites
- 3,081 (92%) influenza RT-PCR negative
- 252 (8%) influenza RT-PCR positive

Cases enrolled by (sub)type, N=234 (18 pending)

B/Victoria 17%
B/Yamagata 24%
H1N1pdm09 48%
H3N2 11%

- H1N1pdm09 (113)
- H3N2 (25)
- B/Yamagata (56)
- B/Victoria (40)

• **FIGURE 11–7** Slides for Exercise 4

12

Résumés and
Job Applications

 Quick Tips

Finding a job is itself a job. Sites such as CareerBuilder (www.careerbuilder.com), Indeed (www.indeed.com), and Monster (www.monster.com) offer important resources to support your search, but you can't be timid or passive. You must search actively and assertively for opportunities.

You will likely experience both excitement and disappointment in your search. It will be exciting to imagine yourself working for different companies and living your life in different cities. It will also be disappointing if you don't immediately get your dream job. The search process is often a roller coaster of emotions—highs and lows.

Keep in mind that a variety of factors entirely aside from your credentials are involved in every hiring decision. If you aren't offered a job for which you applied, don't take it personally and don't be discouraged. The key to finding a satisfying job is perseverance.

The Correspondence of the Job Search

A persuasive letter of application (or *cover letter*) and a clear and organized résumé won't guarantee that you get a job, but it will make it more likely. Also essential is that you make sure that nothing in your online profiles or postings at social media sites such as Facebook, LinkedIn, and Twitter will raise questions in a potential employer's mind about your suitability for a job. Verify especially that details mentioned in your letter of application or listed on your résumé are identical to the information in your online profiles (e.g., names of previous employers, dates of previous employment). Keep in mind that potential employers will almost certainly check your online profiles and postings because it is quick and easy to do so: how you characterize yourself there won't necessarily win you a job but could stop you from getting a job or from keeping a job.

Letter of application. Plan the mechanics of your letter of application carefully. Use a standard font like Arial or Times, and don't use italics or bold: you want the employer to focus on your abilities instead of typographical variations in your letter. Make sure you have no errors in spelling, punctuation, or grammar: you must come across as competent and meticulous. Be brief but not telegraphic. Keep the letter to a single page unless you have extensive pertinent experience to emphasize and explain.

More and more organizations direct job candidates to apply online for positions. If this is the case, you will go online to the organization's job site, enter your name and contact information, identify which job you are applying for, and upload your letter and résumé in the specified format. If no format is specified, submit your materials in one of the widely used formats (e.g., as a doc file, docx file, pdf file, or rtf file) so that it might be easily accessed by the recipient. Make the file read-only so that nothing will be later inserted or deleted.

If you are applying for jobs in the traditional manner, however, make sure to buy high-quality white bond paper and envelopes. This is no time to skimp. Don't send a letter or résumé that has been duplicated. Each copy of your letter must be individually addressed, printed, and signed. Accompany each letter with your résumé.

Pay attention to the style of the letter and the résumé that accompanies it. The tone you want in your letter is one of self-confidence. You must avoid both arrogance and timidity. You must sound interested and eager but not servile or desperate. Don't give the impression that you must have the job, but don't seem indifferent about getting it.

When describing your accomplishments in the letter and résumé, use action verbs. They make your writing concise and specific, and they make you seem dynamic and energetic. For example, instead of reporting that you worked as a salesclerk, explain that you maintained inventories, promoted merchandise, prepared displays, implemented new procedures, and supervised and trained new clerks. Here's a sampling of such action verbs:

administer	edit	oversee
analyze	evaluate	plan
conduct	exhibit	produce
create	expand	reduce
cut	improve	reorganize
design	manage	support
develop	operate	promote
direct	organize	write

You cannot avoid the use of *I* in a letter of application, but take the *you* attitude as much as you can. Emphasize what you can do for the prospective employer—how your getting this job will benefit the company. The letter of application is not the place to be worried about salary and benefits. Above all, be mature and dignified. Write a well-organized, informative letter that highlights the skills you have that the company desires in its employees.

The beginning. Start by explaining that you are applying for a job and by identifying the specific job for which you are applying. Avoid tricky and flashy approaches (e.g., "Are you looking for a sales clerk who will make your cash registers ring? I am your ringing job candidate.") If you can do so legitimately, a bit of name-dropping is a good beginning. Use this technique only if you have permission and if the name you drop will mean something to the prospective employer—for example:

> Dear Mr. García:
>
> Professor Theresa Ricco of Nebraska State University's Department of Biochemistry has suggested that I apply for a research position in the virology division of your company. In June, I will receive my bachelor of science degree in biochemistry from NSU. I have also worked as a research assistant in Dr. Ricco's lab for two years.

Remember that you are trying to arouse immediate interest about yourself in the potential employer.

Sometimes the best approach is a simple statement about the job you seek, accompanied by a description of something in your experience that makes you particularly fit for the job, as in this example:

> Your opening for a researcher in the virology division has come to my attention. In June of this year I will graduate from Nebraska State University with a bachelor of science degree in biochemistry. I have also worked part-time for two years in a cell and molecular research laboratory at NSU. I believe that both my education and my work experience qualify me for this position.

Be specific about the job you want. Quite often, if the job you want is not open, the employer may offer you an alternative one. But employers are not impressed with general statements such as, "I'm willing and able to work at any job you may have open in research, production, or sales." Instead of indicating flexibility, such a claim usually implies that your skills and interests are unfocused—that you would be adequate at several things but truly exceptional at nothing.

In addition, make it clear in this opening that you know something about the company—its history, its achievements, its special projects, its reputation in the industry—anything that will demonstrate that you did your research and picked this company to apply to because something about it was impressive, something about it made you think you could make a key contribution to its success.

The body. In the body of your letter highlight selected items from your education and experience that show your qualifications for the job you seek. Remember always that you are trying to show the employer how nicely you will fit into the job and the organization.

For this section, you need to know what things employers value the most. In evaluating recent college graduates, employers look closely at the major, academic performance, work experience, awards and honors, and extracurricular activities. They also consider recommendations, standardized test scores, military experience, and community service.

Try to include information that employers typically consider important, but emphasize those areas in which you are especially noteworthy. For example, if your grades are good, mention them prominently; otherwise, focus on the more impressive areas of your academic record (e.g., courses in your major, internship experiences, research projects). Speak to the employer's interests, and at the same time highlight your own accomplishments. Show how it would be to the employer's advantage to hire you. The following paragraph does all these things and is an excellent example of the *you* attitude in action:

> The advertisement for this position indicated that extensive interaction with international colleagues, especially from South America, would be expected. I would like to note that I studied Spanish and Portuguese in college and have a high fluency in both languages. I also visited Argentina and Brazil two years ago for a six-week intensive summer session. I am familiar with the major cities as well as the history and key traditions. I would thus be a confident and articulate representative of the company.

Be specific about your accomplishments; otherwise, it will sound like bragging. It is much better to write "I was president of my senior class" instead of "I am a natural leader." "I was president" is a piece of evidence; "I am a natural leader" is a conclusion. Your aim is to give employers the evidence that will lead them to the right conclusions about you.

One tip about job experience: the best experience relates to the job you seek, but mention any job experience, even if it is unrelated. Employers believe that a student who has worked at a job will probably be more mature and reliable than one who has not: you know what's involved in having a job and what's expected of you. If you have worked at a job, you are more likely to possess effective work habits that will translate to your new position.

Don't forget hobbies or skills that might relate to the job (e.g., you are training for your pilot's license while applying for engineering positions in the aviation industry). You're trying to establish that you are interested in, as well as qualified for, the position.

Don't mention salary unless you're answering an advertisement that specifically asks you to discuss your expectations. Keep the *you* attitude. Don't worry about insurance benefits, vacations, and holidays at this point in the process. Keep the prospective employer's interests in the foreground. Your self-interest is taken for granted. If you are offered the job, that's the time to inquire about such details if they are not readily provided to you.

If you are already working, you will emphasize your work experience more than your college experience. Identify your responsibilities and achievements on the job. Never criticize your present employer because you will cause a prospective employer to mistrust you and think of you as inclined to complain about your job.

In the last paragraph of the body, refer the employer to your enclosed résumé. Mention your willingness to supply additional information such as references, letters of recommendation, writing samples, or college transcripts.

The ending. The goal of the letter of application is to be offered an interview with the prospective employer. In your final paragraph, you must request this interview. Your request should be neither apologetic nor aggressive. Simply indicate that you are available for an interview at the employer's convenience, and give any special instructions needed for reaching you. If the prospective employer is in a distant city, indicate (if possible) a convenient time and location that you might meet with a representative of the company, such as the upcoming convention of a professional association. If the employer is really interested, you may be invited to visit the company at its expense.

The complete letter. Figure 12–1 shows a complete letter of application. The beginning of the letter shows that the writer has been interested enough in the company to investigate it. The desired job is specifically mentioned. The middle portion highlights the writer's pertinent studies and work experience relating directly to the job she is seeking. The ending makes an interview convenient for the employer to arrange.

Keep in mind that a hiring official skims your letter and résumé in about 30 seconds. Only if you grab his or her interest in that brief time are you likely to be interviewed for the job.

The résumé. A résumé provides your prospective employer with a convenient summary of your education and experience. As in the letter of application, good grammar, correct spelling, neatness, and brevity—ideally, only one page—are of major importance in your résumé.

The two most widely used types are chronological and functional résumés. Each has advantages and disadvantages.

Chronological résumés. The advantages of a chronological résumé (Figure 12–2) are that it is traditional and acceptable. If your education and experience show a steady progression toward the career you seek, the chronological résumé portrays that progression well. Its disadvantages are that your special capabilities or accomplishments may sometimes get lost in the chronological detail and any holes in your employment or educational history show up clearly.

Put your address at the top. Include your e-mail address and your telephone number (including the area code). If you have a fax number, include it.

For most students, educational information should be placed before work experience. People with extensive work experience, however, may choose to put it before their educational information.

Letter of application

300 East 5th Street
Long Beach, CA 90802
January 23, 2018

Mr. Richard Hernandez
Office of Information Technology
Price Insurance Corporation
2163 Artesia Drive
Denver, CO 80201

Dear. Mr. Hernandez:

Identifies specific job and demonstrates knowledge of company

I am writing to apply for the position of computer security analyst listed in the March issue of *IT Professionals*. Price Insurance has a growing reputation for innovative security solutions, and I believe my education and experience would contribute to your efforts.

Highlights education

I am graduating in May of this year from the University of California at Long Beach with a bachelor's degree in Computer Science. I have completed courses in software verification and validation, fault-tolerant computer systems, web-based software systems, analysis of algorithms, communication networks, database management systems, and human-computer interaction.

Highlights work experience

In addition, I have a year of experience as a programmer and database manager at a local internet service provider. I know how to take a project from application of principles through risk analysis, resource management, and adaptation of improvement techniques, to delivery and billing. I have also worked in computer support, developing both my technical skills and my ability to interact courteously and effectively with customers.

Refers to résumé

You will find more detailed information about my education and work experience in the résumé enclosed with this letter. I could also supply descriptions of the courses I have taken, copies of the projects completed in those courses, and a complete list of my job duties at each place of employment.

Requests interview

In April, I will attend the regional meeting of the Society for IT Professionals in St. Louis. Would it be possible for me to visit with a representative of Price Insurance at that time?

Sincerely yours,

Gillian Woods

Gillian Woods

Enclosure

● **FIGURE 12–1** Complete Letter of Application

Lawrence T. Morgan
ltmorgan@tamu.edu

Local Address: 1221 Locust Street, College Station, TX 77840 (713-515-8601)
Permanent Address: 2305 18th Street, Lubbock, TX, 79401 (806-543-6039)

EDUCATION

Texas A&M University, College Station TX
Bachelor of Science Degree
Double Major: Biomedical Science and Entomology
Graduated: December 2017, GPR: 3.3/4.0

EXPERIENCE

History of Medicine Study Abroad, Germany, Summer 2017
- Shadowed and Observed a triple bypass surgery, North Rhine Westfalia Heart and Diabetes Center
- Visited German veterinary cloning facilities in Bayer Healthcare Target Research Facilities, Chemical Research Facilities, and fermentation facilities
- Participated in research on "Nazi Medicine"

St. Joseph's Hospital, Bryan TX, Day Surgery Tower Volunteer, August 2014–present
- Assisted patients and their families while at the hospital.
- Aided doctors, nurses, and staff in various duties.

International Student Volunteers (ISV) Conservation Project, Summer 2016
- Traveled to Australia to work with a team on trail building in national forests.
- Joined in planting over 1,000 trees, sages, and grasses, and removing invasive species from devastated swamp lands.

ACTIVITIES

Sigma Nu National Fraternity
- Executive Council
  ○ Candidate Marshall–Fall 2016
  ○ Lieutenant Commander (Vice President) Spring 2015–Spring 2016
  ○ Sentinel and Risk Reduction Fall 2014–Spring 2015
- Head of Philanthropy Committee 2015-2016, Recruitment Committee, Academic Chair–Spring 2015, Social Chair – Fall 2014, House Manager–Spring 2014, Candidate Class Social Chair–Fall 2013

Biomedical Science Association (BSA) Fall 2013 – present
Texas A&M Pre-Med Society Fall 2015

AWARDS

Sigma Nu Fraternity
- Commander's Award, Fall 2016–Spring 2017
- Volunteer of the Year, Fall, 2016–Spring 2017
- Scholar of the Year, Fall 2015–Spring 2016
- Big Brother of the Year, Fall 2014–Spring 2015

Study Abroad Fellowship & International Education Scholarship
National Scholar's Honor Society–inducted Fall 2016

January 2018

• **FIGURE 12–2** Chronological Résumé

List the colleges or universities you have attended in reverse chronological order. Do not list your high school. Give your major and date (or expected date) of graduation. Do not list courses, but list anything that is out of the ordinary, such as honors, special projects, and emphases in addition to the major. Extracurricular activities also go here if they are pertinent to the job you seek.

As you did with your educational experience, put your work experience in re verse chronological order. To save space and to avoid the repetition of *I* through-out the résumé, use phrases rather than complete sentences. As in the letter of application, emphasize the experiences that show you in the best light for the job you seek. Use nouns and active verbs in your descriptions. You may include less important jobs (such as from high school), but summarize these briefly unless they are especially pertinent to the position you seek. You would probably put college internships and work-study programs here, though you might choose to put them under education. If you have military experience, put it here. Give the highest rank you held, list service schools you attended, and describe your duties. Make a special effort to show how your military experience relates to the civilian work you seek.

You may wish to provide personal information. Personal information can be a subtle way to point out your desirable qualities and could build rapport with a prospective employer during interviews if he or she shares your interests. Recent travels indicate a broadening of knowledge and probably a willingness to travel. Hobbies listed may relate to the work sought. Participation in sports, drama, or community activities indicates that you enjoy working with people and are civic-minded. Cultural activities indicate you are a person of wide interests. Avoid mentioning activities that could be considered controversial or dangerous. If you are proficient in a language aside from English, definitely mention this: more and more organizations are looking for employees with the potential to support inter-national business activity.

If you choose to indicate that you are married or have a family, you might also emphasize that you are willing to relocate. Don't discuss your health unless it is to describe it as excellent.

You have a choice with references. You can list several references with ad-dresses and phone numbers or simply put in a line like "References available upon request." Both methods have an advantage and a disadvantage. If you pro-vide references, a potential employer can contact them immediately, but you use up precious space that might be better used for more information about yourself. Conversely, if you don't provide the reference information, you save the space but put an additional step between potential employers and informa-tion they may need in order to make a decision. (If one of your references is a prestigious individual in the discipline or industry, be sure to list your references on the résumé.)

In any case, do have at least three references available. Choose from among your college teachers and past employers—people who know you well and will say positive things about you. Get their permission, of course. Also, it's a smart idea to send them a copy of your résumé. If you can't call on them personally, send them a letter that requests permission to use them as a reference, reminds them of their association with you, and sets a time for their reply:

Dear Dr. Singh:

In May of this year, I'll graduate from Waterville Polytechnic Institute with a BS in metallurgical engineering. I'm getting ready to look for a full-time job. If you believe that you know enough about my abilities to give me a good recommendation, I would like your permission to use you as a reference.

As a reminder, during the summers of 2012 and 2013, I worked as a laboratory technician in your testing facility at Waterville. They were highly instructive summers for me and reinforced my college studies.

I want to start sending my résumé out to some potential employers by March 1 and would be grateful for your reply by that time. I've enclosed a copy of my résumé so that you can see in detail what I've been doing.

Thanks for all your help in the past.

Sincerely,

At the bottom of the résumé, insert a dateline—the month and year in which you completed the résumé.

Functional résumés. A main advantage of the functional résumé (Figure 12–3) is that it allows you to highlight the experiences and abilities that show you to your greatest advantage. Extracurricular experiences show up particularly well in a functional résumé. The major disadvantage of this format is the difficulty, for the first-time reader, of deciphering the progression of your jobs and education.

The address portion of the functional résumé is the same as that of the chronological résumé. After the address, you may include a job objective line if you like. A job objective entry specifies the kind of work you want to do and sometimes the industry or service area in which you want to do it, like this:

Work in computer security.

or like this:

Work in computer security for a medical insurance corporation.

Place the job objective entry immediately after the address, and align it with the rest of the entries (as shown in Figure 12–3).

For education, simply give the school from which you received your degree, your major, and your date of graduation. The body of the résumé is essentially a classification. You sort your experiences—educational, business, extracurricular—into categories that reveal capabilities related to the jobs you seek. Remember that in addition to professional skills, employers want good communication skills and good interpersonal skills. Possible categories are technical, professional, team building, communication, research, sales, production, administration, and consulting.

Lawrence T. Morgan

ltmorgan@tamu.edu

Local Address: 1221 Locust Street, College Station, TX 77840 (713-515-8601)
Permanent Address: 2305 18th Street, Lubbock, TX 79401 (806-543-6039)

EDUCATION	**Texas A&M University, College Station TX**
	Bachelor of Science Degree
	Double Major: Biomedical Science and Entomology
	Graduated: December 2017, GPR: 3.3/4.0

LABORATORY QUALIFICATIONS

Collegiate Science Coursework

- General Chemistry w/ lab I & II
- General Biology w/ lab I & II
- Physics w/ lab I & II
- Organic Chemistry w/ lab I & II
- Biochemistry
- Microbiology w/ lab
- Basic knowledge of diseases

Collegiate Entomology

- Identifying Species of Insects
- Collection of Insects
- Forensic Entomology
- Medical Entomology w/ lab
- Veterinary Entomology w/ lab
- General Entomology w/ lab
- Insect Morphology w/ lab

Other Skills

- Microsoft Office – Word, Excel, PowerPoint
- Adobe – Photoshop, Premiere, Illustrator

EXPERIENCE

The History of Medicine Study Abroad, Germany, Summer 2017

- Shadowed and observed a triple bypass surgery, North Rhine Westfalia Heart and Diabetes Center

• **FIGURE 12–3** Functional Résumé

Lawrence T. Morgan 2

- Visited German veterinary cloning facilities in Bayer Healthcare Target Research Facilities, Chemical Research Facilities, and fermentation facilities
- Participated in research on "Nazi Medicine"

St. Joseph's Hospital, Bryan TX, Day Surgery Tower Volunteer, August 2014 – present

- Assisted patients and their families while at the hospital.
- Aided doctors, nurses, and staff in various duties.

International Student Volunteers (ISV) Conservation Project, Summer 2016

- Traveled to Australia to work with a team on trail building in national forests.
- Joined in planting over 1,000 trees, sages, and grasses, and removing invasive species from devastated swamp lands.

ACTIVITIES
- Sigma Nu National Fraternity
- Coastal Conservation Association
- Biomedical Science Association (BSA) Fall 2013 – present
- Texas A&M Pre-Med Society Fall 2015

AWARDS

Sigma Nu Fraternity

- Commander's Award, Fall 2016 – Spring 2017. The chapter president recognizes one exceptional individual within the chapter.
- Volunteer of the Year, Fall 2016 – Spring 2017. The Fraternity recognizes the individual who has demonstrated excellence in the area of Leadership.
- Scholar of the Year, Fall 20125– Spring 2016. The Fraternity recognizes the individual who has demonstrated excellence in the area of scholarship.
- Big Brother of the Year, Fall 2014– Spring 2015. The Fraternity recognizes an exceptional Big Brother in the chapter.

Study Abroad Fellowship and International Education Scholarship

National Scholar's Honor Society – inducted Fall 2016

References Furnished Upon Request

January 2018

• **FIGURE 12–3** *Continued*

To prepare a functional résumé, start by brainstorming. List some categories that you think might nicely describe your experiences. Brainstorm further by listing your experiences in each of those categories. Next, select the categories and experiences that make you a particularly compelling job candidate. Remember, you don't have to display everything you've ever done, just those things that might strike a potential employer as valuable. Finish the functional résumé with a brief reverse chronological work history and insert a dateline, as in the chronological résumé.

For example, Lawrence's chronological résumé (Figure 12–2) is impressive, but he realized it would not help him in applying for a research assistant position in entomology that had been advertised on campus. He developed a functional résumé (Figure 12–3) and submitted it to the entomology professor who needed the research assistant. The functional résumé emphasized what Lawrence knew about entomology and research, which was exactly what the professor wanted, and Lawrence got the job.

Ordinarily, you will prepare your résumé as a traditional formatted document, but you might instead be asked to fill in fields at a job site online that will automatically generate your résumé according to a standardized template.

Traditional formatted résumés. As Figures 12–2 and 12–3 illustrate, in a traditional résumé you use variations in type and spacing to emphasize and organize information. Make the résumé easy to read—leave generous margins and white space. Use distinctive headings and subheadings. You might use a 12-point type like Arial for headings and a 10-point type like Times for the text. Be sparing, however, with the typographical variation: you want to organize the page of information visually without pulling the reader's attention away from your abilities. Ordinarily, résumés use a single-column format with headings above the pertinent information for each category: this single-column format accommodates screen-reading devices for people with limited vision as well as scanning to a database and processing for keywords.

If you are printing your résumé, use a high-quality bond paper in white or off-white. Your letter and résumé should be on matching paper. If you are attaching the résumé as an electronic file to an e-mail message or uploading it to a job site, again make sure you submit it as a read-only file in a widely used format (e.g., doc, docx, pdf, or rtf) for easy access by the recipient.

Who is the recipient of your letters and résumés? When answering an advertisement, you should follow whatever instructions are given there. You might be directed, for example, to a company-wide job site to submit your application electronically. Otherwise, direct the materials, if at all possible, to the person in the organization who directly supervises the position for which you are applying. This person often has the power to hire for the position. Your research into the company may turn up the name you need. If not, don't hesitate to call the company and ask for a name and title. Whatever you do, write to a specific individual by name.

Sometimes, of course, you may gain an interview without having submitted a letter of application—for example, when recruiters come to your campus. Bring a résumé with you, and give it to the interviewer at the start of the interview. The résumé gives the interviewer a starting point for questions and often helps to focus the interview on your qualifications for the job.

Online-generated résumés. More and more companies allow you to submit résumés as documents generated electronically by filling in the fields on a series of pages at the site. The fields solicit the details of your address, credentials, and objectives and allow you to identify the job or jobs for which you are applying. This information is automatically compiled in a standardized format for access by hiring managers. Because you can't organize the information yourself to emphasize your knowledge and experience, using keywords to highlight your suitability for the job will be critical. Think carefully about the words that would likely be used to find the ideal candidate for the position. Make sure you include these words in describing your skills, your previous jobs, and your academic training.

Follow-up letters. Write *follow-up letters* (1) if after 2 weeks you have received no answer to your letter of application, (2) after an interview, (3) if a company fails to offer you a job, and (4) to accept or refuse a job.

No answer. If a company doesn't acknowledge receipt of your original letter of application within 2 weeks, write again with a gracious inquiry such as the following:

Dear Mr. Liao:

On April 12 I applied for a position with your company. I have not heard from you, so perhaps my original letter and résumé have been misplaced. I enclose copies of them.

If you have already reached a decision regarding my application or if there has been a delay in filling the position, please let me know at your earliest opportunity.

I look forward to hearing from you.

Sincerely yours,

After an interview. Within a day's time, follow up your interview with a letter. Such a letter draws favorable attention to your as a person who understands business courtesy and good communication practice. Express your appreciation for the interview. Emphasize any of your qualifications that seemed to be especially important to the interviewer. Express your willingness to live with any special conditions of employment, such as relocation. Make clear that you want the job and feel qualified to do it. If you include a specific question in your letter, it may hasten a reply. Your letter might look like this one:

Dear Ms. Cooper:

Thank you for speaking with me last Tuesday about the computer security position you have open.

Working in computer security relates well to my experience and interests. The job you have available is one I am qualified to do. I am working this semester on a research report that examines the legal and ethical issues regarding electronic retrieval of patient records in public hospitals. May I send you a copy next week when it is completed?

I understand that the position would include working alternating weekends. This requirement presents no difficulty for me.

I look forward to hearing from you soon.

Sincerely yours,

After being refused a job. When a company refuses you a job, good tactics dictate that you acknowledge the refusal. Express thanks for the time spent with you, and state your regret that no opening exists at this time. If you like, express the hope of being considered in the future. You never know—a new job might be available later. In any case, you want to maintain a good reputation with this employer and its representatives. These are people working in the same industry as you: you may encounter them at professional conferences or meetings of the local chapter of your professional association. They may later come to work at your company as your colleague, your supervisor, or your subordinate. The few minutes you devote now to a courteous reply will create a lasting impression on them of your extraordinary professionalism.

Accepting or refusing a job. Writing an acceptance letter presents few problems. Be brief. Thank the employer for the job offer, and accept the job. Determine the day and time you will report for work, and express pleasure at the prospect of joining the organization and working with your new colleagues. A good letter of acceptance might read as follows:

Dear Mr. Simmons:

Thank you for offering me a job as research assistant with your firm. I happily accept. I will report to work on July 1 as you have requested.

I look forward to working with Price Industries and particularly to the opportunity of doing research with Dr. Martinez.

Sincerely yours,

Writing a letter of refusal can be difficult. Be as gracious as possible. Be brief but not so brief as to suggest rudeness or indifference. Make it clear that you appreciate the offer. If you can, give a reason for your refusal. The employer who has spent time and money on interviewing you and corresponding with you deserves these courtesies. And, of course, your own self-interest is involved. In the future, you may wish to work for this organization even though you must decline the opportunity at this time. A good letter of refusal might look like this one:

Dear Ms. Okujeni:

I enjoyed my visit to the research department of your company. I would very much have liked to work with the people I met there. I thank you for offering me the opportunity to do so.

After much serious thought, I have decided that the research opportunities offered to me in another job are closer to the interests I developed at Virginia State University. Therefore, I have accepted that positon and regret that I cannot accept your kind offer.

I appreciate the great courtesy and thoughtfulness that you and your associates extended to me during the application and interview process. The individual you do hire will be blessed with exceptional colleagues.

Sincerely yours,

Your Social Media Profile

Your profile pages on social media sites such as Facebook and LinkedIn are also important factors in your job search. Your profile must reinforce your résumé and letter of application by indicating that you are serious about your field and would be a cordial and productive colleague. Prospective employers will almost always check your social media pages before inviting you for a job interview.

On your social media sites, share information about resources pertinent to your field or profession such as websites, webinars, or software. Discuss presentations at conferences that have informed or changed your thinking on a subject. Mention upcoming meetings and conventions that you also expect to be instructive.

Cite articles in journals, magazines, and newspapers that you have been reading. Offer your analysis and evaluation of this information, especially about its practical applications for your field.

Build a reputation for knowledge and generosity by responding promptly with careful and thoughtful postings to requests for information related to your field or profession. Participate in surveys and interviews, assist with research projects, and contribute to the creation of policies and procedures.

Make it obvious that you enjoy working with others by always being polite and positive in your messages, especially in responding to comments and answering questions. Avoid the temptation to answer insulting or spiteful messages with equally discourteous rejoinders. Prove that you have the maturity to manage negative situations with patience and discipline. Ignore hostile wording and address the key point of a question or comment.

Your profile might also include a variety of other activities with which you are engaged—civic, social, athletic, political, or religious. Exercise your voice judiciously and make sure you communicate consistently as a responsible individual who brings credit to the people and organizations with which you are affiliated—who is liked and admired for your wisdom, diligence, integrity, courage, and friendliness.

Interviewing

The immediate goal of all your preparation and letter and résumé writing is an interview with a potential employer.

The interview. If you have prepared properly, you should show up at the interview knowing almost everything about the organization, including the major milestones in its history as well as its chief locations, important products and services, financial situation, and mission.

You should have the following with you: your résumé; a portfolio of your work, if appropriate; pen and notebook; a list of your references; and your business card.

The interviewer will probably start with a bit of social conversation to put you at ease before proceeding to questions aimed at assessing your skills and interests and how you might be of value to the organization. You might anticipate questions such as the following:

- What can you tell me about yourself?
- What are your strengths and weaknesses?
- What do you want to be doing 5 years from now?
- Why would you want to work for us?
- Why should we hire you?
- What accomplishment are you most proud of?
- What is the biggest problem you've encountered and how did you solve it?
- If a colleague complained to you about his or her boss, how would you manage that situation?
- If a client said/did . . . , how would you manage that situation?

To the question "What can you tell me about yourself?" the interviewer really doesn't expect an extended life history. This question provides you the opportunity to talk about your work, educational experiences, and skills. Try to relate your skills and experience to the needs of the organization. Be sure to mention your people skills and communication skills because these are essential to nearly every professional job.

In your answer to this and other questions, be specific in your examples. If you say something like "I know how to budget for projects" immediately back it up with the details of a particular experience. Focus on offering the specific evidence that will lead the interviewer to the right conclusions about you. These specifics are more memorable than unsupported claims and will help you stick in the interviewer's mind.

In answering questions about your strengths and weaknesses, be honest but don't identify weaknesses that could eliminate you from consideration. Admit to weaknesses that would have no impact on the specific job for which you are interviewing: for example, if the job has no managerial duties, it would be okay to acknowledge that you are still developing your managerial skills.

To the question "What do you want to be doing 5 years from now?" a good answer would identify a position with greater responsibilities at the same company, especially if the job for which you are interviewing would be a suitable stepping stone.

The question "Why do you want to work for us?" allows you to display your knowledge of the organization. In answering this question, you should again emphasize that the skills you have to offer fit the company's needs.

Hypothetical questions examine your problem-solving skills. Here is your chance to demonstrate how you would serve the organization's interests with wisdom, discretion, and integrity.

In the final portion of the interview you will typically be given a chance to ask some questions of your own. It's a good time to get more details about the job or the working environment. Ask about the organization's goals. "What is the company most proud of?" is a good question. Or "What are you looking for in the candidate for this position?" "What will you want him or her to accomplish in the first year on the job?" Don't ask naive questions—questions you ought to know the answers to already from your research—like the size of the company or the number of employees. Don't ask questions just to ask questions. The interview is a good time for you to find out if you really want to work for this organization. Not every organization is going to be a good fit for what you have to offer and what you want to do. Avoid questions about salary and benefits unless the interviewer has raised this subject.

If you really want to work for the organization, make that clear before the interview ends. But don't allow your willingness to appear as desperation. Companies don't hire desperate people. At some point in the interview, be sure to get the interviewer's name (spelled correctly), title, address, phone number, and e-mail address. You'll need them for later correspondence. When the interviewer thanks you for coming to the interview, thank him or her for seeing you and leave. Don't drag the interview out when it's clearly over or linger at the door.

If you are invited for a site visit as a finalist for the position, you will find yourself continuously interviewed—from the minute you arrive until the minute you leave—by a variety of potential colleagues and supervisors during lunches or dinners, while waiting for elevators, while walking from office to office, or while driving to and from the airport. For this series of informal interviews and conversations—each of which could contribute to the hiring decision—prepare 1-minute answers to each of the formal interview questions: that is, abbreviate your usual

3-minute or 5-minute answer in a cogent emphasis on your key selling points. If time allows, you could always elaborate, but if the conversation is interrupted or the subject abruptly changes, you will at least have made your point and proven yourself articulate.

Negotiation. Interviewers rarely raise the subject of salary and benefits until they either see you as a good prospect or are sure they want to hire you. If they offer you the job, the negotiation is sometimes conducted by others in a separate interview.

Sometimes the negotiator may offer you a salary. At other times, you may be asked to name a salary. Now is the time to put to good use the information you may have received through your networking activities. Or check the online job-finding services or specialized services such as Salary.com (www.salary.com) for their estimates of earnings for different positions in different industries.

Your research in these sources will give you not a specific salary but a salary range. If asked to name a salary, do not ask for the bottom of the range. Ask for as near the top as you believe is reasonable given your education and experience. The negotiator will respect you the more for knowing what you are worth. However, balance the compensation package—vacations, pension plans, healthcare, educational opportunities, and so forth—against the salary. Some compensation packages are worth a good deal of money and may allow you to accept a lower salary.

The location of the job is also a critical factor to consider. Online services such as Salary.com offer cost-of-living calculators that assess the comparative prices for housing, food, clothing, transportation, and basic services in different locations. For example, $100,000 in New York would be the equivalent of $64,000 in Chicago, $58,000 in Denver, or $78,000 in Los Angeles.

Before and after the interview. If you have not participated in job interviews before, practice with several friends—two of you as interviewer and interviewee and the remainder as observers. Ask the observers to look for strengths and weaknesses in your answers, diction, grammar, and body language and to give you a candid appraisal. Practice until you feel comfortable with the process.

After you finish each interview, write down your impressions as soon as possible. How did your clothes compare to the interviewer's? Were there unexpected questions? How good were your answers? What did you learn about the organization? What did you learn about a specific job or jobs? Did anything make you uncomfortable about the organization? Do you think you would fit in there? By the next day, get a thank-you note (by letter or e-mail) off to the interviewer.

Job Search Checklist

The Letter and Résumé

Planning

☐ Your letter of application:

- What position are you applying for?
- How did you learn about this position?
- Why are you qualified for this position?
- What interests you about this company?
- What will you do for the organization?
- How will the employer reach you?

☐ Your résumé:

- Do you have all the necessary details about your education and experience (e.g., dates, job descriptions, schools, majors, degrees, extracurricular activities)?
- Which résumé format will suit your experience and abilities, chronological or functional?
- Do you have permission to use the names of at least three people as references?

Revising

☐ Your letter of application:

- Do you seem self-confident but not arrogant or boastful?
- Does your letter show how you could be valuable to the employer?
- Does your letter reflect interest in a specific job?
- Have you emphasized the education and experience that suit you for this job?
- Have you made it clear you would like to be interviewed? Have you made it easy for the employer to arrange the interview?
- Is your letter completely free of grammatical and spelling errors? Is it designed for easy reading and skimming?

☐ Your résumé:

- Have you picked the résumé type that suits your experiences and qualifications?
- Have you listed your education and job experience in reverse chronological order?
- Have you used active verbs and appropriate keywords to describe your education and experience?
- Is your résumé completely free of grammatical and spelling errors? Is it designed for easy reading and skimming?

The Interview

Planning

☐ Have you completed your research about the organization? Are you familiar with its history, goals, products, locations, and reputation in the industry and region?

☐ Have you practiced your answers to likely questions? Do you have good questions to ask the interviewer?

Reviewing

☐ Did you answer all the questions effectively? Which questions could you have answered better?

☐ Were you asked questions that you didn't expect?

☐ How did the interviewer answer your questions? Did he or she seem to think you asked good questions?

☐ Do you think you will be offered the job? Why or why not?

☐ Do you think you are a good fit for this organization? Why or why not?

EXERCISES

1. Investigate a potential employer of majors in your field. Locate information available from the employer's website as well as online job-finding services. Are all the profiles of this employer basically similar? Do you notice important differences? Locate additional information available in newspapers, magazines, and government documents about this employer. Do a search on Twitter for comments about this employer and its products and services. What is the public's perception? What is its reputation in the industry? Interview employees or managers for their perspectives on the advantages and disadvantages of working for this employer. Compile the findings of your research in a slide presentation for majors in your field.

2. Given the findings of your investigation of a major employer in your field, list the characteristics of this employer's ideal job candidate. What would be the ideal job candidate's education and experience? What would be his or her special skills, dispositions, and achievements? How are you similar to this ideal job candidate? How do you differ?

3. Given the findings of your investigation of a major employer in your field and your inventory of the ideal job candidate, tailor your typical letter of job application and résumé to the needs and interests of this employer. What changes will you make in your usual version of each document? How will these changes increase the likelihood of your being interviewed for a job?

4. If a friend of yours were applying for a position as a technical writer at Google, which of the following two résumés (see Figures 12–4 and 12–5) would you advise she include with the letter of application? If you were given 15 more minutes to make either résumé more effective, what changes would you make? How would your advice differ if your friend were applying for a job as a technical writer at Ford Motor Company or General Electric?

Jessica Perry
19309 Magnolia Avenue,
Long Beach CA 90802
562-555-4433
jessica.perry@uclb.edu

Employment Objective	Position as a Technical Communicator, with opportunities to utilize my experience and skills in technical editing and document design
Education *University of California* *at Long Beach*	B.A., Technical Communication, May 2017 GPA: 3.5. GPA in Major: 3.75

Employment Experience

Crisp Communications, *Long Beach, CA* *2016-2017*	• Compose, edit, and design a variety of documents for clients of this consulting firm, including brochures, newsletters, reports and manuals. • Supervise small projects. • Coordinate project staff and schedule. • Maintain billing records.
Mercury Computers, *Los Angeles, CA* *2015-2016*	• Provided customer support for hardware and software products. • Researched customer problems, identified appropriate solutions, and replied to customers. • Served average of 50 customers daily.
Special Skills	• Proficient in all Adobe applications • Read/write/speak Spanish
Scholastic Honors	• Sigma Pi Delta Honor Society, 2017 • R.T. Brookings Scholarship Recipient, 2016 • Dean's Academic Honor Roll, 2015–2017
Professional Memberships	• IEEE Professional Communication Society • Society for Technical Communication (STC)
References	Available upon request from the Career Planning and Placement Center, University of California, Meredith Hall 17, Long Beach, CA 90815.

August 2017

• **FIGURE 12–4** Document for Exercise 4

Jessica Perry

19309 Magnolia Avenue,
Long Beach, CA 90802
562-555-4433
jessica.perry@uclb.edu

Employment Objective:	Position as a Technical Communicator specializing in technical editing and document design
Education	University of California, Long Beach, CA B.A., Technical Communication, May 2017
Communication Skills	• Compose, edit, and design a variety of documents, including brochures, newsletters, reports and manuals. • Create projects using all Adobe applications • Provide customer support for hardware and software products • Research customer problems, identify appropriate solutions, and reply to customers by telephone or e-mail message. • Read/write/speak Spanish • Graduating with Honors: GPA: 3.5; GPA in Major: 3.75
Project Management	• Supervise small document design projects from initial meeting with client through publication and delivery • Coordinate project staff and schedule • Maintain billing records.
Professional Involvement	• Sigma Pi Delta Honor Society • Association for Computing Machinery (ACM) • IEEE Professional Communication Society • Society for Technical Communication (STC)
Work Experience	• Crisp Communications, Long Beach, CA, 2016–2017, document design • Mercury Computers, Los Angeles, CA, 2015–2016, customer support
References	Available upon request from the Career Planning and Placement Center, University of California, Meredith Hall 17, Long Beach, CA 90815.

August 2017

● **FIGURE 12–5** Document for Exercise 4

Appendix A

Brief Guide to Grammar, Punctuation, and Usage

ab ◆ ABBREVIATIONS

Every field and organization uses standard and specialized terms that are abbreviated for convenience and conciseness.

First, determine whether your audience is familiar enough with the unabbreviated term to allow you to use it without defining it. Second, determine whether your audience is familiar enough with the abbreviation for you to use it without spelling it out.

If you decide that an abbreviation is appropriate and must be explained, place the abbreviation in parentheses following your first use of the unabbreviated term. Thereafter, use the abbreviation by itself. If necessary, after the parentheses, provide a definition of the term. Make the definition as brief as possible but as detailed as necessary for your purpose and audience:

> Fluid catalytic cracking (FCC) changes crude oil to gasoline by breaking the long-chain molecules that are characteristic of hydrocarbon liquids. The process involves exposing the oil to a special chemical agent under high temperature and pressure. FCC is the key conversion process at oil refineries.

Use Latin abbreviations like *i.e.* (that is) and *e.g.* (for example) in parenthetical explanations or in tables and figures if space is limited.

Check the appropriate style guide or publication manual in your field for specific guidelines on abbreviations.

apos ◆ APOSTROPHES

Use apostrophes to indicate the possessive case of nouns (e.g., singular: supervisor's; plural: supervisors') and the missing letters in contractions (would've: would have).

With coordinated nouns, use apostrophes to make all possessive in order to indicate individual ownership. To indicate joint ownership, make only the final noun possessive.

- Maria's and Timothy's reports are late. (i.e., The report written by Maria and the report written by Timothy are late.)
- Maria and Timothy's reports are late. (i.e., The reports written together by Maria and Timothy are late.)

Also use apostrophes to indicate the possessive case of indefinite pronouns ending in -one or -body as well as *either, neither, another,* or *other:*

- Everyone's electric bill is incorrect because of a computer error.
- The costly error probably raised somebody's chances of getting fired.
- I asked both supervisors and either's approval is enough to proceed.
- I asked both supervisors but neither's approval was forthcoming.
- We admired each other's patience and persistence.
- He preferred another's answer to mine.

acro ◆ ACRONYMS

Acronyms are names for objects or entities formed from a combination of the initial letters of the words in a title or phrase. Unlike abbreviations (e.g., UN for United Nations), acronyms are pronounced as words (e.g., UNICEF for United Nations International Children's Emergency Fund).

First, determine whether your audience is familiar enough with the acronym to allow you to use it without showing the phrase from which it is derived. Second, determine whether the phrase from which it is derived is itself a sufficient definition or explanation of the acronym.

If you decide that an acronym is appropriate, place the phrase from which it is derived in parentheses following your first use of the acronym. If necessary, supply additional explanation immediately following. Thereafter, use the acronym by itself:

> The mysterious illness was identified as SARS (severe acute respiratory syndrome), a disease similar to pneumonia but viral in nature and thus impervious to antibiotics. SARS killed 299 people in Hong Kong in a 2003 epidemic.

brackets ◆ BRACKETS

Use brackets inside quotations or parentheses to insert a clarifying note or comment:

- "This is the highest level of recognition in the organization [1% of almost 50,000] and certifies exceptional professional achievement."
- If you would like more information about this opportunity, please call my office (M–F 8–5 [ET], 866-555-7423).

cap ◆ CAPITALIZATION

Capitalize months but not seasons (e.g., April was unusually dry this spring).

Capitalize geographic areas but not directions (e.g., The blight started in the Southwest but quickly traveled north and east).

Capitalize titles preceding names but not following or in isolation:

- The invitation is from President Juanita Ramirez.
- Juanita Ramirez, president of the company, arrived earlier this morning.
- The president of the company is Juanita Ramirez.

colon ◆ COLON

Use a colon to introduce quotations, lists, or supporting statements:

- Dr. Raju's assertion was unequivocal: "This test has no predictive value, and it is unethical to advertise it as though it does."
- The company has five international locations: Cairo, Hong Kong, Mumbai, Paris, and Rio de Janeiro.
- The instructor explained the maieutic method: that is, the practice of teaching by asking questions that stimulate thinking and elicit new ideas.

Never use a colon immediately following a verb or a preposition:

WRONG: The five stores are located in: Cairo, Hong Kong, Mumbai, Paris, and Rio de Janeiro.
WRONG: The locations of the five stores are: Cairo, Hong Kong, Mumbai, Paris, and Rio de Janeiro.
RIGHT: The stores are in five locations: Cairo, Hong Kong, Mumbai, Paris, and Rio de Janeiro.
RIGHT: The stores are located in the following five cities: Cairo, Hong Kong, Mumbai, Paris, and Rio de Janeiro.

Use a colon to separate two independent clauses if the second clause explains or amplifies the first:

- The ability to find information is important but insufficient: we must also develop skills in synthesizing and analyzing information.
- The new map uses a different color for each section of the zoo: for example, green for the forest animals and brown for the desert animals.

Use a colon following the salutation of a business letter.

Dear Ms. Hopkins:

c ◆ COMMA

Use a comma before a coordinating conjunction (*and, but, or, nor, for, yet*) that joins two independent clauses:
- The investigators made their recommendation, and the supervisor supported their judgment.
- The investigators made their recommendation, but the supervisor questioned their judgment.

Use a comma after an introductory word, phrase, or clause:

- Unfortunately, the experiment did not yield statistically significant findings.
- In appreciation of their 10 years of service to the hospital, the president awarded each of the nurses a $1,000 prize.
- After you inspect the office space and verify that its condition is satisfactory, you must complete and sign a lease agreement.

Use commas to separate nonrestrictive modifiers from the remainder of the sentence.

RESTRICTIVE: Cotton farming that is conducted without artificial irrigation lowers the cost of production but raises the risk of a poor crop.
NONRESTRICTIVE: Cotton farming, which typically involves the use of fertilizers and defoliants as well as herbicides and pesticides, has high costs for the environment.

A *restrictive modifier* is necessary to the meaning of the sentence. The writer has thus restricted the subject of cotton farming to only that which is conducted without artificial irrigation. The writer here is making a single claim: cotton farming without artificial irrigation has lower costs but higher risks.

A *nonrestrictive modifier* supplies additional information about the subject without restricting it. The writer here is making two claims: cotton farming usually has high costs for the environment, and cotton farming usually involves fertilizers, defoliants, pesticides, and herbicides.

Use commas to separate items in a series of words, phrases, or clauses.

- This facility produces ethylene oxide, ethylene dichloride, and polyethylene.
- Ethylene contributes to the ripening of fruit, the opening of flowers, and the shedding of leaves.
- Florists complain about the impact of ethylene exposure on the life of their flowers, grocers worry about high-ethylene products like bananas, and shippers focus their efforts on good temperature controls to minimize the chemical's release.

Use commas to separate dates, geographical locations, and titles:

- On Tuesday, June 21, 2017, the case was closed.
- The artifact was shipped from Palermo to New York, transported by train to Chicago, and delivered by bicycle messenger to the museum.
- The new members of the grievance committee are Isidore Agboka, Ph.D., and Keisha Pierce, J.D.

dm ♦ DANGLING MODIFIER

A dangling modifier occurs whenever a modifying word or phrase is used without a suitable noun for it to modify:

WRONG: Fearing the infectious disease might jump species, the turkeys were quarantined.
RIGHT: Fearing the infectious disease might jump species, the veterinarian quarantined the turkeys.

dash ◆ DASH

The dash serves the same function as parentheses (i.e., separating a tangential or explanatory comment from the remainder of the sentence), but the dash does so with greater emphasis—like a shout instead of a whisper.

ell ◆ ELLIPSIS

Use three spaced periods to indicate that words have been omitted from a quotation. If the quotation is at the end of a sentence, use four periods (i.e., three spaced periods for the ellipsis and a period for the sentence).

- "On the job, most readers . . . will feel no obligation to read what you write unless your messages are useful to them in making decisions or taking action."
- "On the job, most readers, coming from a variety of professional and cultural backgrounds, will feel no obligation to read what you write unless your messages are useful to them. . . ."

If the quotation is obviously incomplete, skip the ellipsis.

- According to the inspector, the action was "deliberate and malicious."

exc ◆ EXCLAMATION POINT

Avoid using exclamation points in the writing of reports, letters, memos, and e-mail messages so that you never come across as easily excited or agitated. If writing instructions, use exclamation points as necessary with cautions and warnings.

frag ◆ FRAGMENT

A *sentence fragment* is a phrase or a clause punctuated as though it were a complete sentence. This error is usually fixed by connecting the fragment to the preceding sentence.

WRONG: The test results were exactly as we expected. Although we were hoping for better news.
RIGHT: The test results were exactly as we expected, although we were hoping for better news.
WRONG: The recommendations were made by the special committee. Which included the city's forensic scientist and medical examiner.
RIGHT: The recommendations were made by the special committee, which included the city's forensic scientist and medical examiner.

WRONG: The building manager decided to change the security policy. Surprising all of us at the meeting.
RIGHT: The building manager decided to change the security policy, surprising all of us at the meeting.

hyphen ♦ HYPHEN

Use a hyphen with the following suffixes and prefixes:

- -elect: mayor-elect
- -in-law: brother-in-law
- all-: all-around athlete
- ex-: ex-president
- quasi-: quasi-empirical study
- self-: self-confident

Also use hyphens with prefixes preceding capitalized words (e.g., pre-Olympics competition), prefixes preceding numbers (e.g., post-9/11 restrictions), and prefixes preceding acronyms and abbreviations (e.g., anti-HIV medication).
In addition, use a hyphen between words joined together to modify a word:

- The campaign was a time-intensive effort.
- She introduced the jazz-inspired designs at the music festival.
- He makes six-, eight-, and twelve-string guitars.
- She is a well-connected representative for the organization.

ital ♦ ITALICIZATION

Italicize titles of books, journals, magazines, plays, films, radio and television programs, sculptures and paintings, ships, aircraft, and spacecraft.

Italicize words that are considered foreign to the English language.

Italicize Latin words for genus and species (e.g., The zoological designation for the American toad is *Bufo americanus*).

Italicize words, letters, and numbers identified as words, letters, and numbers.

The words *compose* and *comprise* have opposite meanings.
His middle initial is *G*.
She scored *75* on the final exam.

Italicize letters used as statistical symbols:

- *M* (i.e., mean)
- *N* (i.e., number in the population studied)
- *p* (i.e., probability)
- *SD* (i.e., standard deviation)

mm   ◆   MISPLACED MODIFIER

A *misplaced modifier* is a word, phrase, or clause that is in the wrong position in a sentence; as a consequence, it modifies a word that the writer never intended it to and creates ambiguity.

> WRONG: The computer was repaired almost at no cost to us.
> RIGHT: The computer was repaired at almost no cost to us (*or* The computer was almost repaired but at no cost to us).
> WRONG: She tried to find a job for two years.
> RIGHT: She tried for two years to find a job (*or* She tried to find a two-year job).
> WRONG: A great museum in the city that he visits all the time is the Art Exchange on 3rd and Elm.
> RIGHT: A great museum that he visits all the time in the city is the Art Exchange on 3rd and Elm (*or* In the city that he visits all the time, a great museum is the Art Exchange on 3rd and Elm).

np/ag   ◆   NOUN–PRONOUN AGREEMENT

In all formal communication, collective nouns such as *each, everyone, either, neither, anybody, somebody, everybody*, and *no one* use singular pronouns and singular verbs:

> FORMAL: Each has his or her copy.
> FORMAL: Everybody was given his or her ticket.
> For informal communications, the plural pronoun is widely used:
> INFORMAL: Each has their copy.
> INFORMAL: Everybody was given their ticket.

A better solution is to revise the wording:

- All have their copies.
- Each has a copy.
- All were given their tickets.
- Everybody was given a ticket.

The same issue arises with nouns used in a generic sense, such as the word *employee* in the following:

> FORMAL: The employee must complete his or her training within one week of being hired.
> INFORMAL: The employee must complete their training within one week of being hired.

Again, a better solution is to revise the wording:

- All employees must complete their training within one week of being hired.
- The employee must complete the training within one week of being hired.

With collective nouns such as *team*, *group*, *class*, or *committee*, use the singular pronouns *it/its* if you are referring to the unit and the plural pronouns *they/their/them* if you are referring to the constituents.

- The committee is taking its time to find job candidates it believes are qualified.
- The security squad knows that they must keep their uniforms clean and in good repair.

num ♦ NUMBERS

Ordinarily, use words for numbers from zero to nine and numerals for numbers 10 and higher.

In a series of numbers, if any one of the numbers is 10 or higher, use all numerals:

- The collision damaged seven vehicles and injured four passengers.
- The collision damaged 7 vehicles and injured 16 passengers.

Never start a sentence with a numeral: either use words for the number or revise the sentence to shift the number from the opening position.

- Twenty-five economists were involved in the project.
- The project involved 25 economists.

With two adjacent numbers, use words for one and numerals for the other in order to avoid confusion.

- He ordered twelve 50-liter canisters.
- The list includes 15 eight-digit passwords.

Always use numerals for dates, addresses, exact time, exact sums of money, measurements, and cross-references.

- 1 October 2018 or October 1, 2018
- 2305 18th Street
- 2:00 P.M. (but two o'clock)
- $7,889.97 (but about eight thousand dollars)
- 54 km
- 26 g
- see page 89
- see Figure 4

paral ♦ PARALLELISM

Use the same grammatical structure for items in a series:

WRONG: A technical communicator must be skilled in writing proposals, publication management, usability tests, and designing websites.

RIGHT: A technical communicator must be skilled in writing proposals, managing publications, running usability tests, and designing websites.
RIGHT: A technical communicator must be skilled in proposal writing, publication management, usability testing, and website design.

paren ◆ PARENTHESES

Use parentheses to separate a tangential or explanatory comment.

Never put any mark of punctuation before an opening parenthesis, but put any required marks of punctuation inside the parentheses and after the closing parenthesis:

- Use parentheses to separate a tangential or explanatory comment (including definitions, clarifications, examples, quotations, or statistics) from the remainder of a sentence.
- Latin words are used to designate genus and species in zoological classification (e.g., *Bufo americanus* for the American toad). (Note that scientific names are always displayed in italics.)

pron ◆ PRONOUN

Use the subjective case of a pronoun (*I, he, she, we, they*) if it serves as the subject of a clause:

- He and I will write this report.
- He visited the research institute yesterday, even though she and I made the same trip a week ago.

Use the objective case of a pronoun (*me, him, her, us, them*) if it is the object of a verb or the object of a preposition:

- The president praised them and us for working together to find a solution.
- The grievance committee interviewed him and her at separate times.
- The detective reported the results of the investigation to them and me.
- This exhibition is a great opportunity for her and me as emerging artists.

quot ◆ QUOTATION MARKS

Use quotation marks for brief quotations:

- As Simon Miller has noted, "The only solution is to limit the population of the species."
- According to Gloria Martinez, "Trillions of nurdles are made annually in factories on every continent. This growing epidemic of plastic pellets is a looming disaster for the world's environment."

For quotations of four or more lines, skip the quotation marks and indent and single space the quotation.

Use quotation marks around titles of articles from journals and magazines:

- The article "Polymers are Eternal" was originally published by Eduardo Quesada in 1989.
- Emily Kolosov's "The Geologic Heritage in Dispute" claims that crop diversity is diminishing worldwide.

Always put commas and periods inside the quotation marks:

- According to Emily Kolosov's "The Geologic Heritage in Dispute," crop diversity is dwindling worldwide.
- Eduardo Quesada is the author of "Polymers are Eternal."

Put all other marks of punctuation outside the quotation marks unless they are part of the quotation.

- In 1989, Eduardo Quesada published his article "Polymers are Eternal": it was greeted with derision by the plastics industry.
- Emily Kolosov's "The Geologic Heritage in Dispute" asked the question, "What are the avoidable contributors to this growing environmental crisis?"

run-on ♦ RUN-ON SENTENCE

A *run-on sentence* is two independent clauses joined with only a comma or with no punctuation at all.

Two independent clauses may be punctuated as two separate sentences (each with a period) or may be joined with a semicolon, a colon, or a comma and a conjunction (*and, but, for, nor,* or *yet*):

> WRONG: The instructions manual has legal and ethical implications, it must assure safe and efficient operation of the product and minimize liability for the manufacturer.
> RIGHT: The instructions manual has ethical and legal implications. It must assure safe and efficient operation of the product and minimize liability in case of damage or injury.
> RIGHT: The instructions manual has ethical and legal implications; specifically, it must assure safe and efficient operation of the product and minimize liability in case of damage or injury.
> RIGHT: The instructions manual has ethical and legal implications, and thus it must assure safe and efficient operation of the product and minimize liability in case of damage or injury.

A run-on sentence may also be repaired by changing one or the other of the independent clauses to a dependent clause or phrase:

RIGHT: The instructions manual has ethical and legal implications because it must assure safe and efficient operation of the product and minimize liability in case of damage or injury.

RIGHT: The instructions manual has ethical and legal implications, making it necessary to assure safe and efficient operation of the product while minimizing liability in case of damage or injury.

semi   ◆   SEMICOLON

Use a semicolon to join related independent clauses (e.g., Victims of this disease exhibit symptoms of pneumonia; however, unlike typical pneumonia, their disease is viral and unresponsive to antibiotics.).

Use a semicolon to separate items in a series if the items are internally punctuated with commas (e.g., The budget for new equipment included a security camera for each of the 12 building exits; projectors, computers, and screens for all 18 classrooms; and one desk, one desk chair, two guest chairs, three bookshelves, and one filing cabinet for each of the 35 faculty and administrative offices.).

av/ag   ◆   SUBJECT–VERB AGREEMENT

Note that the words *each*, *either*, and *neither* take a singular verb regardless of intervening words:

- Each is qualified.
- Each of the students who applied for the supervisor position is qualified.
- Either is qualified.
- Either of the students who were interviewed yesterday by the hiring committee is qualified.
- Neither is qualified.
- Neither of the students who applied for the supervisor position is qualified.

In a coordinated subject joined by *or* or *nor*, make the verb agree with the closer noun or pronoun:

- The original painting or digital reproductions are always on display.
- Digital reproductions or the original painting is always on display.
- Neither the original painting nor digital reproductions are on display during the building's renovation.
- Neither digital reproductions nor the original painting is on display during the building's renovation.

In a coordinated subject joined by *and*, the verb is always plural:

- The original painting and digital reproductions are available for comparative analysis.

- Digital reproductions and the original painting are available for comparative analysis.

With collective nouns such as *team*, *group*, *class*, and *committee*, use a singular verb if you are referring to the unit and a plural verb if you are referring to the constituents.

- The engineering team is flying to the construction site.
- The engineering team are flying together to the construction site.

WW ♦ WRONG WORD

The following words are often confused and misused. Here are examples of correct usage:

I **accept** your decision.
She liked all the candidates we interviewed **except** the third.

His teaching experience will **affect** [influence] his decisions about this project.
Every delay on the bridge project will **effect** [cause] a loss in sales and profits for local businesses.

The proposal was **already** submitted.
The proposals were **all ready** to be submitted.

The engineers were **all together** [united] in their support of the policy.
The engineers were **altogether** [completely] supportive of the policy.

I don't make paper copies of the progress reports **anymore**.
I don't need **any more** paper copies of the progress reports.

She visited **awhile** with each of the clients.
She visited for **a while** with each of the clients.

The engineers believe that **canvas** is the right material for this product.
I will **canvass** the engineers for their opinions about this product.

He made sure to **cite** each source of information.
He made sure to inspect the **site** of the accident.

Advertising through social media **complemented** the television commercials.
The supervisor **complimented** the students and praised their efficiency.

A biologist, a psychologist, and a sociologist **compose** the hiring committee.
The hiring committee **comprises** a a biologist, a psychologist, and a sociologist.

She is **continually** [repeatedly] making changes to the proposal.
She worked **continuously** [without interruption] to finish the proposal.

The **council** [committee] of advisors made two recommendations.
He thanked his sister for her **counsel** [advice].

She was always **discreet** [prudent] in discussing clients.
The two issues are equally important but **discrete** [separate].

The **eminent** scientists arrived early for their presentation.
The arrival of the scientists is **imminent**.

Their **everyday** practice includes a morning review of local news.
She is in the office **every day** by 7 o'clock.

The train station is six kilometers **farther** [distance].
We will discuss this case **further** [degree].

She invited **fewer** [number] students to this week's meeting.
She is devoting **less** time [amount] to meeting with students.

The hospital is moving **forward** with its new president.
The report includes a **foreword** by the president of the hospital.

Your supervisor **implied** [suggested] that you would be promoted.
Your supervisor **inferred** [concluded] from your record of success that you deserved a promotion.

It's [it is] a historic district of the city.
Its history is impressive.

The walls were covered in **lead** paint.
The architect **led** us on a walking tour of the new building.

Medical research has identified the **principal** cause of this disease.
The **principal** of my loan is shrinking.
We operate the business on the **principle** of sustainability.

She located the source of the **quotation**.
He will **quote** leading scientists in this report.

The police officer talked **respectfully** to the victim about the crime.
The book and film awards were announced on April 2 and 3, **respectively**.

I enjoy riding my **stationary** bicycle.
We printed the letters on the client's **stationery**.

The proposal that **they're** writing is due on Friday.
I visited **their** office earlier this week.
She is taking a taxi to **there** this morning.

The **weather** is usually a safe topic of conversation.
He wants to know **whether** he could leave early today.

She never said **whose** job is at risk.
I know **who's** in the meeting.

I noticed that **you're** writing a feasibility study.
I admire **your** writing ability.

Appendix B

Using Sources of Information

Clear, credible, pertinent, and sufficient information is the basis of all effective technical communication. To deliver the right information at the right time, you must develop the six skills of information literacy:

1. *Locate sufficient sources of information*: Given your audience and purpose, you must determine the nature and scope of potential information sources: interviews, surveys, books, articles, blogs, websites, videos, and government documents.

2. *Examine sources for pertinent information*: After you have compiled the potential sources of information, you must review each for information related to your subject. This effort includes skimming tables of contents and lists of illustrations, searching for keywords, and navigating a subject index.

3. *Evaluate sources for credibility*: You must judge the accuracy of the information in your sources. Ordinarily, the more reliable the source or the more sources offering the same information, the more likely it is you have information that deserves to be trusted. No source is infallible, but credible sources typically identify their sources of information, have a track record of offering valid and reliable information, or publish their information only after rigorous review by editors and subject specialists.

4. *Summarize information in sources clearly and correctly*: You will have to make sure that your abridged version of the information in a source is neither misleading nor mystifying. If you are unsure of the right wording, consider integrating quotations judiciously.

5. *Synthesize information across sources*: You must piece together the information from each of your sources so that it creates a coherent picture of your subject. This job includes explaining the relationships among all the pieces of pertinent information that you have located (e.g., cause/effect, chronological, comparison/contrast) and making sure that your assertions about the relationships are supported by evidence and clarified by examples.

6. *Cite sources of information appropriately*: According to your audience and purpose as well as the practices of your organization, choose a formal or informal method of identifying your information sources. In the majority of documents that you write (e-mail messages, blog postings, letters, and memos), informal citation will be satisfactory (e.g., "According to the book *Risk Analysis for Civil Engineers* . . ." or "The video at www.worldnews.com/112014/solarpower.html makes clear that . . ."). In proposals and reports using multiple sources of information, especially if circulated outside your organization, a

formal system of documentation will be necessary to assist readers in weighing the merits of your research and keeping track of your sources; that is, in addition to specifying which sources supplied which information, a formal system makes evident the number of sources you used, the variety of your sources, and the frequency with which you relied on each source.

Applications such as EndNote (www.endnote.com), RefWorks (www.refworks .com), and Zotero (www.zotero.org) will assist you with managing, citing, and annotating your sources, but your conscientious efforts are critical to developing the skills of information literacy. No application is a substitute for your cautious judgment about information that is clear, credible, pertinent, and sufficient.

The APA System

The APA (American Psychological Association) system of documentation cites sources parenthetically using the author's last name (or authoring organization or title if no author is identified) and the year of publication, separated by a comma. If emphasizing or quoting a specific passage, also cite specific pages using the abbreviation p. or pp. as necessary:

> With its officious language and its emphasis on damage to buildings and vehicles, this accident report "silences the horror of human suffering and the loss of human life" (Sauer, 1993, p. 72).

In a list titled "References" and organized alphabetically by author's last name (or by authoring organization or by title if no author is identified), detail your sources. If available, include the Digital Object Identifier (DOI), a unique alphanumeric designator of the source's fixed location on the Internet. If the source was accessed online but has no DOI, include the URL. If the online source is subject to change (e.g., wiki, website), include the date of retrieval.

Omit from the References all information sources to which your readers have no access (e.g., private e-mail messages, telephone conversations, and interviews): identify these exclusive sources parenthetically in the text as "personal communication":

> Dr. Harris acknowledges that the evidence to support this theory is insufficient but growing (personal communication, April 2, 2017).

Book:

Appadurai, M. (2017). *Ethics and efficiencies of international corporations* (2nd ed.). New York, NY: Dolphin.

Collection:

Williams, M. F., & Pimentel, O. (Eds.). (2013). *Communicating race, ethnicity, and identity in technical communication*. Amityville, NY: Baywood.

Essay or article in a book:

Still, J. L., Carter, B. T., & García, M. V. (2016). The risk and rhetoric of statistical illustrations. In M. M. Staggers & W. B. Cook (Eds.), *Visual communication: Investigations in theory and practice* (pp. 119–131). Chicago, IL: Radius.

Article in a professional journal:

Sauer, B. (1993). Sense and sensibility in technical documentation: How feminist interpretation strategies can save lives in the nation's mines. *Journal of Business and Technical Communication, 7*, 63–83.

Zoetewey, M. W. (2013). The rhetoric of free: Open source software and technical communication during economic downturns. *Technical Communication Quarterly, 23*, 323–342. doi:10.1080/10572252.2013.794090

Article in a monthly publication:

Cook, A. (2016, October). Risk communication and vaccines. *Medicine Monthly, 23*, 49–53.

Article in a daily newspaper, anonymous:

Bike-friendly culture inspires business community (2017, January 14). *The Long Beach Daily*. Retrieved from http://www.lbdaily/com/011417/bike.html

Government publication:

United States Department of Transportation National Highway Traffic Safety Administration. (2012). *Blueprint for Ending Distracted Driving* (DOT HS 811629). Washington, DC: US Government Printing Office. Retrieved from http://www.distraction.gov/blueprint-2012.pdf

Article in an online news source:

Luo, L. (2015, December 17). Broccoli and the brain. *INN.com*. Retrieved from http://www.inn.com/2015/SCIENCE/12/17/broccoli/index.html

Page of a website:

Doctors Without Borders. (2017). *Medical issues*. Retrieved October 11, 2017, from http://www.doctorswithoutborders.org/news/issues.cfm

Online multimedia source:

Luo, L. (2015, December 17). Broccoli and the brain. [Video]. *INN.com*. Retrieved from http://www.inn.com/2015/SCIENCE/12/17/broccoli/index.html#innSTCVideo

Posting to e-mail distribution list:

Vasquez, E. (2017, September 24). Research ethics question. [Email list message]. Retrieved from http://attw.interversity.org/read/messages?id100170924

Posting to a blog or bulletin board:

Taylor, M. R. (2017, August 26). Defining boundaries for caffeine in today's marketplace [Blog post]. Retrieved from http://blogs.fda.gov/fdavoice/index.php/2017/08/defining-boundaries-for-caffeine-in-todays-marketplace/

Tweet:

Rickard, L. [LydiaRickard]. (2018, January 12). FDA regulations promote innovation in devices intended to treat obesity [Tweet]. Retrieved from http://twitter.com/lydiarickard/status/192837465001

The Chicago System

The Chicago (Chicago Manual of Style) notes system of documentation includes a numerical list of citations (i.e., notes) in the order cited as well as a list of sources (i.e., bibliography) in alphabetical order by author's last name (or by authoring organization or by title if no author is identified). For online sources, include the Digital Object Identifier (DOI), a unique alphanumeric designator of the source's fixed location on the Internet. If no DOI is available, include the URL.

For each source cited, insert a superscript number in the text corresponding to the appropriate source in the numerical list:

> With its officious language and its emphasis on damage to buildings and vehicles, this accident report "silences the horror of human suffering and the loss of human life."[2]

Notes:

1. Morarji Appadurai, *Ethics and Efficiencies of International Corporations*, 2nd ed. (New York: Dolphin, 2017), 158.
2. Beverly Sauer, "Sense and Sensibility in Technical Documentation: How Feminist Interpretation Strategies Can Save Lives in the Nation's Mines," *Journal of Business and Technical Communication* 7, no. 1 (1993): 72.
3. Appadurai, *Ethics and Efficiencies of International Corporations*, 184.
4. Sauer, "Sense and Sensibility," 75.

Following are a variety of sources as each would be displayed in numerical notes and in the alphabetical bibliography:

Book:

1. Morarji Appadurai, *Ethics and Efficiencies of International Corporations*, 2nd ed. (New York: Dolphin, 2017), 158.

Appadurai, Morarji. *Ethics and Efficiencies of International Corporations*. 2nd ed. New York: Dolphin, 2017.

Collection:

2. Miriam F. Williams and Octavio Pimentel, eds., *Communicating Race, Ethnicity, and Identity in Technical Communication* (Amityville, NY: Baywood, 2013), ii–iii.

Williams, Miriam F., and Octavio Pimentel, eds. *Communicating Race, Ethnicity, and Identity in Technical Communication*. Amityville, NY: Baywood, 2013.

Essay or article in a book:

3. Joyce L. Still, Brian T. Carter, and Maria V. Garcia, "The Risk and Rhetoric of Statistical Illustrations," in *Visual Communication: Investigations in Theory and Practice*, ed. Mary M. Staggers and William B. Cook (Chicago: Radius, 2016), 124.

Still, Joyce L., Brian T. Carter, and Maria V. García. "The Risk and Rhetoric of Statistical Illustrations." In *Visual Communication: Investigations in Theory and Practice*, edited by Mary M. Staggers and William B. Cook, 119–131. Chicago: Radius, 2016.

Article in a professional journal:

4. Beverly Sauer, "Sense and Sensibility in Technical Documentation: How Feminist Interpretation Strategies Can Save Lives in the Nation's Mines," *Journal of Business and Technical Communication* 7, no. 1 (1993): 72.

Sauer, Beverly. "Sense and Sensibility in Technical Documentation: How Feminist Interpretation Strategies Can Save Lives in the Nation's Mines." *Journal of Business and Technical Communication* 7, no. 1 (1993): 63–83.

5. Meredith W. Zoetewey, "The Rhetoric of Free: Open Source Software and Technical Communication During Economic Downturns," *Technical Communication Quarterly* 23, no. 4 (2013): 323–342, doi:10.1080/10572252.2013.794090.

Zoetewey, Meredith W. "The Rhetoric of Free: Open Source Software and Technical Communication During Economic Downturns." *Technical Communication Quarterly* 23, no. 4 (2013): 323–342. doi:10.1080/10572252.2013.794090.

Article in a monthly publication:

6. Andrew Cook, "Risk Communication and Vaccines," *Medicine Monthly*, October 2016, 51.

Cook, Andrew. "Risk Communication and Vaccines." *Medicine Monthly*, October 2016: 49–53.

Article in a daily newspaper, anonymous:

7. "Bike-Friendly Culture Inspires Business Community," *The Long Beach Daily*, January 14, 2017, http://www.lbdaily/com/011417/bike.html.

"Bike-Friendly Culture Inspires Business Community." *The Long Beach Daily*, January 14, 2017. http://www.lbdaily/com/011417/bike.html.

Government publication:

8. US Department of Transportation, National Highway Traffic Safety Administration, *Blueprint for Ending Distracted Driving*, DOT HS 811 629 (Washington, DC: US Government Printing Office, 2012), 19, http://www.distraction.gov/blueprint-2012.pdf.

US Department of Transportation, National Highway Traffic Safety Administration. *Blueprint for Ending Distracted Driving*. DOT HS 811 629. Washington, DC: US Government Printing Office, 2012. http://www.distraction.gov/blueprint-2012.pdf.

Article in an online news source:

9. Leilei Luo, "Broccoli and the Brain," *INN.com*, December 17, 2015, http://www.inn.com/2015/SCIENCE/12/17/broccoli/index.html.

Luo, Leilei. "Broccoli and the Brain," *INN.com*, December 17 2015. http://www.inn.com/2015/SCIENCE/12/17/broccoli/index.html.

Page of a website:

10. "Medical Issues," *Doctors Without Borders*, accessed October 11, 2017, http://www.doctorswithoutborders.org/news/issues.cfm.

Doctors Without Borders. "Medical Issues." Accessed October 11, 2017. http://www.doctorswithoutborders.org/news/issues.cfm.

Online multimedia source:

11. Leilei Luo, "Broccoli and the Brain," Video. *INN.com*, December 17 2015, http://www.inn.com/2015/SCIENCE/12/17/broccoli/index.html#innSTCVideo.

Luo, Leilei. "Broccoli and the Brain," Video. *INN.com*, December 17 2015. http://www.inn.com/2015/SCIENCE/12/17/broccoli/index.html#innSTCVideo.

Posting to e-mail distribution list:

12. Eduardo Vasquez, "Research Ethics Question," *ATTW-L* (e-mail distribution list), September 24, 2017, http://attw.interversity.org/read/messages?id100170924.

Vasquez, Eduardo. "Research Ethics Question." *ATTW-L* (e-mail distribution list). September 24, 2017. http://attw.interversity.org/read/messages?id100170924.

Posting to a blog or bulletin board:

13. Michael R. Taylor, "Defining Boundaries for Caffeine in Today's Marketplace," *FDA Voice* (blog), August 26, 2017, http://blogs.fda.gov/fdavoice/index.php/2017/08/defining-boundaries-for-caffeine-in-todays-marketplace/.

Taylor, Michael R. "Defining Boundaries for Caffeine in Today's Marketplace." *FDA Voice* (blog). August 26, 2017. http://blogs.fda.gov/fdavoice/index.php/2017/08/defining-boundaries-for-caffeine-in-todays-marketplace/.

Tweet:

14. Lydia Rickard, "FDA regulations promote innovation in devices intended to treat obesity," *Twitter*, January 12, 2018, http://twitter.com/lydiarickard/status/192837465001.

Rickard, Lydia. "FDA regulations promote innovation in devices intended to treat obesity." *Twitter*. January 12, 2018. http://twitter.com/lydiarickard/status/192837465001.

The IEEE System

In the IEEE (Institute of Electrical and Electronics Engineers) system of documentation, you designate the first source that you cite in the text as [1] (displayed in brackets), the second source you cite as [2], and so on. If you cite your first source again later in the text, it is still [1]. If you quote from your source, also cite the specific page or pages, using the abbreviation p. for a single page and pp. for two or more pages and with a comma to separate the number of the reference from the cited pages (e.g., [3, pp. 41–42]).

If your sources include personal e-mail messages, telephone conversations, or interviews, designate these with a bracketed number and identify as "private communication" with the month and year of contact.

The list of sources itself is titled "References": include all the sources of information that you numbered in the text and list them in numerical order.

Book:
[1] M. Appadurai. *Ethics and Efficiencies of International Corporations*, 2nd ed. New York: Dolphin, 2017.

Collection:
[2] M. F. Williams, and Octavio Pimentel, Eds. *Communicating Race, Ethnicity, and Identity in Technical Communication*. Amityville, NY: Baywood, 2013.

Essay or article in a book:
[3] Still, Joyce L., Brian T. Carter, and Maria V. García. "The risk and rhetoric of statistical illustrations.," *Visual Communication: Investigations in Theory and Practice*, M. M. Staggers and W. B. Cook, Eds. Chicago: Radius, 2016, pp. 119–131.

Article in a professional journal:
[4] B. Sauer, "Sense and sensibility in technical documentation: How feminist interpretation strategies can save lives in the nation's mines," *J Bus. and Tech. Commun.*, vol. 7, no. 1, pp. 63–83, 1993.
[5] M.W. Zoetewey, "The rhetoric of free: Open source software and technical communication during economic downturns," *Tech. Commun. Quart.*, vol 23, no. 4, pp. 323–42, 2013. doi:10.1080/10572252.2013.794090

Article in a monthly publication:
[6] A. Cook, "Risk communication and vaccines," *Med. Monthly*, pp. 49–53, Oct. 2016.

Article in a daily newspaper:
[7] "Bike-friendly culture inspires business community." *Long Beach Daily*, Jan. 14, 2017.

Government publication:
[8] U.S. Dept. of Transportation, Nat. Highway Traffic Safety Admin. *Blueprint for Ending Distracted Driving*. DOT HS 811 629. Washington, DC: GPO, 2012. Available: http://www.distraction.gov/blueprint-2012.pdf.

Article in an online news source:
[9] L. Luo, "Broccoli and the brain." INN.com. International News Network. Dec. 17 2015. Available: http://www.inn.com/2015/SCIENCE/12/17/broccoli/index.html.

Page of a website:

[10] Doctors Without Borders. (2017, Oct. 11). *Medical issues.* Available: http://www.doctorswithoutborders.org/news/issues.cfm.

Online multimedia source:

[11] L. Luo, "Broccoli and the brain" [Video] INN.com. Dec. 17, 2015. Available: http://www.inn.com/2015/SCIENCE/12/17/broccoli/index.html#innSTCVideo.

Posting to an e-mail distribution list:

[12] E. Vasquez (2017 Sept. 24) *Research ethics question* [ATTW-L Distribution List Message]. Available: http://attw.interversity.org/read/messages?id100170924

Posting to a blog or bulletin board:

[13] M. R. Taylor, FDA Voice. (2017, Aug. 26) *Defining boundaries for caffeine in today's marketplace* [Blog]. Available: http://blogs.fda.gov/fdavoice/index.php/2017/08/defining-boundaries-for-caffeine-in-todays-marketplace/

Tweet:

[14] L. Rickard (lydiarickard). (2018, Jan. 12). *FDA regulations promote innovation in devices intended to treat obesity* [Tweet]. Available: http://twitter.com/lydiarickard/status/192837465001

Appendix C

Report for Study and Analysis

At ABC University, premedical majors may pursue an internship at a hospital during two summer terms. At the end of their internship, they write a formal report on a disease or condition that they believe to have major importance to medicine. Internship allows them to spend some of their time addressing a major research/medical topic. They have to first propose a topic at the end of their first summer internship, then submit a progress report during the next year, and then submit a final report at the end of their second summer internship.

Hannah Starke chose Alzheimer's as she saw the ravages of this disease during her internship.

MEMORANDUM

To: Elizabeth Tebeaux
From: Hannah Starke
Date: 17 April 2017
Subject: Transmittal of the Final Report on the Mechanics and Treatment of Alzheimer's Disease

Attached to this memo is my final report on Alzheimer's disease. This report provides an overview of the current state of research on Alzheimer's disease on the following topics:

- Alzheimer's effect on the affected individual
- Current physiological hypotheses on disease development
- Risk factors for developing the disease
- Preventative measures
- Current treatment options and their effectiveness

Alzheimer's slowly destroys the brain until the individual is left with no memories and an inability to function independently. Alzheimer's is a terrible fate. I have watched two people I love dearly succumb to this disease, and I can imagine no worse a fate. Their helplessness, confusion, and loss of self rip your heart in two, and I hope that at some point in my career, I will help those who suffer from this disease or prevent them from developing it at all.

The purpose of this report is to inform the reader of all the basic mechanisms that affect Alzheimer's development in individuals. In addition to that, I want to make sure people know that despite the fact that Alzheimer's cannot be cured or reversed in any way, individuals can take several actions to reduce their risk of disease development.

Thank you, Dr. Tebeaux, for your instruction, guidance, and complete dedication to your students. Being in your class reminds me of how it feels to be in the presence of someone who loves what she does and, more importantly, cares more about the success of those she teaches. Also, thank you to the library staff for demonstrating how to use the extensive database network available and fielding all questions I asked them.

Sincerely,

Hannah Starke

Final Report on the Mechanics and Treatment of Alzheimer's Disease

17 April 2017

Hannah Starke	Date

Abstract

This report serves to provide the reader with an overview of the current research on risk factors, disease development, and treatment options for Alzheimer's disease. In addition to overall summaries of the listed topics, this report highlights the possible future development associated with the research developments. The report itself contains a summary, works cited list, and original proposal documents.

Keywords:

Alzheimer's, Amyloid cascade hypothesis, Mitochondrial cascade hypothesis, Glucocorticoid hypothesis, Neurofibrillary tangles, Amyloid β protein

Final Report on the Mechanics and Treatment of Alzheimer's Disease

Table of Contents

List of Figures

Executive Summary

Report Purpose

I hope to educate the reader on all aspects of Alzheimer's disease and steps that he or she can take to try to reduce the risk of developing the disease. The incidence of Alzheimer's has increased in recent years. Actions that people take early in life can greatly influence their risk of developing a disease that changes everything. Understanding the actions you can take to decrease your chances of developing this disease may help you avoid it.

Research Methods

To collect information for the report, I first used medical help sites such as WebMD and the official websites of the Alzheimer's Association and National Institute of Aging to gain basic understanding of Alzheimer's disease and its progression. Next I used the TAMU library databases and free online journals such as *PLoS One* to gain a greater understanding of the mechanisms and technical problems associated with Alzheimer's. Despite continued efforts, I was unable to communicate with any neurological specialists at Baylor College of Medicine or any other institution.

Findings

Alzheimer's begins by attacking and killing brain cells. The brain damage causes horrible memory loss, personality change, and motor skill degeneration. An individual can begin having internal physiological signs of the disease two decades before any noticeable symptoms. Brain scans can detect these early developments, but most of the time detection and preventative measures come too late. Once the disease starts, nothing can stop its progression.

We do not know the exact mechanism of disease formation. As of now, no effective treatments exist to stop disease progression, but research continues to investigate the catalytic events responsible for the disease. With a more comprehensive understanding of what initiates disease formation, scientists can develop more effective treatments. These treatments work by targeting the central problem and stopping the disease before it begins.

Anyone can develop this disease. Genetic predispositions do exist and contribute to the likelihood of development, but genetics is not the deciding factor. Researchers have found several preventative measures that at-risk individuals can take to reduce the risk or slow down the progression of Alzheimer's. Several of these methods include dieting, exercise, and education.

Conclusions

My research and personal experiences have led me to firmly believe that Alzheimer's is one of the worst diseases a person can have. Everyone needs to have a general understanding of the risk factors associated with this disease and the small measures that one can take to reduce the risk of developing this disease. Avoiding any development and catching it early are the two most important steps. As someone who has several of the risk factors associated with developing Alzheimer's, I can assure you that I will do everything reported here. In future years, research advances will lead to effective treatments aimed at eliminating the progression of the disease.

Final Report on the Mechanics and Treatment of Alzheimer's Disease

Introduction?

Alzheimer's is the most common form of dementia, a blanket term for age-associated cognitive decline. Progression of Alzheimer's disease leads to decreases in memory storage, cognitive function, and basic task performance. To a family, Alzheimer's takes a loved one, but to the affected, the disease takes everything. Anyone who has ever watched a loved one struggle with this disease needs no reminder or explanation of the toll it takes on everyone involved.

With the aging of the baby boomer generation comes an increase in the number of individuals entering the high-risk age group. In 2010, almost 5 million individuals suffered from mild to severe Alzheimer's, and by 2050, that number is expected to rise to almost 14 million [1]. The disease takes another toll as well. In this year alone, the United States will spend about $259 billion on the treatment and care of Alzheimer's [2].

Due to these enormous medical and economic impacts on the nation, institutions have increased research to accelerate the creation of effective preventative treatments. Understanding how the disease begins and progresses helps guide this new research. Knowing how the disease advances allows for treatments aimed at stopping the disease before it begins. Through research scientists have developed three main hypotheses explaining how the disease develops. Each one of these hypotheses describes a different physiological mechanism responsible for the initial disease development.

No matter the catalyst, a few neurological changes characterize Alzheimer's:

- Protein aggregations in the brain called "plaques"
- Fibrous networks within brain cells
- Breaking down of synaptic (communicative) connections in the brain
- Neuron (brain cell) death

The progression of these neurological alterations causes the cognitive, psychological, and behavioral symptoms associated with Alzheimer's.

To date, no treatment effectively stops the progression of the disease, but individuals can still reduce their risk of disease development through a few simple actions. A healthy lifestyle, aerobic exercise, and continued educational stimulation all correlate with a reduced risk of Alzheimer's.

This report outlines the following aspects of Alzheimer's:

• General progression
• Developmental hypotheses
• Associated risk factors
• Treatment options available

The most important point to take away from this report is that anyone can develop Alzheimer's disease. Genetic predispositions play a role in who develops this disease, but many other factors contribute. Despite the fact that everyone is at risk, individuals can improve their chances of not developing the disease in several ways.

Alzheimer's Disease

What Is Alzheimer's Disease?

Alzheimer's is a neurodegenerative disease facilitated by neuronal destruction. As the destruction progresses, the affected individual experiences increased memory loss, decreased cognitive function, personality change, and, eventually, inability to care for oneself. The characeristcs of Alzheimer's, the buildup of protein plaques in the brain and fibrous networks (neurofibrillary tangles or NTs) in the neurons, cause cell destruction (Figure 1). Plaques of amyloid β (Aβ) protein build up in the spaces between neurons and cause cell death mainly by destroying cellular communication sites called "synapses." Synapse destruction inhibits message and nutrient transfer between neurons. Scientists are unsure if these plaques are the cause or a by-product of the disease, as discussed later in this report. Within neurons NTs occur when a protein named "tau" begins interacting uncharaceristically with itself [1]. Most likely, Aβ facilitates this unhealthy interaction.

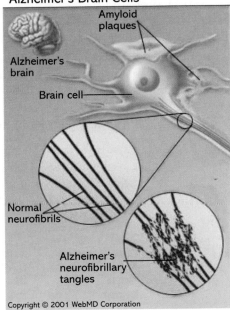

Alzheimer's Brain Cells

Amyloid plaques

Alzheimer's brain

Brain cell

Normal neurofibrils

Alzheimer's neurofibrillary tangles

Copyright © 2001 WebMD Corporation

Figure 1 Neurofibrillary Tangles and Amyloid Plaques. Courtesy of: The Doctor Weighs In http://www.thedoctorweighsin.com/alzheimers-disease-is-there-daylight-at-the-end-of-the-tunnel-or-is-it-an-oncoming-train/

The first site of deterioration in the brain is the hippocampus, the area responsible for memory creation and storage [1]. Accordingly, memory loss is the first symptom of Alzheimer's. Brain damage continues beyond the hippocampus into sites responsible for judgment, personality, speech, and basic motor function. The widespread destruction in the brain leads to progressively less functional individuals (Figure 2). Doctors characterize a person's disease severity through a ranking system. The seven stages of Alzheimer's describe the relative impairment as the disease progresses and affects different areas of the brain [3].

- Stage 1: no signs of disease development in the brain but no cognitive impairment
- Stage 2: very mild cognitive impairment—memory loss of small facts
- Stage 3: mild cognitive impairment—more noticeable memory loss detectable by others, especially short-term memory loss
- Stage 4: moderate cognitive decline—higher cognitive function deterioration begins
- Stage 5: moderately severe cognitive decline—long-term memory deterioration and loss of judgment, moderate assistance necessary
- Stage 6: severe cognitive decline—personality change and constant assistance needed
- Stage 7: very severe cognitive decline—inability to interpret surroundings, loss of motor movement and muscle coordination, eventual death

Due to the extensive brain damage, Alzheimer's eventually kills. The degeneration of the brain leads to impaired ability to perform basic motor skills including bathing, talking, eating, swallowing, and controlling bladder and bowel elimination. Without these basic skills, the affected individual becomes susceptible to

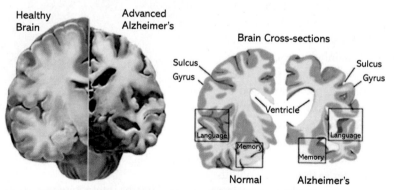

Figure 2 Normal Brain vs. Alzheimer's Brain. Courtesy of: Highlight Health
http://www.highlighthealth.com/neurological-disorders/amyloid-deposits-in-cognitively-normal-people-may-predict-risk-for-alzheimers-disease/

contracting a deadly infection, most commonly pneumonia. If infection does not kill the individual first, the disease will eventually progress to the point where the affected person can no longer breathe without mechanical assistance [4].

Why Is Alzheimer's Disease Important?

Today, Alzheimer's is the 6th leading cause of death in the United States. It is the only one in the top 10 which cannot be prevented, cured, or significantly slowed [2]. The detrimental effects of Alzheimer's reach beyond the loss of life however. For families, caring for an affected individual takes great emotional and psychological tolls. Watching someone you love forget who you are and lose every aspect of his or her personality are two of the most painful human experiences.

For the country, a great economic burden associated with caring and curing the disease exists. The economic toll comes from two sources. First, many family members give up careers to provide constant care for affected loved ones. Caring for an individual with Alzheimer's requires money for medication, doctors' visits, and general care. The loss of wages leads to economic instability for the family [5]. Second, the United States spends an enormous amount of money on caring for those with Alzheimer's. The Alzheimer's Association expects the United States to spend $1.1 trillion in 2050 on Alzheimer's care [1]. As seen in Figure 3, the majority of money is spent on Medicare and Medicaid programs, which facilitate nursing home enrollment and long-term care. This incredible amount of money adds to the ever-worsening economic state of the country.

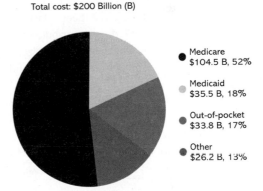

Total cost: $200 Billion (B)

Medicare $104.5 B, 52%

Medicaid $35.5 B, 18%

Out-of-pocket $33.8 B, 17%

Other $26.2 B, 13%

Figure 3 Total US Spending on Alzheimer's, 2012. Courtesy of the Alzheimer's Association [2].

Lastly, for the generation following the affected individual, the genetic link to Alzheimer's takes a cognitive toll. You do not need to have memory loss to be affected by Alzheimer's. You can see physiological findings consistent with Alzheimer's on a magnetic resonance image two decades before any symptoms arise. Having genetic risk factors of Alzheimer's alters cognition even in children. Children who expressed the gene *ApoE4* (discussed later) and had a family history of Alzheimer's scored significantly lower on cognitive tests than those lacking either risk factor [6]. Also, some people who develop Alzheimer's report having reduced memorization or cognitive skills throughout their lives. Finding effective prevention treatments could reduce all three of these detrimental effects.

Physiological Hypotheses

Several hypotheses exist to explain Alzheimer's development, each based on different physiological mechanisms leading to the morphological changes characteristic of Alzheimer's. These hypotheses differ in their explanation of the first step of disease development but not the physical changes of neural anatomy. The following sections highlight the basics of each of these hypotheses and the possible treatment importance of each.

Amyloid Cascade Hypothesis

Since 1992, the amyloid cascade hypothesis (ACH) has become the most widely accepted hypothesis of Alzheimer's development. The ACH proposes that the deposition of Aβ protein, the major component in protein plaques, precedes the development of NTs and neuron death [7]. The theory starts with the synthesis of excess Aβ. Several factors, including genetic ones, contribute to the synthesis of a protein called "APP," which later forms Aβ. This newly synthesized protein aggregates outside the cell or breaks into smaller fragments, enters the cell, and alters cellular processes (Figure 4) [8]. In addition to its role in synaptic degradation discussed earlier, Aβ initiates events in the cell, including tau alteration and mitochondrial dysfunction, by interacting with receptors on the cells. The Aβ protein's interaction with extracellular receptors activates enzymatic proteins within the cell responsible for morphological changes. Scientists have not yet identified the exact mechanism for mitochondrial dysfunction, but isolated Aβ alters the protein tau in a way that contributes to the development of NTs seen in neurons [9].

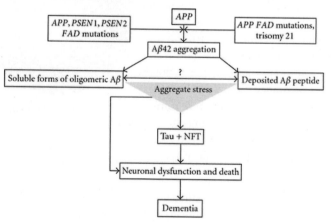

Figure 4 Amyloid Cascade Hypothesis. Courtesy of C. Reitz [8].

Most novel treatments revolve around this hypothetical progression. The treatments target regulatory mechanisms of Aβ through receptor interactions or destruction of free Aβ. Scientists recently discovered a breakthrough treatment utilizing protein antibodies specific for a related protein. I discuss the details later in this report.

Mitochondrial Cascade Hypothesis

The mitochondrial cascade hypothesis (MCH) challenged the ACH by proposing that mitochondrial dysfunction precedes Aβ protein aggregation [10]. Mitochondria generate usable energy (in the form of adenosine triphosphate) necessary for normal cellular function. Neurons require a lot of energy for communication, transport, and growth. The brain only constitutes about 2% of a person's body weight but requires at least 20% of all energy generated. The DNA which resides in the mitochondria regulates function through protein synthesis. If mutations exist in DNA regions encoding vital proteins, the mitochondria will produce malfunctioning proteins which decrease energy production. Proteins outside the mitochondria can alter activity as much as genetics. The Aβ precursor APP blocks nutrient transport channels leading into the mitochondrial matrix, hindering energy production.

There exists an activity threshold required to maintain normal cellular function. When activity drops below the threshold, energy generation decreases and neuronal dynamics deteriorate (Figure 5). Mitochondrial dysfunction causes neuron death by

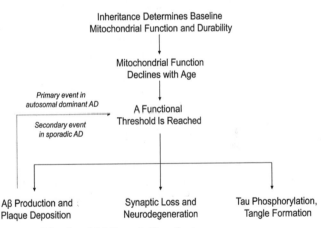

Figure 5 Mitochondrial Cascade Hypothesis.
Courtesy of Swerdlow, R.H. [10].

disrupting synaptic connections long before plaques begin developing; neuron death marks the beginning of Alzheimer's. In addition to cell death, a decrease in mitochondrial activity allows an increase in Aβ production and tau modification in neurons. The glucocorticoid cascade hypothesis argues that the increase in Aβ does not start the cascade, as argued by the ACH; it merely contributes to continued degeneration.

Treatment options pinpointing mitochondrial dysfunction are difficult to test and implement. Altering mitochondrial activity affects every part of the body. Recent research shows that some of the dysfunctions found in neurons affect other cells in the body, most notably platelets. Though the following theory has not been tested extensively, it may propel future studies: Alzheimer's may not be just a neurological disease. Dysfunction in other cells and tissues may contribute to disease development in unknown ways. Once identified, more extensive intracellular relationships may provide researchers targets for experimental treatments.

Glucocorticoid Cascade Hypothesis

The glucocorticoid cascade hypothesis (GCH), the last of the major hypotheses, proposes that long-term depression and exposure to high levels of stress hormones (glucocorticoids) initiate hippocampal damage. In a normal body, stress hormones regulate their own release by binding to receptors in the brain and signaling for the cessation of hormone release. However, long-term elevated stress hormone levels cause destruction of the regulatory receptors, thereby causing hypersecretion. General hormone insensitivity also accompanies prolonged exposure to high cortisol and contributes to increased glucocorticoid circulation. Great increases in hormone concentrations in the brain cause damage to neurons through a variety of mechanisms. The primary mechanism involves the inhibition of glucose uptake into brain cells, which causes brain damage [11].

In addition to direct cell death, high levels of cortisol, the primary stress hormone in humans, accelerate disease progression by contributing to the toxicity and deposition of Aβ and tau proteins in the brain [12]. Hormone mechanisms are incredibly complex, so the exact manner in which they contribute to Aβ and tau deposition is not fully understood. In addition to effectively reducing cortisol levels in the blood, treatment options based on this hypothesis would target maintaining those receptors responsible for muting the stress response.

Risk Factors

Who Is at Risk?

The Alzheimer's Association reports that one in eight individuals above the age of 65 develops the disease and that about half of individuals 85 and older develop it [1]. Several genetic and health factors contribute to Alzheimer's (discussed in the following sections), but everyone is at risk for developing Alzheimer's. As a person ages, neurological decline occurs. The process occurs naturally with aging. Think of it this way: you can physically see the effects of muscle degradation in the elderly, and this degradation occurs due to protein breakdown in tissues over a lifetime. Just like muscles, internal processes depend on proteins, which also degrade. All the internal processes which keep you alive get worn out just like muscles; you just cannot see the effects on the outside. If degradation proceeds far enough, the same kind of damage which occurs in Alzheimer's happens.

Genetic Factors

Mitochondrial DNA carries genes which determine baseline function and durability [10]. When the mitochondria stop working, cell death ensues. Every person inherits mitochondria and mitochondrial DNA from his or her mother. Out of individuals with affected parents, those who had affected mothers had a greater risk of developing Alzheimer's than those who had affected fathers. Whether or not females are more susceptible to Alzheimer's than males has not been investigated thoroughly, but some initial findings do suggest gender linkage [13]. The bottom line is that individuals with mothers affected with Alzheimer's should familiarize themselves with preventative measures. The genetic link cannot be avoided.

Genes found in the human genome (as opposed to the mitochondrial genome) also contribute to disease development. Individuals carrying "the Fatso gene," a gene linked to obesity, appear to have a higher risk of developing Alzheimer's. This increase in susceptibility is because the presence of this gene appears to contribute to brain atrophy, or withering, a contributing factor in disease development [14]. This genetic linkage poses a problem in the United States since an estimated 43% of people of western European descent carry this gene [15]. Other forms of obesity also contribute to Alzheimer's development. Most likely this occurs due to the poor overall health associated with obesity.

Another genomic linkage comes from the encoding of apoproteins, proteins which circulate around the body and deliver lipids to tissues. One specific form of apoprotein E, ApoE4, binds to receptors on brain cells and initiates the generation of Aβ protein. The ApoE4 form promotes this generation more than the other forms E2 and E3 [16]. High concentrations of E4 cause a great increase in Aβ production and buildup. Individuals who inherited the gene for ApoE4 from one parent are three times more likely to develop Alzheimer's than normal individuals, and individuals who inherited it from both parents are ten times more likely to develop Alzheimer's [17]. The presence of ApoE4 appears to correlate most closely with Alzheimer's development.

Early Life Risk Factors

Starting from birth, several factors influence a person's likelihood to develop Alzheimer's. Obesity, chronic stress, vascular disease, head trauma, and other factors increase someone's risk for developing the disease. This section will break down a few of the more prevalent contributing factors.

First, individuals who have had any kind of brain trauma have a much higher risk of developing neurodegenerative diseases including Alzheimer's. Many of these brain injuries cause irreversible brain damage by directly destroying neurons. These small clusters of dead neurons are easy targets for protein buildup and continued degeneration. Athletes, war veterans, and car accident victims with associated head trauma all have an increased risk of developing a neurodegenerative disease [18]. Many adolescent athletes have suffered at least one concussion during their athletic careers. Without realizing it, they increased their risk of developing neurodegenerative diseases with every hard hit taken. Though much of the time people cannot avoid accidents, avoiding brain injuries all together greatly reduced their risk for Alzheimer's.

As described before with the GCH, chronically depressed individuals are more at risk than others to develop Alzheimer's. Prolonged exposure to cortisol and other stress hormones causes neuron death and Aβ production in the hippocampus. Chronic depression arising during adolescence and continuing through much of adult life contributes to disease formation [12]. The longer an individual has suffered from depression, the more susceptible he or she becomes.

Individuals with vascular disease, such as high blood pressure, high cholesterol, and diabetes, have a higher risk for developing Alzheimer's than others. Some studies show that cardiovascular disease is the second most important indicator for Alzheimer's [19]. The vascular destruction that accompanies high blood pressure affects the vasculature in the brain as much as the rest of the body. Decreased blood flow to the brain causes a decrease in nutrient delivery. A decrease in nutrient delivery will eventually lead to mitochondrial dysfunction and cell death.

Lastly, lower education levels and less cognitively driven careers both seem to correlate with higher incidence of Alzheimer's. Learning, memorizing, and processing require the formation of numerous new connections between brain cells. The increased cognitive reserve resists damage [20]. It would take longer to degrade the theoretical increase in intracellular connections, delaying the progression of the disease. Though statistical studies have shown this correlation, it is difficult to prove this anatomical theory due to the inability to cut into a brain and investigate neuronal synapses.

Treatment Options

Alzheimer's has no cure. Once the disease develops, you cannot stop or reverse physiological changes. Once the catalytic event has occurred, it cannot be undone. However, several treatment options exist to help slow down disease progression or reduce the risk of initial development. The following sections dissect several of these options.

Disease Prevention

Since heart and vascular diseases contribute heavily to the development of Alzheimer's, individuals must maintain a healthy heart and vascular system. Staying healthy, avoiding diabetes, not smoking, and lowering blood pressure all promote healthy heart and vascular health and exponentially reduce the risk of developing Alzheimer's [21]. The characteristic vascular degeneration and buildup of cholesterol and other fatty components which accompany vascular disease cause reduced blood flow to the brain. This reduction correlates strongly to Alzheimer's development in elderly individuals. Maintaining a healthy lifestyle greatly reduces an individual's risk of Alzheimer's.

Though exercise is an important component of healthy living, it needs its own section. I cannot overstate the importance of exercising to maintain cognitive health. Numerous studies have found that exercise, most importantly exercise during the teenage years, decreases cognitive decline in the elderly. The probable mechanism is threefold: (1) exercise increases overall health, (2) exercise increases neuronal connections, and (3) exercise increases the concentration of a certain protein in the brain responsible for hippocampal growth and healthy brain function in adults [1]. A study investigating the effect of exercise on plaque formation in mice showed a marked decrease in plaque buildup in the brain [22]. Exercise during all stages of life helps improve every aspect of health in an individual. If you learn nothing else from this report, learn this: never stop exercising.

In addition to maintaining a healthy lifestyle, performing cognitive exercises can improve mental ability and decrease deterioration. Scientists debate the effectiveness of cognitive exercises. As discussed before, cognitive processes require new intracellular connections. Continued educational pursuits in any capacity could help prevent or delay the onset of Alzheimer's.

Over-the-Counter Remedies

Since Alzheimer's gained prominence in the medical world, people have tried to find ways to reduce occurrence with natural, inexpensive remedies. Dietary supplements and so-called medical foods have swept the Alzheimer's world. These alternatives to medical intervention appeal to those individuals wary of medicine or incapable of purchasing prescriptions.

Several companies claim that some dietary supplements will delay the onset of Alzheimer's. Some of these supplements pose no inherent risk to an individual, but some do. The following have not been found to have any significant effect on the development of Alzheimer's [23]:

- Caprylic acid and coconut oil
- Coenzyme Q
- Coral calcium
- *Ginkgo biloba*
- Huperzine A
- Omega-3 fatty acids
- Phosphatidylserine
- Tramiprosate
- Vitamin E [24]

Since dietary supplements are not subjected to the testing required by the US Food and Drug Administration, they can often do more harm than good. Several of these medicinal alternatives contain biological contaminants, interfere with prescription medications, or cause an overdose if taken too liberally. Compounds like omega-3 fatty acids have general health benefits, but they do not necessarily reduce an individual's risk of developing Alzheimer's [23]. The effectiveness of vitamin E drives one of the largest debates in Alzheimer's treatment. Its overall antioxidant and neurological properties lead physicians to insist that it slows cognitive decline. For that reason, vitamin E often supplements Alzheimer's treatment plans [1, 23].

Though most of the advertised supplements have no effect on disease prevention, some may. Scientists have found that high doses of vitamin B reduced the amount of brain atrophy in elderly individuals with mild cognitive impairment [25]. Vitamin D also appears to slow the rate of brain atrophy. Individuals deficient in vitamin D experience faster rates of decline than those with no deficiency [1]. Since Alzheimer's often correlates with heart disease, supplements which reduce cardiovascular disease could reduce Alzheimer's occurrence [25].

Medical Intervention

No cure exists, and there is no way to reverse changes once they begin; however, some medications exist for treating Alzheimer's symptoms. First, physicians prescribe medications which treat the behavioral symptoms of Alzheimer's. These include regular stimulants and tranquilizers prescribed for healthy individuals. Second, some patients qualify for medications to treat the neurological symptoms and slow down disease progression. These medications fall into one of two categories: cholinesterase inhibitors and *N*-methyl-D-aspartate (NMDA) receptor antagonists. (See Appendix C for specific name brand drug details.)

Cholinesterase inhibitors act to decrease the enzyme responsible for breaking down acetylcholine, a chemical messenger important in memory, judgment, and other thought processes. The effects of these drugs are small, but they can delay or slow symptom worsening for a period of time (typically no longer than one year). Physicians frequently prescribe cholinesterase inhibitors for the treatment of mild to moderate Alzheimer's [26]. Unfortunately, these drugs only work on about half of the medicated individuals, another indicator of disease complexity.

The NMDA receptor antagonists work by blocking NMDA receptors in the brain. They work by regulating the activity of glutamate, a chemical involved in information processing, storage, and retrieval. Glutamate plays an essential role in learning and memory by triggering NMDA receptors to allow a controlled amount of calcium to flow into a nerve cell. Excess glutamate overstimulates these receptors, allowing too much calcium to enter the cell. Too much calcium causes disruption of cellular processes and cell death. The NMDA receptor antagonists are used more frequently for the treatment of severe Alzheimer's where overstimulation of these receptors can be seen [26]. Again, these drugs do not help every prescribed individual and do not prevent disease progression. They provide short-term relief from some of the worst symptoms.

In March 2012, researchers published a study investigating a new treatment targeting Aβ protein. This new treatment utilized antibodies for a protein called "DKK1," a closely related protein to Aβ. The results showed a clear decrease in synaptic toxicity of Aβ and cessation of disease progression [27]. Scientists used rat models to test this treatment option; but with the promising results, continued research could potentiate an effective treatment option.

Summary and Conclusions

Alzheimer's is a terrible disease, and anyone can develop it. Unlike other neurological diseases, like Huntington's disease, genetic predispositions are not the necessary condition. Old age, early life injuries, heart disease, and many other, common factors exponentially increase your likelihood of developing the disease.

As the ever-growing population ages, the importance of finding treatment options increases. In the coming years, the prevalence of Alzheimer's will drastically increase, leading to larger economic and medical burdens in the United States (Figure 6). The general population does not know how to identify risk factors or successfully live a life that could ward off Alzheimer's. Individuals can get genetic testing to determine if they have genetic predispositions like ApoE4 or not. Lack of knowledge could theoretically further increase disease incidence. As this disease gains recognition, people need to be informed about the possible implications and avoidance techniques.

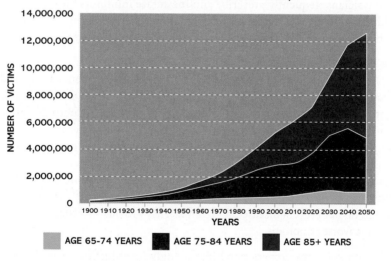

PREVALENCE OF ALZHEIMER'S DISEASE
(BY DECADES IN U.S.A. FROM 1900–2050)

This graph portrays how many Americans over the age of 65 are currently affected by Alzheimer's, and a projection of how many more will become affected with it as time passes.
w3.ouhsc.edu

Figure 6 Increase in Alzheimer's through 2050. Courtesy of Stanford University http://folding.stanford.edu/English/FAQ-Diseases

As research continues, scientists find key mechanisms to attack for more effective treatments. Like with most diseases, early detection and treatment are pivotal. Since right now no effect treatments exist to combat early stages of the disease, the most important thing you can do is prevent any development. Simple, healthier lifestyle changes can reduce the likelihood of developing Alzheimer's.

Works Cited

1. US Department of Health and Health Services. *2013–2014 Alzheimer's Disease Progress Report: Insights and Challenges.* National Institutes of Health, 2014, https://www.nia.nih.gov/alzheimers/publication/2013–2014-alzheimers-disease-progress-report.

2. Alzheimer's Association. *2017 Alzheimer's Disease Facts and Figures.* Alzheimer's Association, 2017, https://www.alz.org/documents_custom/2017-facts-and-figures.pdf.

3. "Seven Stages of Alzheimer's." Alzheimer's Association, 2012, http://alz.org/alzheimers_disease_stages_of_alzheimers.asp.

4. "How Does Alzheimer's Kill?" *Slate Magazine* (April 30, 2001), http://www.slate.com/articles/news_and_politics/explainer/2001/04/how_does_alzheimers_kill.

5. Ernst RL, Hay JW. "The US Economic and Social Costs of Alzheimer's Disease Revisited." *American Journal of Public Health* 84, no. 8 (1994): 1261–1264.

6. Bloss CS, Delis DC, Salmon DP, Bondi MW. "Decreased Cognition in Children with Risk Factors for Alzheimer's Disease." *Biological Psychiatry* 64, no. 10 (2008): 904–906.

7. Hardy JA, Higgins GA. "Alzheimer's Disease: The Amyloid Cascade Hypothesis." *Science* 256, no. 5054 (1992): 184–185.

8. Reitz C. "Alzheimer's Disease and the Amyloid Cascade Hypothesis: A Critical Review." *International Journal of Alzheimer's* (2012).

9. Jin M, Shepardson N, Yang T, et al. "Soluble Amyloid Beta-Protein Dimers Isolated from Alzheimer Cortex Directly Induce Tau Hyperphosphorylation and Neuritic Degeneration." *Proceedings of the National Academy of Sciences USA* 108 (2011): 5819–5824.

10. Swerdlow RH. "Mitochondria and Cell Bioenergetics: Increasingly Recognized Components and a Possible Etiologic Cause of Alzheimer's Disease." *Antioxidants and Redox Signaling* 16, no. 12 (2012): 1434–1455.

11. Sapolsky RM, Krey LC, McEwen BS. "The Neuroendocrinology of Stress and Aging: The Glucocorticoid Cascade Hypothesis." *Science of Aging Knowledge Environment* 38 (2002).

12. Aznar S, Knudsen GM. "Depression and Alzheimer's Disease: Is Stress the Initiating Factor in a Common Neuropathological Cascade?" *Journal of Alzheimer's Disease* 23, no. 2 (2011): 177–193.

13. Edland SD, Silverman JM, Peskind ER, Tsuang D, Wijsman E, Morns JC. "Increased Risk of Dementia in Mothers of Alzheimer's Disease Cases: Evidence for Maternal Inheritance." *Neurology* 4 (1996): 254–256.

14. Ho AJ. "A Commonly Carried Allele of the Obesity-Related FTO Gene Is Associated with Reduced Brain Volume in the Healthy Elderly." *Proceedings of the National Academy of Sciences USA* 107, no. 18 (2010): 8404–8409.

15. James F. " 'Fatso Gene' Linked to Alzheimer's." National Public Radio, April 19, 2010, http://www.npr.org/blogs/thetwo-way/2010/04/fatso_gene_linked_to_alzheimer.html?ps=rs.

16. He X, Cooley K, Chung CHY, Dashti N, Tang J. "Apolipoprotein Receptor 2 and X11α/β Mediate Apolipoprotein E–Induced Endocytosis of Amyloid β Precursor Protein and β-Secretase, Leading to Amyloid β Production." *Journal of Neuroscience* 27, no. 15 (2007): 4052–4060.

17. Oklahoma Medical Research Foundation. "Why ApoE4 Increases Alzheimer's Risk." Wellcome Trust. October 4, 2007, http://genome.wellcome.ac.uk/doc_WTX038956.html.

18. "Brain Injury May More than Double Dementia Risk in Older Veterans." Alzheimer's Association, July 18, 2011, http://www.alz.org/aaic/monday_1230amCT_news_release_brain_injury.asp.

19. Borenstein AR, Wu Y, Mortimer JA, Schellenberg GD, McCormick WC, Bowen JD, McCurry A, Larson EB. "Developmental and Vascular Risk Factors for Alzheimer's Disease." *Neurobiology of Aging* 26, no. 3 (2005): 325–334.

20. American Academy of Neurology. "Low Education Level Linked to Alzheimer's, Study Shows." ScienceDaily, October 1, 2007, http://www.sciencedaily.com/releases/2007/10/071001172855.htm.

21. Raymond R. "Evidence Lacking to Support Alzheimer's Prevention." National Public Radio, April 28, 2010, http://www.npr.org/blogs/health/2010/04/alzheimers_prevention_exercise.html?ps=rs.

22. Lunde A. "Preventing Alzheimer's: Exercise Still Best Bet." Mayo Clinic, March 25, 2008, http://www.mayoclinic.com/health/alzheimers/MY00002.

23. "Alternative Treatments." Alzheimer's Association, 2012, http://www.alz.org/alzheimers_disease_alternative_treatments.asp.

24. Butler R. "Can Vitamins Prevent Alzheimer's?" Everyday Health, November 20, 2006, http://www.everydayhealth.com/specialists/senior_aging/butler/qa/can-vitamins-prevent-alzheimers/index.aspx.

25. Russell P. "Vitamin B May Help Prevent Mental Decline in Elderly." WebMD.com, September 9, 2010, http://www.webmd.com/brain/news/20100909/vitamin-b-may-help-preventmental-decline-elderly.

26. Morris J. "Cognitive Symptoms." Alzheimer's Association, 2012, http://www.alz.org/professionals_and_researchers_cognitive_symptoms.asp.

27. Purro SA, Dickens EM, Salinas PC. "The Secreted Wnt Antagonist Dickkopf-1 Is Required for Amyloid β Mediated Synaptic Loss." *Journal of Neuroscience* 32, no. 10 (2012): 3492–3498.

28. "Alzheimer's Disease Medications Fact Sheet." National Institute on Aging, December 13, 2011, http://www.nia.nih.gov/alzheimers/publication/alzheimers-disease-medications-fact-sheet.

APPENDIX A: Proposal

MEMORANDUM

DATE:	February 14, 2017
TO:	Dr. Elizabeth Tebeaux
FROM:	Hannah Starke
SUBJECT:	**Proposal for Report on Alzheimer's Disease**

Project Summary

The term *dementia* describes a wide range of diseases which cause destruction of brain cells and eventual performance impairment. Alzheimer's disease is the most well known of these disorders. Alzheimer's is a neurodegenerative disease which impairs one's ability to remember, think, and perform basic tasks; and the disease leads to death. This debilitating disease changes every aspect of the lives of the afflicted individual and his or her loved ones. Though not fully known, causes for the disease have been linked to some genetic and environmental factors. The time is now to understand this disease because with an ever aging population comes the possibility for a large increase in occurrence of Alzheimer's disease.

I will consult online medical journals and consumer health articles to gain a better understanding of this disease. People have a hard time comprehending the disease in its most basic sense, and that misunderstanding is due to improper communication between the general public and those researching and treating the disease. I hope to present the information in a concise, readable manner so that readers will understand the risks and what they can do to try to prevent the disease.

This project has two main objectives:

- Investigate the biological processes and genetic factors responsible for the development of Alzheimer's disease
- Discuss recent medical and homeopathic remedies found for the prevention and treatment of Alzheimer's

Introduction and Rationale

It is estimated that today 2.4 to 5.1 million people in the United States alone suffer from Alzheimer's disease. The economic and psychological tolls that this disease has on those afflicted and those who care for the afflicted are astronomical, and those issues will only increase since recent studies estimate that the number of cases will triple by 2050.

Just as the economic impact of the disease increases every year, research on treatment options has as well. The primary at-risk age group is 65 and older, and with the baby boomer population quickly entering into that age group, research has greatly increased to try to answer questions regarding the mechanics of the disease and to prevent its development in the population.

Alzheimer's disease plays a large role in my life. My maternal grandmother was diagnosed with the disease when I was four years old, and by the time I was six she had no recollection of who I was. I remember my grandmother asking my mom, "Who is that little girl standing by the door?" every time we visited. In December 1998, Alzheimer's took her life. With recent research showing that there may be a genetic link flowing through the maternal bloodline, the time is now for me to begin to understand the steps that could be taken to reduce my risk of developing this disease. In addition, this past December, a woman so close to me that I call her my mother was diagnosed with the same disease, and she seems to be fading fast. I am a junior in college on my way to medical school, and yet she asks me and those around me if I like finally being a high school student. I have gone through caring for someone with this disease, and I never want someone I love to feel the way that I have when I realize someone I care so deeply about cannot remember who I am. I seek to shine light on this disease and all the ways that people can try to prevent it.

Research Methods

In order to begin proper research, basic information about the disease needs to be understood. There are many intricacies associated with this disease, so I must find sources that expect no prior knowledge. The best sources used in order to achieve this come from the Alzheimer's Association, the National Institute on Aging, and MedlinePlus. These sources require no prior technical knowledge on behalf of the reader in order to understand.

To gain further understanding of the mechanics of the disease, I will consult more technical journals. Databases such as the National Institute of Health's PubMed and EBSCO Medline will link me to journals such as *PLoS ONE*, the *New England Journal of Medicine*, *Neurology*, *Nature*, and other advanced scientific journals. This will provide me with an intimate familiarity of how the disease acts on the body and the biomechanics of treatment options. Gaining a more intricate understanding of the disease will allow me to synthesize all necessary data.

I hope to complete interviews with several physicians at Baylor College of Medicine in Houston. They will be able to shed light on the most recent developments and possible implications from those findings.

Tentative Outline for the Report

I. Summary/Abstract
II. Introduction
    a. What is Alzheimer's and why is it important?
    b. What the reader needs to learn from this report
III. Rationale for report
    a. What the recent research is showing as the physiological issues associated with the progression of the disease
    b. Why the general population needs to be well informed on Alzheimer's
        i. Who is at risk
IV. Risk factors for the disease
    a. Genetic factors
    b. Child development
V. Current treatment options
    a. Homeopathic remedies
    b. Preventative measures
    c. Medical intervention methods and their effectiveness
VI. Summary and Conclusion

Initial Bibliography

Abbott A. "Dementia: A Problem for Our Age." *Nature* 475, no. 7355 (2011): S2–S4.

"About Alzheimer's Disease: Treatment." National Institute on Aging, 2011. http://www.nia.nih.gov/alzheimers/topics/treatment.

"Alternative Treatments." Alzheimer's Association, 2012, http://www.alz.org/alzheimers_disease_alternative_treatments.asp.

Bloss CS, Delis DC, Salmon DP, Bondi MW. "Decreased Cognition in Children with Risk Factors for Alzheimer's Disease." *Biological Psychiatry* 64, no. 10 (2008): 904–906.

Honea RA, Swerdlow RH, Vidoni ED, Burns JM. "Progressive Regional Atrophy in Normal Adults with a Maternal History of Alzheimer Disease." *Neurology* 76, no. 9 (2011): 822–829.

Jin M, Shepardson N, Yang T, et al. "Soluble Amyloid Beta-Protein Dimers Isolated from Alzheimer Cortex Directly Induce Tau Hyperphosphorylation and

Neuritic Degeneration." *Proceedings of the National Academy of Sciences USA* 108 (2011): 5819–5824.

Jucker M, Walker LC. "Pathogenic Protein Seeding in Alzheimer's Disease and Other Neurodegenerative Disorders." *Annals of Neurology* 70 (2011): 532–540.

Maruszak A, Żekanowski C. "Mitochondrial Dysfunction and Alzheimer's Disease." *Progress in Neuro-Psychopharmacology and Biological Psychiatry* 35, no. 2 (2011): 320–330.

US Department of Health and Health Services. *2013-2014 Alzheimer's Disease Progress Report: Insights and Challenges.* National Institutes of Health, 2014, https://www.nia.nih.gov/alzheimers/publication/2013-2014-alzheimers-disease-progress-report.

APPENDIX B: Progress Report

MEMORANDUM

To:	Elizabeth Tebeaux
From:	Hannah Starke
Date:	22 March 2017
Subject:	**Progress on Semester Project**

Summary

With my report, I hope to educate readers as much as possible on the mechanics, physiological hypotheses, risk factors, and treatment options of Alzheimer's disease (AD). I have found amazing, unbelievably interesting information on AD and everything associated with it. Since I have a lot of personal connections with the disease, I hope to educate myself as much as the reader as to how I can try to prevent the development of the disease.

Everyone has risk for developing AD. Some researchers say that around half of individuals 85 and older have AD, and this number will only continue to rise in the coming decades. Many of the neurological developments associated with AD are the same as natural processes associated with aging, but different characteristics distinguish AD from natural aging.

Many different hypotheses exist to explain how AD develops, each one with a key mechanical malfunction in the brain. All of these hypotheses provide a plausible explanation of the progression of AD. I will investigate three main hypotheses. In addition to the mechanical issues associated with AD, I have found several studies discussing risk factors associated with developing AD, many of which I had never heard. Early life experiences could carry serious implications for mental health and cognitive decline.

Though no cure exists, physicians use a variety of treatments to halt the progression of symptoms. Individuals can take a variety of steps to decrease the likelihood of developing the disease.

I have found amazing sources, but much work remains. I have created a great basis from which I can work, but I still have to find the real details and fill in the gaps.

Work Completed

Introduction/Background Information

- *2013–2014 Alzheimer's Disease Progress Report: Insights and Challenges*
  - This report provides a great, concise overview of the current status and basic information about AD and some basic research associated with it.
  - This is a great source. It will be used for sure.
  - US Department of Health and Health Services. *2013–2014 Alzheimer's Disease Progress Report: Insights and Challenges*. National Institutes of Health, 2014, https://www.nia.nih.gov/alzheimers/publication/2013-2014-alzheimers-disease-progress-report/.

- "Understanding Alzheimer's Disease: What You Need to Know."
  - The National Institute on Aging created this booklet to provide caregivers information on what to expect as loved ones develop AD. It highlights basic knowledge, beginning signs, treatment options, and other information that affected individuals and caregivers must know.
  - This is a good source; it provides basic information which may or may not help. Whether or not I use this source remains to be seen.
  - "Understanding Alzheimer's Disease: What you need to know." National Institute on Aging, June 2011, http://nia.nih.gov/sites/default/files/UnderstandingADfinal.pdf

- "Decreased Cognition in Children with Risk Factors for Alzheimer's Disease."
  - This article describes a very early finding, so the reader must take caution. However, the study found that children who possess two major risk factors for AD, a positive family history and the presence of the allele ApoE4, had decreased performance on cognitive testing compared to peers who had one or none of the risk factors. Risk factors for AD could cause decreased cognition in young people.
  - The researchers make an interesting discovery linking childhood development to possible disease development in the future. I will discuss this point.
  - Bloss CS, Delis DC, Salmon DP, Bondi MW. "Decreased Cognition in Children with Risk Factors for Alzheimer's Disease." *Biological Psychiatry* 64, no. 10 (2008): 904–906.

Physiological Hypotheses

Amyloid Cascade Hypothesis

- "Soluble Amyloid Beta-Protein Dimers Isolated from Alzheimer Cortex Directly Induce Tau Hyperphosphorylation and Neuritic Degeneration."
  - This study provides an extremely detailed look at a very specific morphological change in neuronal components found in the brain. Doctors use the formation of amyloid β-protein plaques in the brain as a diagnostic tool. This study took that information to another level and looked at the link between these proteins and neuronal degeneration.
  - I will most likely use this report. I need to find supplementing materials to help explain the technical details, but the article itself makes good points and provides possible explanations.
  - Jin M, Shepardson N, Yang T, et al. "Soluble Amyloid Beta-Protein Dimers Isolated from Alzheimer Cortex Directly Induce Tau Hyperphosphorylation and Neuritic Degeneration." *Proceedings of the National Academy of Sciences USA* 108 (2011): 5819–5824.

- "Alzheimer's Disease: The Amyloid Cascade Hypothesis."
  - This review provides a simple overview of what the amyloid cascade hypothesis states. It also connects it to events which could have occurred earlier in life. Very well done. Will certainly be used.
  - Hardy JA, Higgins GA. "Alzheimer's Disease: The Amyloid Cascade Hypothesis." *Science* 256, no. 5054 (1992): 184–185.

Mitochondrial Cascade Hypothesis

- "Mitochondrial Dysfunction and Alzheimer's Disease."
  - This article highlights a lesser-known hypothesis of AD progression called the "mitochondrial cascade hypothesis." This hypothesis postulates that mitochondrial degeneration plays a large role in the progression of AD through a lack of respiration and energy production. In addition to that, mitochondrial baseline function and durability may be genetically determined. This article outlines a variety of enzymes whose activity correlates with AD development.
  - This article will be used. It provides an amazing overview of a more complicated, more recent hypothesis for neurodegeneration.
  - Maruszak A, Zekanowski C. "Mitochondrial Dysfunction and Alzheimer's Disease." *Progress in Neuro-Psychopharmacology and Biological Psychiatry* 32, no. 5 (2011): 320–330.

Glucocorticoid Cascade Hypothesis

- "Depression and Alzheimer's Disease: Is Stress the Initiating Factor in a Common Neuropathological Cascade?"
    - This article suggests that there exists an important link between depressive disorders earlier in life and the development of AD later in life.
    - I had never heard of this linkage before, so this article spurred my desire to look further into this connection. I will use this and other articles on this topic.
    - Aznar S, Knudsen GM. "Depression and Alzheimer's Disease: Is Stress the Initiating Factor in a Common Neuropathological Cascade?" *Journal of Alzheimer's Disease* 23, no. 2 (2011): 177–193.

- "The Neuroendrocrinology of Stress and Aging: The Glucocorticoid Cascade Hypothesis."
    - Overview of the glucocorticoid hypothesis. Good.
    - Sapolsky RM, Krey LC, McEwen BS. "The Neuroendocrinology of Stress and Aging: The Glucocorticoid Cascade Hypothesis." *Science of Aging Knowledge Environment* 38 (2002).

Risk Factors

Genetic Factors

- " 'Fatso Gene' Linked to Alzheimer's."
    - This short report provided great, basic information on a possible correlation between a gene deemed the "fatso gene," linked to the condition of obesity, and susceptibility to develop Alzheimer's. This could be a major problem because scientists estimate that one in three Americans have this gene.
    - This is a slightly different article, and even though it provides little information, the small fact that it does provide is incredibly interesting. It will provoke more research on the topic.
    - James, Frank. " 'Fatso Gene' Linked to Alzheimer's." National Public Radio, April 29, 2010, http://www.npr.org/blogs/thetwo-way/2010/04/fatso_gene_linked_to_alzheimer.html?ps=rs

- "Increased Risk of Dementia in Mothers of Alzheimer's Disease Cases: Evidence for Maternal Inheritance."
    - This study found increased risk of developing AD when an individual had an affected mother.
    - Findings were simple but proved important. I will use this source unless I find one with more conclusive, comprehensive results.

○ Edland SD, Silverman JM, Peskind ER, Tsuang D, Wijsman E, Morns JC. "Increased Risk of Dementia in Mothers of Alzheimer's Disease Cases: Evidence for Maternal Inheritance." *Neurology* 4 (1996): 254–256.

Early Life Risk Factors

See "Depression and Alzheimer's Disease: Is Stress the Initiating Factor in a Common Neuropathological Cascade?" above

• "Do Early Life Insults Contribute to the Late-Life Development of Parkinson and Alzheimer Diseases?"
  ○ Events that occurred early in life (as early as in the womb) appear to play a role in the development of AD. Also, AD and cardiovascular disease have similar risk factors. Obesity is a big risk factor. And AD may be linked to a unique form of insulin resistance in the brain.
  ○ This article discusses many of the lesser-known risk factors contributing to the development of AD. It is really fascinating.
  ○ Miller DB, O'Callaghan. "Do Early Life Insults Contribute to the Late-Life Development of Parkinson and Alzheimer Diseases?" *Metabolism* 57, suppl. 2 (2008): S44–S49.

• "Developmental and Vascular Risk Factors for Alzheimer's Disease."
  ○ The authors found that the constellation of risk factors depended on ApoE-ε4 status. In those with one or more ε4 alleles, developmental risk factors played an important role, whereas in those without ε4 alleles, vascular risk factors were more predictive of disease. More generally, they found that both vascular and developmental risk factors play a role in the development of AD.
  ○ I will use this article because it investigates other risk factors that people would not know about. People would not think of early life experiences as risk factors in the development of a disease like AD, but this study shows that there is probably a connection.
  ○ Borenstein AR, Wu Y, Mortimer JA, Schellenberg GD, McCormick WC, Bowen JD, McCurry A, Larson EB. "Developmental and Vascular Risk Factors for Alzheimer's Disease." *Neurobiology of Aging* 26, no. 3 (2005): 325–334.

Treatment Options

Preventative Measures/Lifestyle Changes?

- "Evidence Lacking to Support Alzheimer's Prevention."
  - Keep your cardiovascular health up, no smoking, eat a bunch of omega-3 fatty acids, be physically active, and so on.
  - I may or may not use this source. I will look for a more comprehensive source that discusses these points along with others.
  - Raymond, R. "Evidence Lacking to Support Alzheimer's Prevention." National Public Radio, April 28, 2010, http://www.npr.org/blogs/ health/2010/04/alzheimers_prevention_exercise.html?ps=rs.

- "How Exercise Might Help Keep Alzheimer's at Bay."
  - A panel convened by the NIH warned the public that it's not clear whether physical activity and associated exercises can prevent AD or other forms of mental decline. Mental exercise is another possible strategy that seems like a good idea to many AD researchers. After all, it appears to increase connections in the brain and perhaps makes the brain more resilient. Members of the panel found some benefit from omega-3 fatty acids like those in fish. But it found no convincing studies in people that antioxidants like vitamin E could make a difference.
  - The points presented here supplement basic findings. Overall, a strong report.
  - Hamilton, J. "How Exercise Might Help Keep Alzheimer's at Bay." National Public Radio, April 29, 2010, http://www.npr.org/templates/ story/story.php?storyId=126370279.

- "Preventing Alzheimer's: Exercise Still Best Bet."
  - Physical activity appears to inhibit brain changes in mice. Of 6,000 women aged 65+, those who were physically active showed less decline in mental abilities. Mice bred to have AD-like brain plaques and allowed to exercise had fewer plaques than those that were sedentary.
  - This article does highlight the advantages of exercise within a rodent model and a general trend in humans.
  - Lunde, A. "Preventing Alzheimer's: Exercise Still Best Bet." *Mayo Clinic*, March 25, 2008, http://www.mayoclinic.com/health/alzheimers/ MY00002.

Over-the-Counter Remedies

- "Vitamin B May Help Prevent Mental Decline in Elderly."
  - A University of Oxford study found that taking vitamin B tablets every day can reduce the rate of brain atrophy in older people with mild cognitive impairment by as much as half. Some B vitamins—folic acid, vitamin B_6, and vitamin B_{12}—control levels of the amino acid homocysteine in the blood, and high levels of homocysteine are associated with an increased risk of AD.
  - I will use this review in my report. It provides a simple understanding of a specific intervention the public knows.
  - Russell, P. "Vitamin B May Help Prevent Mental Decline in Elderly." WebMD.com, September 9, 2010, http://www.webmd.com/brain/news/20100909/vitamin-b-may-help-preventmental-decline-elderly.

- "Alternative Treatments."
  - This site provides an overview of some of the more well-known alternative treatments for AD as well as possible concerns that one should have before undertaking any treatment. Mainly, this site discusses the theories behind using caprylic acid and coconut oil, coenzyme Q_{10}, coral calcium, *Ginkgo biloba*, huperzine A, omega-3 fatty acids, phosphatidylserine, and Tramiprosate. It does a great job of highlighting whether or not researchers have found each treatment to be effective or not. Most have not been found to have much effect.
  - I will use this source in the report because it is a great overview.
  - "Alternative Treatments." Alzheimer's Association, 2012, http://www.alz.org/alzheimers_disease_alternative_treatments.asp.

- "Can Vitamins Prevent Alzheimer's?"
  - Some research has suggested that vitamin E might help, but a large-scale study published in the *New England Journal of Medicine* in June 2005 showed otherwise. It was the largest such study of vitamin E ever undertaken.
  - Factors that increase the risk of heart disease, such as high blood pressure and hardening of the arteries due to high cholesterol, appear to be related to an increased risk of AD. In fact, autopsies show that one-third of those who had "vascular dementia" (cognitive changes brought on by atherosclerosis of the brain) also had evidence of AD. Therefore, prevention and treatment of these risk factors and conditions may be good preventive medicine. That means keeping blood pressure under control, keeping cholesterol levels low, not smoking, and maintaining a

healthy weight. If diabetes is present, it also means keeping blood sugar controlled as tightly as possible.

- ○ Butler, R. "Can Vitamins Prevent Alzheimer's?" *Everyday Health*, November 20, 2006, http://www.everydayhealth.com/specialists/senior_aging/butler/qa/can-vitamins-prevent-alzheimers/index.aspx.

Medical Intervention

- Cognitive Symptoms
  - ○ The Alzheimer's Association reports on various known, used treatment options. It lays out the pros and cons of the treatments and how they work.
  - ○ The US Food and Drug Administration has approved two classes of drugs to treat cognitive symptoms of AD: cholinesterase inhibitors and NMDA receptor antagonists. Vitamin E supplements are frequently prescribed and have become part of a standard treatment regimen for most people with AD.
  - ○ Morris, J. "Cognitive Symptoms." Alzheimer's Association, 2012, http://www.alz.org/professionals_and_researchers_cognitive_symptoms.asp.

- "About Alzheimer's Disease: Treatment."
  - ○ This review of the basic treatment options provides a good overview and links to outside sources which provide more in-depth knowledge about options. This approach provides a great starting point.
  - ○ "About Alzheimer's Disease: Treatment." National Institute on Aging, 2011, http://www.nia.nih.gov/alzheimers/topics/treatment.

- "Alzheimer's Disease Medications Fact Sheet."
  - ○ This site provides an overview of the medication available for use for different stages of AD. It does a great job of highlighting what the drugs do, who should use them, possible side effects, and much more. Great source.
  - ○ "Alzheimer's Disease Medications Fact Sheet." National Institute on Aging, December 13, 2011, http://www.nia.nih.gov/alzheimers/publication/alzheimers-disease-medications-fact-sheet.

Work Remaining

I have had a few issues getting in touch with many physicians. Due to timing, most have been in and out of the office or out of town. As of now, I have not conducted any interviews, but I assure you, I am still trying.

I will research further on the hypotheses as well. I found different theories toward the end of my research marathon, so I will continue to look into those. I believe they provide an important basis of understanding.

I will need to fill in gaps in research as I begin to write. That will require extra research as the paper progresses.

Changes in the Initial Proposal

I have not made many drastic changes in the paper topic. I changed how I will structure the paper. I made one big addition: the section on hypotheses. Once I started reading, I had to learn more. There exist many different mechanical hypotheses, and I felt it important to investigate and report on them. A general overview cannot present the true intricacies of the disease the way that it needs to be done. It will make the report a little more technical, but if I succeed, it will not seem more technical.

Updated Outline
    I. Summary/Abstract
    II. Introduction
        a. What is Alzheimer's and why is it important
        b. What the reader needs to learn from this report
    III. Rationale for report
    IV. Physiological hypotheses
        a. Amyloid cascade hypothesis
            i. Amyloid protein plaques
        b. Mitochondrial cascade hypothesis
            i. Mitochondrial degradation
        c. Glucocorticoid cascade hypothesis
    V. Risk factors
        a. Who is at risk
            i. Genetic factors
                1. Maternal links
                2. Mitochondrial links
            ii. Early-life risk factors
    VI. Treatment options
        a. Over-the-counter remedies
        b. Preventative measures/lifestyle changes
            i. Exercise
            ii. Avoidance of head trauma
        c. Medical intervention
            i. Methods
            ii. Effectiveness
    VII. Summary and Conclusion

Updated Bibliography

"About Alzheimer's Disease: Treatment." National Institute on Aging, 2011, http://www.nia.nih.gov/alzheimers/topics/treatment.

"Alternative Treatments." Alzheimer's Association, 2012, http://www.alz.org/alzheimers_disease_alternative_treatments.asp.

"Alzheimer's Disease Medications Fact Sheet." National Institute on Aging, December 13, 2011, http://www.nia.nih.gov/alzheimers/publication/alzheimers-disease-medications-fact-sheet.

Aznar S, Knudsen GM. "Depression and Alzheimer's Disease: Is Stress the Initiating Factor in a Common Neuropathological Cascade?" *Journal of Alzheimer's Disease* 23, no. 2 (2011): 177–193.

Bloss CS, Delis DC, Salmon DP, Bondi MW. "Decreased Cognition in Children with Risk Factors for Alzheimer's Disease." *Biological Psychiatry* 64, no. 10 (2008): 904–906.

Borenstein AR, Wu Y, Mortimer JA, Schellenberg GD, McCormick WC, Bowen JD, McCurry A, Larson E.B. "Developmental and Vascular Risk Factors for Alzheimer's Disease." *Neurobiology of Aging* 26, no. 3 (2005): 325–334.

Butler R. "Can Vitamins Prevent Alzheimer's?" Everyday Health, November 20, 2006, http://www.everydayhealth.com/specialists/senior_aging/butler/qa/can-vitamins-prevent-alzheimers/index.aspx.

Edland SD, Silverman JM, Peskind ER, Tsuang D, Wijsman E, Morns JC. "Increased Risk of Dementia in Mothers of Alzheimer's Disease Cases: Evidence for Maternal Inheritance." *Neurology* 4 (1996): 254–256.

Hamilton J. "How Exercise Might Help Keep Alzheimer's at Bay." National Public Radio, April 29, 2010, http://www.npr.org/templates/story/story.php?storyId126370279.

Hardy JA, Higgins GA. "Alzheimer's Disease: The Amyloid Cascade Hypothesis." *Science* 256, no. 5054 (1992): 184–185.

James F. " 'Fatso Gene' Linked to Alzheimer's." National Public Radio, April 19, 2010, http://www.npr.org/blogs/thetwo-way/2010/04/fatso_gene_linked_to_alzheimer.html?ps=rs.

Jin M, Shepardson N, Yang T, et al. "Soluble Amyloid Beta-Protein Dimers Isolated from Alzheimer Cortex Directly Induce Tau Hyperphosphorylation and Neuritic Degeneration." *Proceedings of the National Academy of Sciences USA* 108 (2011): 5819–5824.

Lunde A. "Preventing Alzheimer's: Exercise Still Best Bet." *Mayo Clinic.* 25 March 2008. http://www.mayoclinic.com/health/alzheimers/MY00002.

Maruszak A, Żekanowski C. "Mitochondrial Dysfunction and Alzheimer's Disease." *Progress in Neuro-Psychopharmacology and Biological Psychiatry* 35, no. 2 (2011): 320–330.

Miller DB, O'Callaghan JP. "Do Early-Life Insults Contribute to the Late-Life Development of Parkinson and Alzheimer Diseases?" *Metabolism* 57, suppl. 2 (2008): S44–S49.

Morris J. "Cognitive Symptoms." Alzheimer's Association, 2012, http://www.alz.org/professionals_and_researchers_cognitive_symptoms.asp.

Raymond R. "Evidence Lacking to Support Alzheimer's Prevention." National Public Radio, April 28, 2010, http://www.npr.org/blogs/health/2010/04/alzheimers_prevention_exercise.html?ps=rs.

Russell P. "Vitamin B May Help Prevent Mental Decline in Elderly." WebMD.com, September 9, 2010, http://www.webmd.com/brain/news/20100909/vitamin-b-may-help-preventmental-decline-elderly.

Sapolsky RM, Krey LC, McEwen BS. "The Neuroendocrinology of Stress and Aging: The Glucocorticoid Cascade Hypothesis." *Science of Aging Knowledge Environment* 38 (2002).

US Department of Health and Health Services. *2013–2014 Alzheimer's Disease Progress Report: Insights and Challenges.* National Institutes of Health, 2014, https://www.nia.nih.gov/alzheimers/publication/2013-2014-alzheimers-disease-progress-report/.

APPENDIX C: Medication Chart [28]

Medications to Treat Alzheimer's Disease

This brief summary does not include all information important for patient use and should not be used as a substitute for professional medical advice. Consult the prescribing doctor and read the package insert before using these or any other medications or supplements. Drugs are listed in order of FDA approval, starting with the most recent.

DRUG NAME	DRUG TYPE AND USE	HOW IT WORKS	COMMON SIDE EFFECTS	MANUFACTURER'S RECOMMENDED DOSAGE	FOR MORE INFORMATION
Namenda® (memantine)	N-methyl D-aspartate (NMDA) antagonist prescribed to treat symptoms of moderate to severe Alzheimer's	Blocks the toxic effects associated with excess glutamate and regulates glutamate activation	Dizziness, headache, constipation, confusion	• Tablet: Initial dose of 5 mg once a day • May increase dose to 10 mg/day (5 mg twice a day), 15 mg/day (5 mg and 10 mg as separate doses), and 20 mg/day (10 mg twice a day) at minimum 1-week intervals if well tolerated • Oral solution: same dosage as above • Extended-release tablet: Initial dose of 7 mg once a day; may increase dose to 14 mg/day, 21 mg/day, and 28 mg/day at minimum 1-week intervals if well tolerated	For current information about this drug's safety and use, visit www.namenda.com. Click on "Prescribing Information" to see the drug label.
Razadyne® (galantamine)	Cholinesterase inhibitor prescribed to treat symptoms of mild to moderate Alzheimer's	Prevents the breakdown of acetylcholine and stimulates nicotinic receptors to release more acetylcholine in the brain	Nausea, vomiting, diarrhea, weight loss, loss of appetite	• Tablet: Initial dose of 8 mg/day (4 mg twice a day) • May increase dose to 16 mg/day (8 mg twice a day) and 24 mg/day (12 mg twice a day) at minimum 4-week intervals if well tolerated • Oral solution: same dosage as above • Extended-release capsule: same dosage as above but taken once a day	For current information about this drug's safety and use, visit www.razadyneer.com. Click on "Important Safety Information" to see links to prescribing information.
Exelon® (rivastigmine)	Cholinesterase inhibitor prescribed to treat symptoms of mild to moderate Alzheimer's	Prevents the breakdown of acetylcholine and butyrylcholine (a brain chemical similar to acetylcholine) in the brain	Nausea, vomiting, diarrhea, weight loss, loss of appetite, muscle weakness	• Capsule: Initial dose of 3 mg/day (1.5 mg twice a day) • May increase dose to 6 mg/day (3 mg twice a day), 9 mg (4.5 mg twice a day), and 12 mg/day (6 mg twice a day) at minimum 2-week intervals if well tolerated • Patch: Initial dose of 4.6 mg once a day; may increase to 9.5 mg once a day after minimum of 4 weeks if well tolerated • Oral solution: same dosage as capsule	For current information about this drug's safety and use, visit www.fda.gov/cder. Click on "Drugs@FDA," search for Exelon, and click on drug-name links to see "Label Information."
Aricept® (donepezil)	Cholinesterase inhibitor prescribed to treat symptoms of mild to moderate, and moderate to severe Alzheimer's	Prevents the breakdown of acetylcholine in the brain	Nausea, vomiting, diarrhea	• Tablet: Initial dose of 5 mg once a day • May increase dose to 10 mg/day after 4–6 weeks if well tolerated, then to 23 mg/day after at least 3 months • Orally disintegrating tablet: same dosage as above • 23-mg dose available as brand-name tablet only	For current information about this drug's safety and use, visit www.fda.gov/cder. Click on "Drugs@FDA," search for Aricept, and click on drug-name links to see "Label Information."

Index